Reproducing Kernels and their Applications

International Society for Analysis, Applications and Computation

Volume 3

The titles published in this series are listed at the end of this volume.

Reproducing Kernels and their Applications

Edited by

Saburou Saitoh
Gunma University, Japan

Daniel Alpay
Ben-Gurion University, Israel

Joseph A. Ball
Virginia Tech, U.S.A.

and

Takeo Ohsawa
Nagoya University, Japan

KLUWER ACADEMIC PUBLISHERS
DORDRECHT / BOSTON / LONDON

A C.I.P. Catalogue record for this book is available from the Library of Congress.

ISBN 978-1-4419-4809-0

Published by Kluwer Academic Publishers,
P.O. Box 17, 3300 AA Dordrecht, The Netherlands.

Sold and distributed in North, Central and South America
by Kluwer Academic Publishers,
101 Philip Drive, Norwell, MA 02061, U.S.A.

In all other countries, sold and distributed
by Kluwer Academic Publishers,
P.O. Box 322, 3300 AH Dordrecht, The Netherlands.

Printed on acid-free paper

TABLE OF CONTENTS

PREFACE

The First International Congress of the International Society for Analysis, its Applications and Computations (ISAAC'97) was held at the University of Delaware from 3 to 7 June 1997. As specified in the invitation of the President Professor Robert P. Gilbert of the ISAAC, we organized the session on

Reproducing Kernels and Their Applications.

In our session, we presented 24 engaging talks on topics of current interest to the research community. As suggested and organized by Professor Gilbert, we hereby publish its Proceedings. Rather than restricting the papers to Congress participants, we asked the leading mathematicians in the field of the theory of reproducing kernels to submit papers. However, due to time restrictions and a compulsion to limit the Proceedings a reasonable size, we were unable to obtain a comprehensive treatment of the theory of reproducing kernels. Nevertheless, we hope this Proceedings of the First International Conference on reproducing kernels will become a significant reference volume. Indeed, we believe that the theory of reproducing kernels will stand out as a fundamental and beautiful contribution in mathematical sciences with a broad array of applications to other areas of mathematics and science.

We would like to thank Professor Robert Gilbert for his substantial contributions to the Congress and to our Proceedings. We also express our sincere thanks to the staff of the University of Delaware for their manifold cooperation in organizing the Congress. Finally, we wish to express our sincere thanks to the staff of Kluwer Academic Publishers and Mrs. Noriko Kimura of Gunma University for their help in publishing the Proceedings.

August, 1998

S. Saitoh
D. Alpay
J. Ball
and
T. Ohsawa

1 OPERATOR THEORETICAL CLASSIFICATION OF REPRODUCING KERNEL HILBERT SPACES

Shigeo Akashi

Department of Mathematics,
Faculty of Science,
Niigata University, Japan
akashi@math.sc.niigata-u.ac.jp

Abstract: The homeomorphism problems of subspaces with norms in terms of the ranges of the closed unit ball under compact positive operators are examined. These results will be applied to the operator theoretical classification of reproducing kernel Hilbert spaces.

1 INTRODUCTION

The general theory of reproducing kernel Hilbert spaces was developed by Nathan Aronszajn ([2]) in 1950 and the general theory and its various applications, see the recent book ([7]).

The concept of ϵ-entropy was developed by Andrei N. Kolmogorov ([4]) in 1957 in connection with the 13th problem given by David Hilbert, and he showed that not all continuous functions of three variables are representable as superpositions of continuous functions of two variables. Reese T. Prosser ([6]) applied Kolmogorov's methods to the entropy theoretical classification of compact operators.

In this paper, the homeomorphism problems of subspaces with norms in terms of the ranges of the closed unit ball under compact positive operators are examined. These results will be applied to the classification problems of reproducing kernel Hilbert spaces whose kernels are jointly continuous positive matrices on $[0,1]$.

2 PRELIMINARIES

Throughout this paper, N denotes the set of all positive integers. Let $\mathcal{H}$ be a separable Hilbert space with inner product $<\cdot,\cdot>$ and norm $||\cdot||$.

Let P be an abstract set and k be a complex-valued positive matrix on P in the sense of Aronszajn ([2]), so that, for any finite number of points $\{p_j\}$ of P and for any complex numbers $\{c_j\}$,

$$\sum_i \sum_j \overline{c_i} c_j k(p_i, p_j) \geq 0.$$

Then, there exists a uniquely determined functional Hilbert space $(\mathcal{H}(k), <\cdot,\cdot>_k)$ consisting of functions on P and admitting the reproducing kernel $k(\cdot,\cdot)$ with the properties that

(i) $\quad K(\cdot,q)\in\mathcal{H}(k)$ for any $q\in P$, and

(ii) $\quad <f(\cdot),k(\cdot,q)>_k = f(q)$ for any $q\in P$ and for any $f\in\mathcal{H}(k)$.

For any compact positive operator T on $\mathcal{H}$, there exist the non-increasing sequence of eigenvalues $\{\lambda_i(T); i\in N\}$ and the orthonormal system of eigenvectors $\{e_i(T); i\in N\}$ satisfying

$$Te_i(T) = \lambda_i(T)e_i(T), \qquad i\in N.$$

Then, T can be represented by

$$Tx = \sum_{i=1}^{\infty} \lambda_i(T) < x, e_i(T) > e_i(T), \qquad x\in\mathcal{H}.$$

Here, the exponent of convergence $E(T)$ is defined by

$$E(T) = inf\left\{ r\geq 0; \sum_{i=1}^{\infty} \lambda_i(T)^r < \infty \right\}.$$

Let $card(T,r)$ be the distribution of the eigenvalues of T which is defined by

$$card(T,r) = max\{i\in N; \lambda_i(T) > r\}, \quad r>0.$$

Moreover, let $G(T)$ and $g(T)$ be the upper growth order and the lower growth order of $\{\lambda_i; i\in N\}$ which are defined by

$$G(T) = \limsup_{r\to+0} \frac{card(T,r)}{log(1/r)}$$

and

$$g(T) = \liminf_{r \to +0} \frac{card(T,r)}{log(1/r)},$$

respectively. It is well known that the equality

$$E(T) = G(T)$$

holds ([5]).

For any $\epsilon > 0$ and any relatively compact subset $\mathcal{F}$ of $\mathcal{H}$, an ϵ-covering is defined by a family of open balls with radii ϵ and whose union can cover $\mathcal{F}$, and an ϵ-packing is defined by a family of open balls with centers in $\mathcal{F}$ and radii ϵ whose pairwise intersections are all empty. Here, the ϵ-entropy of $\mathcal{F}$, which is denoted by $S(\mathcal{F}, \epsilon)$, is defined by the base-2 logarithm of the minimum number of elements of any ϵ-covering of $\mathcal{F}$, and the ϵ-capacity of $\mathcal{F}$, which is denoted by $C(\mathcal{F}, \epsilon)$, is defined by the base-2 logarithm of the maximum number of elements of any ϵ-packing of $\mathcal{F}$.

For any positive number ϵ, the ϵ-entropy of T and the ϵ-capacity of T are defined by $S(T(\mathcal{U}), \epsilon)$ and $C(T(\mathcal{U}), \epsilon)$, respectively, where $\mathcal{U}$ is the closed unit ball of $\mathcal{H}$. Then, by Prosser ([6]) and Akashi ([1]), it is known that, for any positive number δ, there exists a positive number ϵ_δ satisfying

$$\left(\frac{1}{\epsilon}\right)^{g(T)-\delta} \leq S(T(U), \epsilon) \leq \left(\frac{1}{\epsilon}\right)^{G(T)+\delta}, \qquad 0 < \epsilon < \epsilon_\delta.$$

3 THE HOMEOMORPHISM PROBLEMS OF SUBSPACES OF A SEPARABLE HILBERT SPACE

In this section, we shall consider the homeomorphism problems of subspaces with norms in terms of the range of the closed unit ball under compact positive operators.

Let T be an injective compact positive operator on $\mathcal{H}$. Then, the range of the closed unit ball under T is represented by

$$T(\mathcal{U}) = \left\{ y \in \overline{linear\{e_i(T); i \in N\}};\ \sum_{i=1}^{\infty} \left| \frac{< y, e_i(T) >}{\lambda_i(T)} \right|^2 \leq 1 \right\}.$$

Here, we define the subspace $\mathcal{F}_T$ of $\mathcal{H}$ by

$$\mathcal{F}_T = \bigcup_{c>0} cT(\mathcal{U}),$$

and define the new norm q_T of Minkowski type on $\mathcal{F}_T$ by

$$q_T(x) = inf\,\{c > 0;\ x \in cT(\mathcal{U})\}.$$

Then we have the following

Theorem 3.1. Let T_1 and T_2 be two injective, compact positive operators on $\mathcal{H}$ satisfying $G(T_1) \neq g(T_2)$. Then, $(\mathcal{F}_{T_1}, q_{T_1})$ is not homomorphically homeomorphic to $(\mathcal{F}_{T_2}, q_{T_2})$, that is, there does not exist a bijective, bicontinuous linear operator W on $\mathcal{H}$ with values in $\mathcal{H}$ such that, $\mathcal{F}_{T_2} = W(\mathcal{F}_{T_1})$ holds and, for some positive constants c and d, the two inequalities

$$q_{T_2}(Wx) \leq c q_{T_1}(x), \quad x \in \mathcal{F}_{T_1},$$

$$q_{T_1}(x) \leq d q_{T_2}(Wx), \quad x \in \mathcal{F}_{T_1}$$

hold.

Proof. Without loss of generality, we can assume $G(T_1) < G(T_2)$. Assume that there exists a bijective, continuous linear operator W which is stated in this theorem. Then, we have

$$\begin{aligned} T_2(\mathcal{U}) &= \{y \in \mathcal{F}_{T_2};\, q_{T_2}(y) \leq 1\} \\ &\subset \{Wx;\, x \in \mathcal{F}_{T_1}, q_{T_1}(x) \leq d\} \\ &= W(dT_1(\mathcal{U})). \end{aligned}$$

But these inclusions imply that, for any positive number ϵ,

$$S(T_2(\mathcal{U}), \epsilon) \leq S(W(dT_1(\mathcal{U})), \epsilon).$$

Moreover, according to the relations betweem $q_{T_1}(\cdot)$ and $q_{T_2}(\cdot)$, we have

$$S(W(T_1(\mathcal{U})), \|W\|\epsilon) \leq S(T_1(\mathcal{U}), \epsilon).$$

Therefore, these two inequalities shows that

$$G(T_2) \leq G(T_1).$$

But this inequality contradicts $G(T_1) < G(T_2)$.

Remarks. In the theory of nuclear spaces ([3]), the theory of compact operators ([8]) plays an important role and especially, examining the homeomorphism problems of subspaces of a separable Hilbert space is exactly equivalent to examining whether a given nuclear space is homomorphically homeomorphic to a certain Hilbert space or not. Therefore, it is important to investigate such homeomorphism problems.

4 THE CLASSIFICATION OF REPRODUCING KERNEL HILBERT SPACES

In this section, we shall apply the result in the previous section to the operator theoretical classification of reproducing kernel Hilbert spaces in $L^2[0, 1]$.

Let k be a jointly continuous, complex-valued positive matrix defined on $[0,1]$, and K denotes the integral kernel operator defined by

$$(Kf)(s) = \int_0^1 k(s,t)f(t)dt, \quad f \in L^2[0,1],\ s \in [0,1].$$

Let $\{\lambda_i(K); i \in N\}$ be the non-increasing sequence of eigenvalues of K and $\{e_i(K,\cdot); i \in N\}$ be the sequence of eigenfunctions of K associated with $\{\lambda_i(K); i \in N\}$. Then, Mercer's theorem [9] shows the equalities

$$lim_{n\to\infty} \int_0^1 \int_0^1 \left| k(s,t) - \sum_{i=1}^{n} \lambda_i(K) e_i(K,s) \overline{e_i(K,t)} \right|^2 dsdt = 0$$

and

$$lim_{n\to\infty} sup_{s,t\in[0,1]} \left| k(s,t) - \sum_{i=1}^{n} \lambda_i(K) e_i(K,s) \overline{e_i(K,t)} \right| = 0.$$

Let $\mathcal{H}(k)$ be the reproducing kernel Hilbert space with inner product $<\cdot,\cdot>_k$ and norm $||\cdot||_k$ whose reproducing kernel is k. Then, we have the following

Lemma 4.1.

$$||k(\cdot,t)||_k = q_{\sqrt{K}}(k(\cdot,t))$$

holds for any $t \in [0,1]$.

Proof. Indeed, for any $t \in [0,1]$, we have

$$\begin{aligned} ||k(\cdot,t)||_k &= \sqrt{< k(\cdot,t), k(\cdot,t) >_k} \\ &= \sqrt{k(t,t)} \\ &= \sqrt{\sum_{i=1}^{\infty} \lambda_i(K) |e_i(K,t)|^2} \\ &= \sqrt{\sum_{i=1}^{\infty} \left| \frac{< e_i(K,\cdot), k(,t) >}{\sqrt{\lambda_i(K)}} \right|^2}. \end{aligned}$$

Let r be any positive number and $\mathcal{U}$ be the closed unit ball of $L^2[0,1]$. Then, $k(\cdot,t) \in r\sqrt{K}(\mathcal{U})$ implies that

$$\sqrt{\sum_{i=1}^{\infty} \left| \frac{< e_i(K,\cdot), k(,t) >}{\sqrt{\lambda_i(K)}} \right|^2} \leq r$$

holds. Therefore, we have

$$\sqrt{\sum_{i=1}^{\infty} \left| \frac{< e_i(K,\cdot), k(,t) >}{\sqrt{\lambda_i(K)}} \right|^2} \leq q_{\sqrt{K}}(k(\cdot,t)).$$

Conversely, for any positive number δ,

$$k(\cdot,t) \in \left(\sqrt{\sum_{i=1}^{\infty} \left| \frac{< e_i(K,\cdot), k(,t) >}{\sqrt{\lambda_i(K)}} \right|^2} + \delta \right) \sqrt{K}(\mathcal{U})$$

holds. Therefore, we have

$$q_{\sqrt{K}}(k(\cdot,t)) \leq \sqrt{\sum_{i=1}^{\infty} \left| \frac{< e_i(K,\cdot), k(,t) >}{\sqrt{\lambda_i(K)}} \right|^2}.$$

These inequalities conclude the proof of this lemma.

Let k_1 (resp. k_2) be a jointly continuous, positive matrix on $[0,1]$ and $\mathcal{H}(k_1)$ (resp. $\mathcal{H}(k_2)$) be the reproducing kernel Hilbert space whose reproducing kernel is k_1 (resp. k_2). Then, we have our main

Theorem 4.2. If $G(K_1) \neq G(K_2)$ holds, then $\mathcal{H}(k_1)$ is not homomorphically homeomorphic to $\mathcal{H}(k_2)$, that is, there does not exist any bijective, bicontinuous linear operator W defined on $L^2[0,1]$ with values in $L^2[0,1]$ such that, $\mathcal{H}(k_2) = W(\mathcal{H}(k_1))$ holds and, for some positive constants c and d,

$$\|Wf\|_{k_2} \leq c\|f\|_{k_1}, \quad f \in \mathcal{H}(k_1),$$

$$\|f\|_{k_1} \leq d\|Wf\|_{k_2}, \quad f \in \mathcal{H}(k_1)$$

hold.

Proof. Since $\{k_1(\cdot,t);\ t \in [0,1]\}$ (resp. $\{k_2(\cdot,t);\ t \in [0,1]\}$) is a dense subset of $\mathcal{H}(k_1)$ (resp. $\mathcal{H}(k_2)$), $(\mathcal{H}(k_1), \|\cdot\|_{k_1})$ (resp. $(\mathcal{H}(k_2), \|\cdot\|_{k_2})$) is homomorphically homeomorphic to $(\mathcal{F}_{\sqrt{K_1}}, q_{\sqrt{K_1}})$ (resp. $(\mathcal{F}_{\sqrt{K_2}}, q_{\sqrt{K_2}})$). Moreover, $(\mathcal{F}_{\sqrt{K_1}}, q_{\sqrt{K_1}})$ is not homomorphically homeomorphic to $(\mathcal{F}_{\sqrt{K_2}}, q_{\sqrt{K_2}})$, because $G(\sqrt{K_1}) \neq G(\sqrt{K_2})$ holds. Therefore, Theorem 3.1. leads us to the conclusion.

Acknowledgment. The author would like to express his hearty thanks to Professor Saburou Saitoh who gave much enlightening and suggestive advice in the course of stimulating discussions.

References

[1] S. Akashi, The asymptotic behavior of ϵ-entropy of a compact positive operator, J. Math. Anal. Appl., 153(1990), 250-257.

[2] N. Aronszajn, Theory of reproducing kernels, Trans. Amer. Math. Soc., 68(1950), 337-405.

[3] I. M. Gel'fand and N. Ya. Vilenkin, Generalized functions, 4(1964), Academic Press, New York.

[4] A. N. Kolmogorov, On the representation of continuous functions of several variables by superpositions of continuous functions of one variable and addition, Dokl., 114(1957), 679-681.

[5] B. Ja. Levin, Distributions of zeros of entire functions, Translation of Mathematical Monographs 5(1964), Amer. Math. Soc., Providence.

[6] R. T. Prosser, The ϵ-entropy and ϵ-capacity of certain time-varying channels, J. Math. Anal. Appl., 16(1966), 553-573.

[7] S. Saitoh, Integral transforms, reproducing kernels and their applications, Pitman Research Notes in Mathematics, 369(1997), Addison-Longman, Harlow.

[8] R. Schatten, Norm ideals of completely continuous operators, Springer Verlag, Berlin, 1970.

[9] K. Yosida, Lectures on differential and integral equations, Interscience Publishers, New York, 1960.

2 HOLOMORPHIC FACTORIZATION OF MATRICES OF POLYNOMIALS

John P. D'Angelo

Dept. of Mathematics
University of Illinois
Urbana IL 61801
USA
jpda@math.uiuc.edu

INTRODUCTION

This paper considers some work done by the author and Catlin [CD1,CD2,CD3] concerning positivity conditions for bihomogeneous polynomials and metrics on bundles over certain complex manifolds. It presents a simpler proof of a special case of the main result in [CD3], providing also a self-contained proof of a generalization of the main result from [CD1]. Some new examples and applications appear here as well. The idea is to use the Bergman kernel function and some operator theory to prove purely algebraic theorems about matrices of polynomials.

The main idea arises from generalizing a classical factorization question. See [Dj] and [RR] for many aspects of factorization of non-negative matrices and operators on Hilbert spaces. Consider a real-analytic matrix-valued function $F(z,\overline{z})$ that is positive semi-definite at each point. Is there a holomorphic matrix-valued function $A(z)$ such that $F(z,\overline{z}) = A(z)^*A(z)$? Here A^* denotes the conjugate transpose of A. In general the answer is no, even when F is a scalar, positive-definite, and its entries are bihomogeneous polynomials. Because such factorizations have many applications, we allow ourselves a generalization; we can first multiply F by powers of a scalar function R, and ask whether we can factor R^dF for sufficiently large d. This is a natural thing to do when one studies proper holomorphic mappings between balls in different dimensions, and one chooses R to be the squared Euclidean norm. See [CD1] for applications. This multiplication also admits an interpretation in terms of metrics on tensor products of Hermitian line bundles.

We write $\langle\zeta, w\rangle$ for the Euclidean Hermitian inner product of ζ and w on any finite-dimensional complex Euclidean space, and $||\zeta||^2$ for the squared norm.

Later we will use subscripts to denote L^2 norms. We write $\mathbf{V}(A)$ to denote the variety defined by the simultaneous vanishing of the component functions of a holomorphic mapping A.

A *bihomogeneous polynomial* on $\mathbf{C}^n$ is a polynomial function $f : \mathbf{C}^n \times \mathbf{C}^n \to \mathbf{C}$ that is homogeneous of the same degree m in each set of variables. We will be considering $f(z, \overline{w})$, which is conjugate-analytic in the second set of variables. The polynomial defined by $f(z, \overline{z})$ is real-valued if and only if $f(z, \overline{w}) = \overline{f(w, \overline{z})}$; we call such an f a bihomogeneous real-valued polynomial on $\mathbf{C}^n$ of degree $2m$.

Suppose that f is a bihomogeneous real-valued polynomial of degree $2m$. By elementary linear algebra, it is possible to find holomorphic polynomial mappings A and B, with finitely many components, that are homogeneous of degree m, and such that

$$f(z, \overline{w}) = \langle A(z), A(w) \rangle - \langle B(z), B(w) \rangle \tag{1}$$

It follows from (1) that

$$f(z, \overline{z}) = ||A(z)||^2 - ||B(z)||^2 \tag{2}$$

Suppose in addition that $f(z, \overline{z}) \geq 0$. We investigate the following questions. Can we choose $B = 0$ in (2), and if we cannot, can we do this for $R^d f$, where R is an appropriate multiplier and d is sufficiently large? Suppose more generally that $F(z, \overline{z})$ is a matrix of bihomogeneous polynomials each of degree $2m$, and that is positive semi-definite at each $z \neq 0$. Can we factor $||z||^{2d} F(z, \overline{z})$ for sufficiently large d?

The following theorem gives a decisive answer in the positive-definite case.

Theorem 1. [Catlin-D'Angelo]. Suppose that f is a bihomogeneous real-valued polynomial on $\mathbf{C}^n$ of degree $2m$. Then f is positive away from the origin if and only there is an integer d and a holomorphic homogeneous polynomial mapping A, whose components span the space of homogeneous polynomials of degree $m + d$, such that

$$||z||^{2d} f(z, \overline{z}) = ||A(z)||^2. \tag{3}$$

Suppose that $F(z, \overline{z})$ is an r by r matrix whose entries are bihomogeneous polynomials of degree $2m$. Then $F(z, \overline{z})$ is positive-definite at each point $z \neq 0$ if and only if there is an integer d and a holomorphic homogeneous polynomial matrix A, whose row vectors span the space of r-tuples of homogeneous polynomials of degree $m + d$, such that

$$||z||^{2d} F(z, \overline{z}) = A(z)^* A(z). \tag{4}$$

Note that (3) is the scalar version of (4). The scalar statement about f was proved in [CD1]. The matrix version is a special case of a general result from [CD3] about Hermitian metrics on bundles over certain complex manifolds.

Because it is a special case, some steps in the proof simplify; its intrinsic interest justifies giving the simpler proof here. In the scalar statement, one can replace the condition that the components of A span, by the condition that $\mathbf{V}(A) = \{0\}$; the exponent required may be smaller. The proof reveals that the stronger condition on A is the natural one.

When $F(z,\overline{z}) = A(z)^*A(z)$, necessarily $F(z,\overline{z})$ is positive semi-definite at each point. A general result such as Theorem 1 cannot hold in the positive semi-definite case, for the following simple reason. Suppose $r = 1$ for simplicity. If the zero set of f is not an analytic variety, then there is no hope to write $||z||^{2d}f(z,\overline{z}) = ||A(z)||^2$, because the zero set of the right side is an analytic variety. A simple example where the zero set of a bihomogeneous polynomial fails to be an analytic variety is given by (5).

$$f(z,\overline{z}) = (|z_1|^2 - |z_2|^2)^2. \tag{5}$$

Some results hold under specific hypotheses in the semi-definite case, but most of this paper considers only the positive-definite case.

We give a complete proof of Theorem 1, relying on the Bergman kernel function for the unit ball. In Theorem 2 we give a simple application to elliptic PDE. In Theorem 3 we reinterpret Theorem 1 in terms of the universal bundle over complex projective space. We also provide some illuminating examples along the way. We close the paper with some brief remarks about factorization theorems proved in the 1970s.

I. HOLOMORPHIC FACTORABILITY

Suppose that we are given an r by r matrix $F(z,\overline{w})$ whose entries $F_{ij}(z,\overline{w})$ are bihomogeneous polynomials of degree $2m$ on $\mathbf{C}^n$. Let $N = N(n,m)$ denote the dimension of the vector space of homogeneous polynomials of degree m in n variables. We write V_d for the vector space of r-tuples of homogeneous polynomials of degree d on $\mathbf{C}^n$.

Definition 1. The r by r matrix of bihomogeneous polynomials $F(z,\overline{w})$ of degree $2m$ is called *holomorphically factorable* if there is an integer s and a matrix $(E_{jk}(z))$, for $j = 1,...,r$ and $k = 1,...,s$, of homogeneous polynomials of degree m such that (6) holds.

$$F_{ij}(z,\overline{w}) = \langle E_j(z), E_i(w)\rangle = \sum_{k=1}^{s} E_{jk}(z)\overline{E_{ik}(w)} \tag{6}$$

Here $E_j(z) = (E_{j1}(z),...,E_{js}(z))$.

Let A be the transpose of E; we observe immediately that Definition 1 implies that

$$F(z,\overline{w}) = A(w)^*A(z). \tag{7}$$

Definition 2. The matrix of bihomogeneous polynomials $F(z,\overline{w})$ of degree $2m$ is called *strictly holomorphically factorable* if it is holomorphically factorable, and in addition the s column vectors of $(E_{jk}(z))$ are a basis for V_d.

Note that the notion of strict holomorphic factorability requires that we choose $s = N$. The first concept allows us to write (7), forcing $F(z,\overline{z})$ to be positive semi-definite. The second concept ensures that $F(z,\overline{z})$ is positive-definite, but it is an even stronger condition. We give a simple example when $r = 1$ to illustrate the difference.

Example. Let $F(z,\overline{w})$ be the one-by-one matrix given by $z_1^2\overline{w_1}^2 + z_2^2\overline{w_2}^2$. Here Definition 1 holds with $E_1(z) = (z_1^2, z_2^2)$. Note that $F(z,\overline{z})$ is positive-definite away from the origin, but the absence of the $z_1 z_2$ cross-term means that Definition 2 fails.

II. The link to operator theory

We will prove Theorem 1 by using some facts about compact operators on a Hilbert space. Because we consider (matrices of) bihomogeneous polynomials and use the Euclidean norm as a multiplier, it suffices to consider the Hilbert space $\mathcal{H}$ of r-tuples of L^2 functions on the unit ball B in complex Euclidean space $\mathbf{C}^n$. We let $\mathcal{A}$ denote the closed subspace of r-tuples of holomorphic functions in $L^2(B)$. We write P for the Bergman projection from $\mathcal{H}$ to $\mathcal{A}$; it is the usual Bergman projection acting on each component.

There is an orthogonal decomposition of $\mathcal{A}$ into the finite-dimensional subspaces V_d. The following result provides the crucial link between matrices whose entries are bihomogeneous polynomials and operator theory on $\mathcal{A}$.

Proposition. Let $F(z,\overline{z}) = (F_{ij}(z,\overline{z}))$ be an r by r matrix of bihomogeneous polynomials of degree $2d$ on $\mathbf{C}^n$. The following are equivalent:

1. The matrix F is *strictly holomorphically factorable.* Thus there is an s by r matrix $E(z)$ of holomorphic homogeneous polynomials whose column vectors give a basis for V_d for each $z \neq 0$. Furthermore, with A the transpose of E, (8) holds.

$$F(z,\overline{w}) = A(w)^* A(z) \tag{8}$$

2. Consider the integral operator $T : V_d \to V_d$ defined by

$$Th(z)_j = \int_B \sum_i F_{ij}(z,\overline{w}) h_i(w) dV(w) \tag{9}$$

Then T is positive on V_d. That is, there is a positive constant c so that for $h \in V_d$,

$$\langle Th, h\rangle_{\mathcal{H}} \geq c||h||^2_{\mathcal{H}} \tag{10}$$

3. Write

$$F_{ij}(z,\overline{w}) = \sum F_{ij\alpha\beta} z^\alpha \overline{w}^\beta \tag{11}$$

for constants $F_{ij\alpha\beta}$. Then the matrix $(F_{ij\alpha\beta})$ is positive-definite on $\mathbf{C}^{Nr} = \mathbf{C}^r \otimes \mathbf{C}^N$; that is, there is a positive constant c' so that

$$\sum F_{ij\alpha\beta}t_i\bar{t}_j s_\alpha \bar{s}_\beta \geq c' \sum |t_i s_\alpha|^2 = c' \sum |t_i|^2 \sum |s_\alpha|^2 \tag{12}$$

Proof. First we show that 1 implies 2. Given $h \in V_d \subset \mathcal{A}$, we write $h = (h_1, ..., h_r)$. We assume that (6) holds and compute the left side of (10), to obtain

$$\langle Th, h\rangle_{\mathcal{H}} = \int\int \sum_{k=1}^{s} \sum_{i,j=1}^{r} \overline{E_{ik}(w)} E_{jk}(z) h_i(w) \overline{h_j(z)} dV(w) dV(z)$$

$$= \sum_{k=1}^{s} | \int \sum_{j} E_{jk}(z)\overline{h_j(z)} dV(z)|^2 = \sum_{k=1}^{s} |\langle A_k, h\rangle_{\mathcal{H}}|^2 \tag{13}$$

The condition of strict holomorphic factorability guarantees that the vectors A_k, the column vectors of E, form a basis for V_d. The last expression in (13) is therefore $\geq c||h||^2_{\mathcal{H}}$. This proves that 1 implies 2.

Next we show that 3 implies 1. Recall that N is the dimension of the space of homogeneous polynomials of degree d in n variables, and that (F_{ij}) is an r by r matrix. If (12) holds, there is a basis $\{E_{i\beta}\}$ of $\mathbf{C}^{Nr}$ so that $F_{ij\alpha\beta} = \langle E_{j\alpha}, E_{i\beta}\rangle$. Plug this in (11) to obtain

$$F_{ij}(z, \overline{w}) = \sum F_{ij\alpha\beta} z^\alpha \overline{w}^\beta = \sum \langle E_{j\alpha}, E_{i\beta}\rangle z^\alpha \overline{w}^\beta \tag{14}$$

Now define $A_j(z)$ by $A_j(z) = \sum E_{j\alpha} z^\alpha$. We see that

$$F_{ij}(z, \overline{w}) = \langle A_j(z), A_i(w)\rangle \tag{15}$$

and hence that 1 holds.

It remains to prove that 2 implies 3. We write $h_i(z) = \sum H_{i\alpha} z^\alpha$, and we plug this into $\langle Th, h\rangle_{\mathcal{H}}$. Recall that distinct monomials are orthogonal, so we may write

$$\langle Th, h\rangle_{\mathcal{H}} = \sum \int\int F_{ij\alpha\beta} z^\alpha \overline{w}^\beta H_{i\mu} w^\mu \overline{H}_{j\nu} \overline{z}^\nu dV(w) dV(z) = \sum F_{ij\alpha\beta} H_{i\beta} \overline{H}_{j\alpha} p_\alpha p_\beta \tag{16}$$

where the positive numbers p_α are equal to $||z^\alpha||^2_{L^2}$.

On the other hand, we have $||h||^2_{\mathcal{H}} = \sum |H_{i\alpha}|^2 p_\alpha$ by a similar calculation. Thus (10) implies that there is a positive constant c so that

$$\sum F_{ij\alpha\beta} p_\alpha p_\beta H_{i\beta} \overline{H}_{j\alpha} \geq c \sum |H_{i\alpha}|^2 p_\alpha \tag{17}$$

Since the p_β are positive numbers, (17) implies that the matrix with entries $F_{ij\alpha\beta}$ is also positive-definite, with a different constant c'. This gives 3. ♠

III. PROOF OF THEOREM 1

Suppose that the entries of F are bihomogeneous polynomials of degree $2m$. Let Q_d be the operator on V_{m+d} whose kernel is given by $\langle z, w\rangle^d F(z, \overline{w})$. In order to prove Theorem 1, Proposition 1 implies that we must find an integer d so that Q_d is positive-definite on V_{m+d}. We observe immediately, that if this holds for some d, then it holds for all larger integers. See [CD1], whose title suggests this stabilization process. Furthermore, the operators Q_d are zero except on V_{m+d}. This suggests considering their sum, weighted by positive constants, on the whole space $\mathcal{A}$.

If we choose the positive constants C_d appropriately, then

$$\sum C_d \langle z, w\rangle^d = \frac{n!}{\pi^n} \frac{1}{(1 - \langle z, w\rangle)^{n+1}} = B(z, w) \tag{18}$$

Here $B(z, w)$ is the Bergman kernel function for the unit ball B in complex Euclidean space $\mathbf{C}^n$. The crucial property of the Bergman kernel function is that it is the integral kernel of the Bergman projection mapping $L^2(B)$ to its closed subspace $A^2(B)$ of holomorphic functions. The kernel function satisfies

$$B(z, w) = \sum_{\alpha} \phi_\alpha(z) \overline{\phi_\alpha(w)}$$

where the collection $\{\phi_\alpha\}$ is any complete orthonormal set for the Hilbert space $A^2(B)$. For the unit ball, one can choose $\phi_\alpha = c_\alpha z^\alpha$, where c_α is a normalizing constant, and z^α denotes the indicated monomial.

Recall that $\mathcal{H}$ denotes the Cartesian product of r copies of $L^2(B)$. Let $P : \mathcal{H} \to \mathcal{A}$ denote the Bergman projection, acting componentwise. Motivated by (18), we let Q denote the integral operator on $\mathcal{H}$ whose kernel is given by $B(z, w)F(z, \overline{w})$. For a scalar function ψ, we let M_ψ denote the operator on $\mathcal{H}$ given by multiplication by ψ. Also M_F denotes matrix multiplication by F. There is no integration involved in these operators. Choose a smooth non-negative function ϕ of compact support that is positive near the origin.

We may write

$$Q = (M_F P + P M_\phi) + (Q - M_F P) - P M_\phi = T_1 + T_2 + T_3 \tag{19}$$

We claim that T_1 is positive, and that T_2 and T_3 are compact. This will show that $Q = S + K$, where S is positive on $\mathcal{A}$ and K is compact.

Lemma 1. T_1 is positive on all of $\mathcal{A}$, and T_3 is compact on $\mathcal{H}$.

Proof. The second statement is immediate, because the integral kernel is smooth everywhere on the ball. The first statement follows because P is a self-adjoint projection. To see this, let $h \in \mathcal{A}$.

$$\langle T_1 h, h\rangle_{\mathcal{H}} = \langle M_F P h + P M_\phi h, h\rangle_{\mathcal{H}} = \langle M_F h, h\rangle_{\mathcal{H}} + \langle M_\phi h, h\rangle_{\mathcal{H}} = \langle M_{F+\phi} h, h\rangle_{\mathcal{H}} \tag{20}$$

Since the multiplier $F + \phi$ is strictly positive-definite at all points, the last expression in (20) is at least $C||h||^2_{\mathcal{H}}$, and the result follows. ♠

Lemma 2. T_2 is compact.

Proof. This follows from Theorem 1 in [CD2], but is elementary in this case, because of the explicit nature of the Bergman kernel. The kernel of $Q - M_F P$ is

$$\frac{n!}{\pi^n} \frac{(F(z,\overline{w}) - F(z,\overline{z}))}{(1-\langle z,w\rangle)^{n+1}}$$

The numerator (a matrix of polynomials) vanishes on the boundary diagonal, where the only singularities of the denominator occur. One can use Young's inequality to verify that T_2 is compact. ♠

We summarize what we have proved so far. The operator Q on $\mathcal{H}$ has kernel given by $B(z,w)F(z,\overline{w})$. By Lemmas 1 and 2, we have written $Q = S + K$, where S is positive on $\mathcal{A}$ and K is compact.

The operator Q vanishes off $\mathcal{A}$, and we have $\mathcal{A} = \oplus V_j$. Write Q_d for the restriction of Q to V_{m+d}. If we show that Q_d is positive on V_{m+d} for sufficiently large d, then an application of Proposition 1 completes the proof of Theorem 1.

Since S is positive, there is $c > 0$ so that $\langle Sh,h\rangle_{\mathcal{H}} \geq c||h||^2_{\mathcal{H}}$. Since K is compact, there is a finite rank operator L such that the operator norm $|||K - L||| < \frac{c}{3}$. See [R]. Write $Q = S + L + (K - L)$ so that

$$\langle Qh,h\rangle_{\mathcal{H}} = \langle Sh,h\rangle_{\mathcal{H}} + \langle Lh,h\rangle_{\mathcal{H}} + \langle (K-L)h,h\rangle_{\mathcal{H}} \tag{21}$$

Using the lower bound on S, and because $|\langle (K-L)h,h\rangle_{\mathcal{H}}| \leq \frac{c}{3}||h||^2_{\mathcal{H}}$, we can write

$$\langle Qh,h\rangle_{\mathcal{H}} \geq c||h||^2_{\mathcal{H}} - \frac{c}{3}||h||^2_{\mathcal{H}} - |\langle Lh,h\rangle_{\mathcal{H}}| \geq \frac{2c}{3}||h||^2_{\mathcal{H}} - |\langle Lh,h\rangle_{\mathcal{H}}| \tag{22}$$

Because L is finite rank, we can choose d_0 sufficiently large such that, for $d \geq d_0$, the restriction of L to V_{m+d} satisfies $|\langle Lh,h\rangle_{\mathcal{H}}| \leq \frac{c}{3}||h||^2_{\mathcal{H}}$ also. Combining this with (22) implies that the restriction of Q to V_{m+d} is positive. By Proposition 1, its kernel $\langle z,w\rangle^d F(z,\overline{w})$ can be written $A(w)^* A(z)$, completing the proof of Theorem 1. ♠

IV. EXAMPLES AND APPLICATIONS

The integer d in Theorem 1 can be arbitarily large even when F has fixed degree. The example $f_c(z,\overline{z}) = |z_1|^4 + c|z_1 z_2|^2 + |z_2|^4$ is positive away from the origin for $c > -2$. By Theorem 1, for each c with $c > -2$, there is a minimum d_c for which (3) holds. It is elementary to show that $d_c \to \infty$ as $c \to -2$. See [CD1].

Because the integer d can be arbitarily large, the holomorphic mapping A from Theorem 1 can have an arbitrarily large number of components. This fact has consequences for proper holomorphic mappings between balls in different dimensions. For example, in [CD1], Theorem 1 is used to prove the following.

Given a holomorphic polynomial $q : \mathbf{C}^n \to \mathbf{C}$ that doesn't vanish on the closed unit ball, there is an integer N and a holomorphic polynomial mapping $p : \mathbf{C}^n \to \mathbf{C}^N$ such that $\frac{p}{q}$ is reduced to lowest terms, and defines a proper map between balls. The integer N can be arbitarily large. This is in sharp contrast to the case when $n = 1$, where the result is trivial, and we can take $N = 1$ as well.

Next we give an application to symbols of differential operators. Let D be a linear partial differential operator on real Euclidean space $\mathbf{R}^{2n}$ of even order $2m$. Recall (See [F] for example) that the principal symbol, or characteristic form, $p(\xi)$ of D governs whether it is elliptic. We suppose that the principal symbol has constant coefficients. Thus $p(\xi) = \sum_{|\alpha|=2m} c_\alpha \xi^\alpha$, and the operator is elliptic precisely when p vanishes only at the origin. If we make the usual identification of $\mathbf{R}^{2n}$ with $\mathbf{C}^n$, then we can express D in terms of the operators $\frac{\partial}{\partial z_j}$ and $\frac{\partial}{\partial \overline{z}_j}$. Using multi-index notation we can then write the principal symbol as

$$\sum_{|\alpha|+|\beta|=2m} c_{\alpha\beta}(\frac{\partial}{\partial z})^\alpha(\frac{\partial}{\partial \overline{z}})^\beta = f(\frac{\partial}{\partial z}, \frac{\partial}{\partial \overline{z}}).$$

In general, f is not bihomogeneous. There are simple simple necessary and sufficient condition for a real-valued polynomial $f(z, \overline{z})$ to be bihomogeneous of degree $2m$. One is that it be both homogeneous of degree $2m$ over $\mathbf{R}$ and invariant under replacing z by $e^{i\theta}z$. (Here $e^{i\theta}$ is a scalar, not an n-tuple). Another is the existence of holomorphic polynomial mappings A and B, each homogeneous of degree m, such that $f(z, \overline{z}) = ||A(z)||^2 - ||B(z)||^2$. An arbitrary real-valued polynomial p can be written as the difference of squared norms of holomorphic polynomials, but the holomorphic polynomials will not be homogeneous of the same degree when p fails to be bihomogeneous. See [D] for uses of such a decomposition.

We say that the partial differential operator D on $\mathbf{R}^{2n}$ is *complex bihomogeneous* if its principal symbol is a bihomogeneous polynomial. In this case we may apply Theorem 1 to obtain the following conclusion. We write Δ for the Laplace operator defined by $\sum \frac{\partial}{\partial z_j}\frac{\partial}{\partial \overline{z}_j}$.

Theorem 2. Let D be a complex bihomogeneous linear partial differential operator. Suppose that p is the absolute value of the principal symbol of D. Let q_d be the absolute value of the principal symbol of $\Delta^d D$. The following are equivalent:

1) D is elliptic (that is, $p(z, \overline{z}) > 0$ for $z \neq 0$).

2) There is an integer d and a positive-definite matrix $(E_{\mu\nu})$ so that q_d satisfies

$$q_d(\frac{\partial}{\partial z}, \frac{\partial}{\partial \overline{z}}) = \sum_{|\mu|=m} \sum_{|\nu|=m} E_{\mu\nu}(\frac{\partial}{\partial z})^\mu(\frac{\partial}{\partial \overline{z}})^\nu.$$

3) There is an integer d so that q_d is a squared norm of a holomorphic differential operator:

$$q_d(\frac{\partial}{\partial z}, \frac{\partial}{\partial \overline{z}}) = ||\sum A_\mu(\frac{\partial}{\partial z})^\mu||^2 = \sum_i |\sum A_{\mu i}(\frac{\partial}{\partial z})^\mu|^2. \tag{23}$$

We assume also that the indicated homogeneous polynomials span V_{m+d}.

4. There is an integer d' so that $q_{d'}$ satisfies (23), and such that $\mathbf{V}(A) = \{0\}$.

Proof. The principal symbol of $\Delta^d D$, evaluated at $(z, \overline{z})$, is $q_d(z, \overline{z}) = ||z||^{2d} p(z, \overline{z})$. The operator D is elliptic precisely when $|p|$ is strictly positive away from the origin. Therefore, by Theorem 1, D is elliptic if and only there is d so that $||z||^{2d} p(z, \overline{z})$ satisfies any of the equivalent conditions of Proposition 1. Equation (10), applied when $r = 1$, is equivalent to the positive-definiteness of $(E_{\mu\nu})$. The strict holomorphic factorability there is equivalent to statement 3 here. Statement 3 obviously implies statement 4, which in turn implies that $p(z, \overline{z})$ is positive away from the origin. Thus the four statements are equivalent. ♠

Remark. This result extends to systems of PDE in a straightforward fashion.

V. REINTERPRETATION OF THEOREM 1

Next we reinterpret Theorem 1 in terms of pullbacks of the universal bundle over Grassman manifolds. See [W] for more details about the universal bundle. Let $\mathbf{G}_{p,N}$ denote the Grassman manifold of p planes in complex N-space. When $p = 1$ we have complex projective space, and we write as usual $\mathbf{P}^{N-1}$ for $\mathbf{G}_{1,N}$. Let $\mathbf{U}_{p,N}$ denote the universal bundle over $\mathbf{G}_{p,N}$. This bundle is sometimes known as the tautological bundle or the stupid bundle; a point in $\mathbf{U}_{p,N}$ is a pair (S, ζ) where S is a p-dimensional subspace of $\mathbf{C}^N$ and $\zeta \in S$.

We let g_0 denote the Euclidean metric on $\mathbf{U}_{p,N}$. In terms of a local frame e of $\mathbf{U}_{p,N}$, we define $g_0(e) = e^* e$. Observe that if T determines a change of frames by acting on the right, then

$$g_0(eT) = (eT)^*(eT) = T^* e^* e T = T^* g_0(e) T.$$

This is the correct transformation law, and hence g_0 defines a Hermitian metric on $\mathbf{U}_{p,N}$. We may consider the matrix representation of g_0, with respect to a local frame. We have $(g_0)_{ij} = \langle e_j, e_i \rangle$ where $\langle , \rangle$ denotes the usual Hermitian inner product on complex Euclidean space $\mathbf{C}^N$, and the vectors e_i for $i = 1, ..., r$ are linearly independent. Note the interchange of indices. We see immediately that g_0 is of the form $A^* A$.

Let L denote the universal line bundle $L = \mathbf{U} = \mathbf{U}_{1,n}$ over complex projective space $\mathbf{P}^{n-1}$. We consider also its m-th tensor power $\mathbf{U}^m$. Let E denote the vector bundle over $\mathbf{P}^{n-1}$ equal to the direct sum of r copies of $\mathbf{U}^m$. On L we use the Euclidean metric, written $||z||^2$, and on E we use the metric determined by a matrix of bihomogeneous polynomials $F(z, \overline{z})$ that is positive-definite for $z \neq 0$. Theorem 1 now admits the following restatement.

Theorem 3. Suppose that L and E are the bundles over $\mathbf{P}^{n-1}$ as described in the previous paragraph, equipped with the given metrics. Then there are integers N and d, so that the bundle $L^d \otimes E$ over $\mathbf{P}^{n-1}$, with metric determined by $||z||^{2d}F(z,\overline{z})$, is the isometric pullback of the vector bundle $\mathbf{U}_{r,N}$, with the Euclidean metric, over the Grassmanian $\mathbf{G}_{r,N}$ via a holomorphic embedding.

The link to bihomogeneous polynomials arises because one can identify homogeneous polynomials of degree m on $\mathbf{C}^n$ with sections of the m-th power H^m of the hyperplane bundle H over $\mathbf{P}^{n-1}$. The bundle H is dual to $\mathbf{U}$. A matrix of bihomogeneous polynomials determines a metric on the direct sum of r copies of $\mathbf{U}$.

The general result in [CD3] considers certain base manifolds M, a line bundle L and a vector bundle E over M, and metrics R and F on them satisfying certain conditions. One of the main ideas is that the metrics be *globalizable*; this generalizes properties of the Euclidean metric on the universal bundle. This property extends the definition of the metric to be a real-analytic function on the total space of the bundle cross itself, and that is Hermitian symmetric. See [CD3] for the precise definition. In some sense a globalizable metric means that the inner product of two vectors can be evaluated at bundle points with different base points. The conclusion of the theorem again guarantees the existence of integers N and d so that $L^d \otimes E$, with metric R^dF, is the isometric pullback of $\mathbf{U}_{r,N}$, with the Euclidean metric, over the Grassmanian $\mathbf{G}_{r,N}$. The proof again relies on the Bergman kernel and facts about compact operators, but it is technically more difficult than the special case considered here.

VI. REMARKS ON CLASSICAL FACTORIZATION

We briefly mention some of the results in [Dj] and [RR]. Djokovic [Dj] considers for example an r by r positive semi-definite matrix $F(\lambda,\mu)$ whose entries are complex-valued homogeneous polynomials of degree $2m$ in the pair of real variables (λ,μ). He proves that one can write $F(\lambda,\mu) = A(\lambda,\mu)^*A(\lambda,\mu)$ where the entries in A are homogeneous polynomials of degree m. Two nice things about this result are that it is not required to multiply F by powers of a scalar function, and F is allowed to be semi-definite. On the other hand, the theorem holds only when the entries depend upon two real variables, the analogue of one complex variable. For us, making $A(z)$ depend holomorphically on z in (4) requires that we work with bihomogneous polynomials. The only bihomogeneous polynomials in one complex variable are constants times $|z|^{2m}$. Hence we could factor this scalar out of the matrix completely, and the analogue of the result in [Dj] becomes trivial in our setting. The idea of multiplying by powers of a scalar factor does not appear in [Dj].

The work in [RR] concerns functions from either $\mathbf{R}$ or the unit circle S^1 that take values in non-negative operators on a Hilbert space. The authors study many aspects of the factorization question in detail, including holomorphic extension to the upper half plane or to the unit disc. One of many results there is that if $P(x)$ is a non-negative operator on a Hilbert space, that is a polynomial of degree $2m$ in the real variable x, then there is an operator $Q(x)$ such that $P(x) = Q(x)^*Q(x)$ and such that Q is a polynomial of degree m

in x. Other results in [RR] are related to an application of Theorem 1 here from [CD1]. Suppose that $f(z,\overline{z})$ is an arbitrary polynomial that is positive on the unit sphere. Then there is a holomorphic polynomial mapping g such that $f(z,\overline{z}) = ||g(z)||^2$ on the unit sphere. In [RR] however positivity questions are considered only for functions depending on one variable.

REFERENCES

[CD1]. David W. Catlin and John P. D'Angelo, *A stabilization theorem for Hermitian forms and applications to holomorphic mappings*, Math Research Letters **3** (1996), 149-166.

[CD2]. David W. Catlin and John P. D'Angelo, *Positivity conditions for bihomogeneous polynomials*, Math Research Letters **4** (1997), 1-13.

[CD3]. David W. Catlin and John P. D'Angelo, *An isometric embedding theorem for holomorphic bundles.* (preprint)

[D]. John P. D'Angelo, *Several complex variables and the geometry of real hypersurfaces*, CRC Press, Boca Raton (1993).

[Dj]. D. Z. Djokovic, *Hermitian matrices over polynomial rings*, J.Algebra **43** (1976), 359-374.

[F]. Gerald B. Folland, *Introduction to Partial Differential Equations*, Princeton University Press (1976).

[RR]. M. Rosenblum and J. Rovnyak, *The factorization problem for nonnegative operator valued functions*, Bulletin A.M.S. **77** (1971), 287-318.

[Ru]. Walter Rudin, *Functional Analysis*, McGraw-Hill, New York (1973).

[W]. Raymond O. Wells, *Differential Analysis on Complex Manifolds*, Prentice-Hall, Englewood Cliffs, New Jersey (1973).

3 BERGMAN-CARLESON MEASURES AND BLOCH FUNCTIONS ON STRONGLY PSEUDOCONVEX DOMAINS

Hitoshi Arai

Mathematical Institute,
Tohoku University, Japan
arai@math.tohoku.ac.jp

INTRODUCTION

In their paper [4], Choa, Kim and Park proved the following characterization of Bloch functions on the unit ball B_n in $\mathbf{C}^n$.

Theorem CKP ([4, Main Theorem]) *For a holomorphic function f on B_n, let*

$$d\nu_f(z) = (1-|z|^2)\left(|\nabla f(z)|^2 - |\mathcal{R}f(z)|^2\right) dV(z),$$

where $|\nabla f(z)|^2 = \sum_{j=1}^n |\partial f(z)/\partial z_j|^2$, $\mathcal{R}f(z) = \sum_{j=1}^n z_j \partial f(z)/\partial z_j$, and dV is the $2n$-dimensional Lebesgue measure on $\mathbf{C}^n$.

Then f is a Bloch function if and only if ν_f is a Bergman-Carleson measure on B_n.

The purpose of this paper is to give a simple proof to the theorem, and moreover generalize it to strongly pseudoconvex domains. We also characterize little Bloch functions on strongly pseudoconvex domains in terms of vanishing Carleson measures.

Let Ω be a strongly pseudoconvex domain in $\mathbf{C}^n$ with C^∞ boundary $\partial\Omega$. In order to describe our theorem, let us mention some notation. Denote by $F(z,\xi)$ the infinitesimal Kobayashi metric at $z \in \Omega$ in the direction $\xi \in \mathbf{C}^n$, that is,

$$F_K(z,\xi) = \inf\{\alpha > 0 : \exists u \in \Delta(\Omega) \text{ with } u(0) = z \text{ and } u'(0) = \xi/\alpha\},$$

where $\Delta(\Omega)$ denotes the set of all holomorphic mappings from the open unit disc Δ to Ω. For a smooth function f on Ω, its modulus of the covariant derivative $Q(f)(z)$ at $z \in \Omega$ is defined by

$$Q(f)(z) = \sup_{0 \neq \xi \in \mathbf{C}^n} \frac{1}{F(z,\xi)} \left(\left| \sum_{j=1}^{n} \frac{\partial f}{\partial z_j}(z)\xi_j \right| + \left| \sum_{j=1}^{n} \frac{\partial f}{\partial \overline{z_j}}(z)\overline{\xi_j} \right| \right).$$

We consider the following measure

$$d\mu_f^p(z) = Q(f)(z)^p dV(z), \tag{1}$$

for $0 < p < \infty$, where dV is the $2n$ dimensional Lebesgue measure on $\mathbf{C}^n$. When Ω is the open unit ball, this measure $d\mu_f^2$ is equivalent to $d\nu_f$ defined in Theorem CKP. Indeed, it is easy to see that there exists a positive constant C such that for all holomorphic functions f on B_n,

$$\begin{aligned} C^{-1}(1-|z|^2)\left(|\nabla f(z)|^2 - |\mathcal{R}f(z)|^2\right) &\leq Q(f)(z)^2 \\ &\leq C(1-|z|^2)\left(|\nabla f(z)|^2 - |\mathcal{R}f(z)|^2\right). \end{aligned}$$

Let us recall the definitions of Bergman-Carleson measure and Bloch functions. Let

$$L_a^2(\Omega) = \left\{ f \in H(\Omega) : ||f||_2 = \left(\int_\Omega |f(z)|^2 dV(z) \right)^{1/2} < \infty \right\},$$

where $H(\Omega)$ is the set of all holomorphic functions on Ω.

Following Luecking [9] we call a measure μ on Ω a Bergman-Carleson measure if

$$||\mu||_c = \sup \left\{ \left(\int_\Omega |h(w)|^2 d\mu(w) \right)^{1/2} : h \in L_a^2(\Omega),\ ||h||_2 = 1 \right\} < \infty.$$

In other words, a measure μ is a Bergman-Carleson measure if and only if the identity mapping I is bounded from $L_a^2(\Omega)$ to the Banach space $L_a^2(\Omega,\mu)$ of all holomorphic functions f on Ω satisfying

$$||f||_{2,\mu} = \left(\int_\Omega |f(z)|^2 d\mu(z) \right)^{1/2} < \infty.$$

Further, if I is compact from $L_a^2(\Omega)$ to $L_a^2(\Omega,\mu)$, we call μ a vanishing Bergman-Carleson measure.

Now, we recall the definition of Bloch function by Krantz and Ma [7]. In [7], they defined holomorphic Bloch functions on Ω. Following their definition, we call a function f on Ω is a pluriharmonic Bloch function if f is pluriharmonic on Ω and $Q(f)$ is bounded on Ω. In addition, f is called a pluriharmonic little Bloch function if f is pluriharmonic and

$$\lim_{z\in\Omega,\delta(z)\to 0} Q(f)(z) = 0,$$

where $\delta(z) = \inf\{|z - \zeta| : \zeta \in \partial\Omega\}$.

Our main theorem is the following

Theorem 1 (Main Theorem) *Suppose f is a pluriharmonic function on Ω.*

(A) *The following statements are mutually equivalent:*

(i) *f is a pluriharmonic Bloch function on Ω*

(ii) *μ_f^p is a Bergman-Carleson measure on Ω for every $0 < p < \infty$.*

(iii) *μ_f^p is a Bergman-Carleson measure on Ω for some $0 < p < \infty$.*

(B) *The following statements are mutually equivalent:*

(i) *f is a pluriharmonic little Bloch function on Ω.*

(ii) *μ_f^p is a vanishing Bergman-Carleson measure on Ω for every $0 < p < \infty$.*

(iii) *μ_f^p is a vanishing Bergman-Carleson measure on Ω for some $0 < p < \infty$.*

As a corollary of Theorem 1 (A) for the case $p = 2$ we have Theorem CKP. Furthermore, Theorem 1 (A) and (B) yield characterizations of (little) Bloch functions on the unit disc obtained by Xiao an Zhong ([12], [13]).

Our proof of Theorem 1 differs from one of Theorem CKP in [4] in detail and methods. Indeed, the proof in [4] is based on the transitivity of the group of automorphisms on the unit ball. On the other hand, our proof is based on the generalized subharmonic property of the modulus of covariant derivatives, which we will prove in this paper. From this property and a theorem by Zhu [14] stated later, it follows Theorem 1.

We note that in [3], other characterizations of Bloch functions were proved by using invariant diffusion process (cf. [2]).

1 GENERALIZED SUBHARMONIC PROPERTY OF $Q(f)^p$

Let $E(z, r)$ be the Kobayashi metric ball with centre $z \in \Omega$ and radius r, that is, $E(z, r) = \{w \in \Omega : \beta(z, w) < r\}$, where $\beta(z, w)$ is the distance between z and w with respect to the infinitesimal Kobayashi metric on Ω.

Let F be a nonnegative function on Ω. As in Zhu [15], we say that F has the generalized subharmonic property if there exist positive constants C and α such that for every $z \in \Omega$,

$$F(z) \leq C \frac{1}{V(E(z,\alpha))} \int_{E(z,\alpha)} F(w) dV(w).$$

To prove Theorem 1, we verify that $Q(f)^p$ has the so-called generalized subharmonic property.

In this section we prove the following

Theorem 2 *Let $0 < p < \infty$. Then there exists a positive constant C_p such that for every pluriharmonic function f on Ω and $z \in \Omega$,*

$$Q(f)(z)^p \le C_p \frac{1}{V(E(z,c_0))} \int_{E(z,c_0)} Q(f)(w)^p dV(w),$$

where c_0 is a positive constant independent of f, z and p.

This theorem asserts that if f is a pluriharmonic function on Ω, then $Q(f)^p$ has the generalized subharmonic property for every $0 < p < \infty$.

To prove Theorem 2, let us recall some facts on boundary behavior of the infinitesimal Kobayashi metric. Let $\delta_0 \in (0,1)$ be a positive constant such that for each $z \in \mathbf{C}^n$ with $\delta(z) < \delta_0$, there is a unique boundary point $b(z) \in \partial\Omega$ with $|z - b(z)| = \delta(z)$. Let $\Omega_0 = \{z \in \Omega : \delta(z) < \delta_0\}$. For $\zeta \in \partial\Omega$, we denote by N_ζ the complex linear space spanned by the inward unit normal ν_ζ at ζ, and by $N_\zeta^\perp$ its orthogonal complement. Then $\mathbf{C}^n = N_\zeta \oplus N_\zeta^\perp$. Let Π_ζ be the orthogonal projection of $\mathbf{C}^n$ to N_ζ and $\Pi_\zeta^\perp$ the orthogonal projection on $\mathbf{C}^n$ to $N_\zeta^\perp$. Since Ω is strongly pseudoconvex, it is now well understand the boundary behavior of infinitesimal Kobayashi metric (cf. Graham [6], Aladro [1], Ma [10]): For insetance, there exists a constant $C_0 > 0$ such that for each $\xi \in \mathbf{C}^n$ and each $z \in \Omega_0$ near to $\partial\Omega$, the following inequality holds ([1]; see also [10]).

$$C_0^{-1}\left(\frac{|\Pi_{b(z)}\xi|}{\delta(z)} + \frac{|\Pi_{b(z)}^\perp \xi|}{\sqrt{\delta(z)}}\right) \le F(z,\xi) \le C_0\left(\frac{|\Pi_{b(z)}\xi|}{\delta(z)} + \frac{|\Pi_{b(z)}^\perp \xi|}{\sqrt{\delta(z)}}\right). \quad (2)$$

In our paper we need the following special case of a result by Aladro [1]:

Lemma 3 *Let c be a constant with $0 < c < 1$. Then there exists positive constant C_1 such that for each $\zeta \in \partial\Omega$, each $\xi \in \mathbf{C}^n$ and each $z \in \Omega_0$ with $|z - \zeta| < c\sqrt{\delta(z)}$, the following inequalities hold*

$$C_1^{-1}\left(\frac{|\Pi_\zeta\xi|}{\delta(z)} + \frac{|\Pi_\zeta^\perp \xi|}{\sqrt{\delta(z)}}\right) \le F(z,\xi) \le C_1\left(\frac{|\Pi_\zeta\xi|}{\delta(z)} + \frac{|\Pi_\zeta^\perp \xi|}{\sqrt{\delta(z)}}\right). \quad (3)$$

Proof. For the reader's convenience, we give a simple proof of the lemma by using the estimate (2). We denote by $C_1, \cdots, C_6$ positive constants depending only on Ω, δ_0 and c. Let $C(\eta,\zeta) = \Pi_\eta - \Pi_\zeta$ and $D(\eta,\zeta) = \Pi_\eta^\perp - \Pi_\zeta^\perp$. Then $||C(\eta,\zeta)|| \le C_2|\eta - \zeta|$ and $||D(\eta,\zeta)|| \le C_3|z-\zeta|$, where $||\cdot||$ means the operator norm. Since $|b(z) - \zeta| < (c+1)\sqrt{\delta(z)}$ by the assumptions on z, we have that

$$|\Pi_{b(z)}\xi| \le |\Pi_\zeta\xi| + |C(b(z),\zeta)\xi| \le |\Pi_\zeta\xi| + C_2(c+1)\sqrt{\delta(z)}|\xi|$$

Similarly we have

$$|\Pi_{b(z)}^\perp\xi| \le |\Pi_\zeta^\perp\xi| + C_3(c+1)\sqrt{\delta(z)}|\xi|.$$

Therefore by (2) we get

$$F(z,\xi) \le C\left(\frac{|\Pi_{b(z)}\xi|}{\delta(z)} + \frac{|\Pi^{\perp}_{b(z)}\xi|}{\sqrt{\delta(z)}}\right) \le C_4\left(\frac{|\Pi_{\zeta}\xi|}{\delta(z)} + \frac{|\Pi^{\perp}_{\zeta}\xi|}{\sqrt{\delta(z)}}\right),$$

To prove the first inequality of (3), let $z' = \delta(z)\nu_\zeta + \zeta$. We may assume $z' \in \Omega_0$. We will prove

$$|z' - b(z)| \le c'\sqrt{\delta(z)}, \tag{4}$$

where c' is a constant depending only on c and Ω. After proving this inequality we have by the similar way as above that

$$F(z',\xi) \le C_5\left(\frac{|\Pi_{b(z)}\xi|}{\delta(z')} + \frac{|\Pi^{\perp}_{b(z)}\xi|}{\sqrt{\delta(z')}}\right).$$

Since $\delta(z') = \delta(z)$, the estimates (2) yield that

$$\begin{aligned} C^{-1}\left(\frac{|\Pi_{\zeta}\xi|}{\delta(z)} + \frac{|\Pi^{\perp}_{\zeta}\xi|}{\sqrt{\delta(z)}}\right) &\le F(z',\xi) \\ &\le C_5\left(\frac{|\Pi_{b(z)}\xi|}{\delta(z')} + \frac{|\Pi^{\perp}_{b(z)}\xi|}{\sqrt{\delta(z')}}\right) \\ &= C_5\left(\frac{|\Pi_{b(z)}\xi|}{\delta(z)} + \frac{|\Pi^{\perp}_{b(z)}\xi|}{\sqrt{\delta(z)}}\right) \\ &\le C_5 C F(z,\xi), \end{aligned}$$

Therefore we gain the lemma.

Now we prove (4). Let $N(b(z),\zeta) = \nu_{b(z)} - \nu_\zeta$. Then $||N(b(z),\zeta)|| \le C_6|b(z) - \zeta|$. Therefore

$$\begin{aligned} z' - b(z) &= \delta(z)\nu_{b(z)} - \delta(z)N(b(z),\zeta) + \zeta - b(z) \\ &= \big(\delta(z)\nu_{b(z)} + b(z)\big) - \zeta - \delta(z)N(b(z),\zeta) + 2(\zeta - b(z)) \\ &= z - \zeta - \delta(z)N(b(z),\zeta) + 2(\zeta - b(z)). \end{aligned}$$

Moreover we have that

$$|\zeta - b(z)| \le |\zeta - z| + |z - b(z)| < c\sqrt{\delta(z)} + \delta(z) < (1+c)\sqrt{\delta(z)}.$$

Thus

$$\begin{aligned} |z' - b(z)| &\le c\sqrt{\delta(z)} + \delta(z)C_6|\zeta - b(z)| + 2|\zeta - b(z)| \\ &\le c\sqrt{\delta(z)} + (1+c)\sqrt{\delta(z)}(2 + C_6\delta(z)) \le c'\sqrt{\delta(z)}. \end{aligned}$$

■

Let $Df(z) = (\partial f(z)/\partial z_j, \cdots, \partial f(z)/\partial z_n)$ and $\bar{D}f(z) = (\partial f(z)/\partial \bar{z}_j, \cdots, \partial f(z)/\partial \bar{z}_n)$. For simplicity of notation, we write, $N_\zeta f(w) = \Pi_\zeta(Df(w))$, $\bar{N}_\zeta f(w) = \Pi_\zeta(\bar{D}f(w))$, $T_\zeta f(w) = \Pi_\zeta^\perp(Df(w))$, and $\bar{T}_\zeta f(w) = \Pi_\zeta^\perp(\bar{D}f(w))$. In addition, for every $w \in \Omega$ and $\zeta \in \partial\Omega$, let

$$M_f(w,\zeta) = \delta(w)\left(|N_\zeta f(w)| + |\bar{N}_\zeta f(w)|\right) + \sqrt{\delta(w)}\left(|T_\zeta f(w)| + |\bar{T}_\zeta f(w)|\right).$$

From Lemma 3 we have the following

Lemma 4 *There exist positive constants C_7, r_0 and δ_1 with $\delta_1 < \delta_0$ such that for every $z \in \Omega_0$ with $\delta(z) < \delta_1$, every $w \in E(z, r_0)$ and every smooth function f on Ω, the following inequalities hold:*

$$C_7^{-1} M_f(w, b(z)) \leq Q(f)(w) \leq C_7 M_f(w, b(z)). \tag{5}$$

Proof. Let $\zeta = b(z)$, and

$$P_z(r_1, r_2) = \{w \in \mathbf{C}^n : |\Pi_{b(z)}(w-z)| < r_1, |\Pi_{b(z)}^\perp(w-z)| < r_2\}, \quad 0 < r_1,\ r_2.$$

Then there exist positive constants a, b, A and B depending only on Ω such that for every Kobayashi metric ball $E(z,r)$ $(r > 0)$,

$$P_z(ar\delta(z), br\sqrt{\delta(z)}) \subset E(z,r) \subset P_z(Ar\delta(z), Br\sqrt{\delta(z)}), \text{ (see [8, Lemma 6]).} \tag{6}$$

Therefore if $(1 + A/2)\delta(z) < \delta_0$ and $Br + (Ar+1)\delta(z)^{1/2} < c$, where c is a constant in Lemma 3, then

$$E(z,r) \subset P_z(Ar\delta(z), Br\sqrt{\delta(z)}) \subset \Omega_0 \cap \{w \in \mathbf{C}^n : |w-\zeta| < c\sqrt{\delta(z)}\}. \tag{7}$$

Hence we can choose r_0 and δ_1 so that

$$E(z, r_0) \subset \Omega_0 \cap \{w \in \mathbf{C}^n : |w - \zeta| < c\sqrt{\delta(z)}\}$$

holds true for every point $z \in \Omega$ with $\delta(z) < \delta_1$. In what follows we consider only such a Kobayashi metric ball $E(z, r_0)$.

If $\xi \in \mathbf{C}^n$ satisfies $|\xi| = 1$ and $\Pi_\zeta^\perp \xi = 0$, then by Lemma 3 we have

$$\begin{aligned} Q(f)(w) &\geq \frac{|Df(w)\cdot\xi|}{F(w,\xi)} + \frac{|\bar{D}f(w)\cdot\xi|}{F(w,\xi)} \\ &\geq C_1^{-1}\frac{|N_\zeta f(w)|}{\delta(w)^{-1}|\Pi_\zeta\xi|} + \frac{|\bar{N}_\zeta f(w)|}{\delta(w)^{-1}|\Pi_\zeta\xi|} \\ &= C_{-1}^{1}\delta(w)\{|N_\zeta f(w)| + |\bar{N}_\zeta f(w)|\}. \end{aligned}$$

On the other hand, if $\xi \in \mathbf{C}^n$ satisfies $|\xi| = 1$ and $\Pi_\zeta \xi = 0$, then by Lemma 3 we have

$$
\begin{aligned}
Q(f)(w) &\geq \sup_{0 \neq \theta \in \mathbf{C}^n} \left\{ \frac{|Df(w) \cdot \theta|}{F(w,\theta)} + \frac{|\bar{D}f(w) \cdot \theta|}{F(w,\theta)} \right\} \\
&\geq C_1^{-1} \left\{ \frac{|T_\zeta f(w)|}{\sqrt{\delta(w)}^{-1} |\Pi_\zeta^\perp \xi|} + \frac{|\bar{T}_\zeta f(w)|}{\sqrt{\delta(w)}^{-1} |\Pi_\zeta^\perp \xi|} \right\} \\
&= C_1^{-1} \left\{ \sqrt{\delta(w)} |T_\zeta f(w)| + \sqrt{\delta(w)} |\bar{T}_\zeta f(w)| \right\}.
\end{aligned}
$$

Therefore we have the first inequality in (5).

For proving the second inequality in (5), we note that for every $\xi \neq 0$,

$$
\begin{aligned}
\frac{|Df(w) \cdot \xi|}{F(w,\xi)} &\leq C_1 \frac{|Df(w) \cdot \Pi_\zeta \xi| + |Df(w) \cdot \Pi_\zeta^\perp \xi|}{\delta(w)^{-1} |\Pi_\zeta \xi| + \delta(w)^{-1/2} |\Pi_\zeta^\perp \xi|} \\
&= C_1 \frac{|N_\zeta f(w) \cdot \Pi_\zeta \xi| + |T_\zeta f(w) \cdot \Pi_\zeta^\perp \xi|}{\delta(w)^{-1} |\Pi_\zeta \xi| + \delta(w)^{-1/2} |\Pi_\zeta^\perp \xi|} \\
&\leq C_1 \left(\delta(w) |N_\zeta f(w)| + \sqrt{\delta(w)} |T_\zeta f(w)| \right).
\end{aligned}
$$

By the same way, we have

$$
\frac{|\bar{D}f(w) \cdot \xi|}{F(w,\xi)} \leq C_1 \left(\delta(w) |\bar{N}_\zeta f(w)| + \sqrt{\delta(w)} |\bar{T}_\zeta f(w)| \right).
$$

Thus we have the desired inequality. ■

Now we are ready to prove Theorem 2.

Proof of Theorem 2. Let $z \in \Omega_0$, and let $\zeta = b(z)$. Consider a coordinate system satisfying that

$$
\zeta = 0, \quad N_\zeta = \{(w_1, 0, \cdots, 0) : w_1 \in \mathbf{C}\}, \quad C_\zeta = \{(0, w') : w' \in \mathbf{C}^{n-1}\}.
$$

Here we have $\nu_\zeta = (1, 0, \cdots, 0)$. Note that in this coordinate we have

$$
N_\zeta f(w) = \frac{\partial f}{\partial z_1}(w), \quad \bar{N}_\zeta f(w) = \frac{\partial f}{\partial \bar{z}_1}(w),
$$

$$
T_\zeta f(w) = \left(\frac{\partial f}{\partial z_j}(w) \right)_{j=2,\cdots,n}, \text{and} \quad \bar{T}_\zeta f(w) = \left(\frac{\partial f}{\partial \bar{z}_j}(w) \right)_{j=2,\cdots,n}.
$$

Suppose $c_0 = r_0$ and $z \in \Omega_0$ with $\delta(z) < \delta_1$. Let $P_z = P_z(ar_0\delta(z), br_0\delta(z))$, where a and b are constants in the inequality (6). Then by (6) we see easily that

$$
C_8^{-1} V(E(z, r_0)) \leq V(P_z) \leq C_8 V(E(z, r_0)),
$$

where C_8 is a positive constant depending only on Ω. Therefore in order to prove Theorem 2 it is sufficient to show that

$$Q(f)(z)^p \le C_9 \frac{1}{V(P_z)} \int_{P_z} Q(f)(w)^p dV(w),$$

where $C_9 > 0$ is a constant independent of f and z. We will prove this inequality.

Note that $\partial f/\partial z_j$ and $\partial f/\partial \bar{z}_j$ ($j = 1, \cdots, n$) are pluriharmonic. Hence by applying [5] to these functions in each variable, we get that there exists a positive constant C_{10} depending only on p satisfying

$$\left|\frac{\partial f}{\partial z_j}(z)\right|^p \le \frac{C_{10}}{V(P_z)} \int_{P_z} \left|\frac{\partial f}{\partial z_j}(w)\right|^p dV(w),$$

and

$$\left|\frac{\partial f}{\partial \bar{z}_j}(z)\right|^p \le \frac{C_{10}}{V(P_z)} \int_{P_z} \left|\frac{\partial f}{\partial \bar{z}_j}(w)\right|^p dV(w).$$

Therefore we obtain by Lemma 4 that

$$\begin{aligned} &Q(f)(z)^p \le C_7 M(z, b(z))^p \\ &\le C_{11} \left\{ \delta(z) \left(\left|\frac{\partial f}{\partial z_1}(z)\right|^p + \left|\frac{\partial f}{\partial \bar{z}_1}(z)\right|^p \right) + \sqrt{\delta(z)} \sum_{j=1}^{n} \left(\left|\frac{\partial f}{\partial z_j}(z)\right|^p + \left|\frac{\partial f}{\partial \bar{z}_j}(z)\right|^p \right) \right\} \\ &\le \frac{C_{12}}{V(P_z)} \int_{P_z} \left\{ \delta(z) \left(\left|\frac{\partial f}{\partial z_1}(w)\right|^p + \left|\frac{\partial f}{\partial \bar{z}_1}(w)\right|^p \right) \right. \\ &\qquad \left. + \sqrt{\delta(z)} \sum_{j=1}^{n} \left(\left|\frac{\partial f}{\partial z_j}(w)\right|^p + \left|\frac{\partial f}{\partial \bar{z}_j}(w)\right|^p \right) \right\} dV(w) =: \text{(I), say,} \end{aligned}$$

where C_{11} and C_{12} are positive constants depending only on Ω, p and C_7. Since for each $w \in P_z$, $\delta(w)$ is compatible to $\delta(z)$, there exists a positive constant C_{13} depending only on p and Ω such that

$$\text{(I)} \le \frac{C_{13}}{V(P_z)} \int_{P_z} M_f(w, b(z))^p dV(w) \le \frac{C_{13} C_7}{V(P_z)} \int_{P_z} Q(f)(w)^p dV(w).$$

(End of the proof of Theorem 2)

2 A THEOREM BY ZHU AND THE PROOF OF THEOREM 1.

In [15, p.124-125], Zhu proved the following theorem:

Theorem Z *Let D be the open unit disc in $\mathbf{C}$, and f a nonnegative function on D. If f has the generalized subharmonic property, then $f(z)dV(z)$ is a Bergman-Carleson measure on D if and only is f is bounded. Moreover, $f(z)dV(z)$ is a vanishing Bergman-Carleson measure on D if and only if $f(z) \to 0$ as $|z| \to 1$.*

This theorem was proved also implicitly in Zhu [14] when D is a bounded symmetric domain in $\mathbf{C}^n$. Furthermore, all inequalities he used in that paper hold true for the case of strongly pseudoconvex domains. Indeed they were proved in [8]. Therefore by combining this fact with Theorem 2 we gain Theorem 1.

3 A NOTE ON TOEPLITZ OPERATORS

By virtue of the systematic study of Toeplitz operator by Zhu (cf. [14] and [15]), Theorem 1 implies the following result on Toeplitz operators with symbol $Q(f)^p$: Let

$$T_{Q(f)^p}g(z) = \int_\Omega K(z,w)Q(f)^p(w)g(w)dV(w),$$

where g is a function on Ω, and $K(z,w)$ is the Bergman kernel on Ω.

From Zhu's theorem ([15, p.125], see also [14]) with the remark in the previous section and Theorem 1 it follows the following

Theorem 5 *Let f be a pluriharmonic function on Ω.*

(A) *The following statements are mutually equivalent:*

(i) *f is a pluriharmonic Bloch function on Ω.*

(ii) *For every $0 < p < \infty$, the Toeplitz operator with symbol $Q(f)(z)^p$ is bounded on $L^2_a(\Omega)$.*

(iii) *For some $0 < p < \infty$, the Toeplitz operator with symbol $Q(f)(z)^p$ is bounded on $L^2_a(\Omega)$.*

(B) *The following statements are mutually equivalent:*

(i) *f is a pluriharmonic little Bloch function on Ω.*

(ii) *For every $0 < p < \infty$, the Toeplitz operator with symbol $Q(f)(z)^p$ is compact on $L^2_a(\Omega)$.*

(iii) *For some $0 < p < \infty$, the Toeplitz operator with symbol $Q(f)(z)^p$ is compact on $L^2_a(\Omega)$.*

References

[1] G. Aladro, The comparability of the Kobayashi approach region and the admissible approach region, Illinois J. of Math. 33 (1989), 42–63.

[2] H. Arai, Degenerate elliptic operators, Hardy spaces and diffusions on strongly pseudoconvex domains, Tohoku Math. J. 46 (1994), 469–498.

[3] H. Arai, Some characterizations of Bloch functions on strongly pseudoconvex domains, Tokyo J. Math., 17 (1994), 373–383.

[4] J. S. Choa, H. O. Kim and Y. Y. Park, A Bergman-Carleson measure characterization of Bloch functions in the unit ball of $\mathbf{C}^n$, Bull. Korean Math. Soc. 29 (1992), 285-293.

[5] C. Fefferman and E. M. Stein, H^p spaces of several variables, Acta Math. 129 (1972), 137–193.

[6] I. Graham, Boundary behavior of the Carathéodory and Kobayashi metrics on strongly pseudoconvex domains in $\mathbf{C}^n$ with smooth boundary, Trans. Amer. Math. Soc. 207 (1975), 219–240.

[7] S. G. Krantz and D. Ma, Bloch functions on strongly pseudoconvex domains, Indiana Univ. Math. J. 37 (1988), 145-163.

[8] H. Li, BMO, VMO and Hankel operators on the Bergman spaces of strongly pseudoconvex domains, J. Funct. Anal. 106 (1992), 375-408.

[9] D.Luecking, A technique for characterizing Carleson measures on Bergman spaces, Proc. Amer. Math. Soc. (1983), 656-660.

[10] D. Ma, On iterates of holomorphic maps, Math. Z. 207 (1991), 417–428.

[11] E. M. Stein, Boundary Behavior of Holomorphic Functions of Several Variables, Math. Notes, Princeton Univ. Press, 1972.

[12] J. Xiao, Carleson measure, atomic decomposition and free interpolation from Bloch space, Ann. Acad. Sci. Ser. A. I. Math. 19 (1994), 175–184.

[13] J. Xiao and L. Zhong, On little Bloch space, its Carleson measure, atomic decomposition and free interpolation, Complex Variables 27 (1995), 175–184.

[14] K. Zhu, Positive Toeplitz operators on weighted Bergman spaces of bounded symmetric domains, J. Operator Theory 20 (1988), 329–357.

[15] K. Zhu, Operator Theory in Function Spaces, Marcel Dekker, New York, 1990.

4 THE ROLE OF THE AHLFORS MAPPING IN THE THEORY OF KERNEL FUNCTIONS IN THE PLANE

Steven R. Bell

Department of Mathematics
Purdue University, USA
bell@math.purdue.edu

Abstract: We describe recent results that establish a close relationship between the Ahlfors mapping function associated to an n-connected domain in the plane and the Bergman and Szegő kernels of the domain. The results show that the Ahlfors mapping plays a role in the multiply connected setting very similar to that of the Riemann mapping in the simply connected case. We also describe how the Ahlfors map is connected to the Poisson and other kernels.

1. INTRODUCTION

The Bergman and Szegő kernel functions associated to a simply connected domain Ω in the complex plane are easily expressed in terms of a single Riemann mapping function associated to the domain. Indeed, if a is a point in Ω and $f_a(z)$ is the Riemann mapping function mapping Ω one-to-one onto the unit disc $D_1(0)$ with $f_a(a) = 0$ and $f_a'(a) > 0$, then the Szegő kernel $S(z, w)$ is given by

$$S(z,w) = \frac{c\,S(z,a)\overline{S(w,a)}}{1 - f_a(z)\overline{f_a(w)}},$$

and the Bergman kernel $K(z, w)$ is given by

$$K(z,w) = \frac{4\pi c^2 S(z,a)^2\overline{S(w,a)^2}}{(1 - f_a(z)\overline{f_a(w)})^2},$$

where $c = 1/S(a, a)$ in both the formulas. The Szegő kernel could be eliminated from the right hand side of these formulas by noting that $f_a'(z) = 2\pi c S(z, a)^2$. However, we shall see that the formulas above have natural generalizations to multiply connected domains.

These identities reveal that, not only can Riemann maps be computed by means of kernel functions, but kernel functions can be computed by means of Riemann maps. In this paper, I shall describe analogous results for n-connected domains that show that the Bergman and Szegő kernels are simple rational combinations of an Ahlfors map and n other basic functions of one variable related to the zeroes of the Ahlfors map.

There are many ways in which the Ahlfors map can be thought of as the "Riemann mapping function for multiply connected domains." Indeed, the Ahlfors map associated to a point a in a multiply connected domain Ω is the unique holomorphic function f mapping Ω into the unit disc that makes $f'(a) > 0$ and as large as possible. Furthermore, the Ahlfors map also has mapping properties analogous to the Riemann map; it maps Ω *onto* the unit disc and maps each boundary curve of Ω one-to-one onto the unit circle (see §2 for more details). We shall see that, from the point of view of kernel functions, the Ahlfors map also takes on the role of a "Riemann map in multiply connected domains."

In the last section of this paper, I describe the relationship between the Ahlfors map and the Poisson kernel, the Garabedian kernel, and the complimentary kernel to the Bergman kernel.

2. THE AHLFORS MAP AND ZEROES OF THE SZEGO KERNEL

We shall study the kernel functions on a finitely connected domain in the plane such that no boundary component reduces to a point. Such a domain can be mapped biholomorphically to a bounded domain Ω with C^∞ smooth boundary, i.e., a bounded domain whose boundary $b\Omega$ is given by finitely many non-intersecting C^∞ simple closed curves.

In order to continue, we must list some basic facts about the kernel functions. Suppose that Ω is a bounded n-connected domain in the plane with C^∞ smooth boundary. Let γ_j, $j = 1, \ldots, n$, denote the n non-intersecting C^∞ simple closed curves which define the boundary of Ω, and suppose that γ_j is parameterized in the standard sense by $z_j(t)$, $0 \le t \le 1$. Let $T(z)$ be the C^∞ function defined on $b\Omega$ such that $T(z)$ is the complex number representing the unit tangent vector at $z \in b\Omega$ pointing in the direction of the standard orientation. This complex unit tangent vector function is characterized by the equation $T(z_j(t)) = z_j'(t)/|z_j'(t)|$.

We shall let $A^\infty(\Omega)$ denote the space of holomorphic functions on Ω that are in $C^\infty(\overline{\Omega})$. The space of complex valued functions on $b\Omega$ that are square integrable with respect to arc length measure ds will be denoted by $L^2(b\Omega)$. We shall let $H^2(b\Omega)$ denote the space of functions in $L^2(b\Omega)$ that represent the L^2 boundary values of holomorphic functions on Ω (as described in [1]) and we shall call $H^2(b\Omega)$ the *Hardy* space. The inner product associated to $L^2(b\Omega)$ shall be written

$$\langle u, v \rangle_{b\Omega} = \int_{b\Omega} u\, \bar{v}\, ds.$$

For each fixed point $a \in \Omega$, the Szegő kernel $S(z, a)$, as a function of z, extends to the boundary to be a function in $A^\infty(\Omega)$. Furthermore, $S(z, a)$ has exactly $(n-1)$ zeroes as a function of z in Ω (counting multiplicities) and does not vanish at any points z in the boundary of Ω. Furthermore, $S(z,w)$ is in $C^\infty((\overline{\Omega}\times\overline{\Omega}) - \{(z,z) : z \in b\Omega\})$ as a function of (z,w).

The *Garabedian kernel* $L(z,a)$ is a kernel related to the Szegő kernel via the identity

$$\frac{1}{i}L(z,a)T(z) = S(a,z) \qquad \text{for } z \in b\Omega \text{ and } a \in \Omega. \tag{2.1}$$

For fixed $a \in \Omega$, the kernel $L(z,a)$ is a holomorphic function of z on $\Omega - \{a\}$ with a simple pole at a with residue $1/(2\pi)$. Furthermore, as a function of z, $L(z,a)$ extends to the boundary and is in the space $C^\infty(\overline{\Omega} - \{a\})$. In fact, $L(z,a)$ extends to be in $C^\infty((\overline{\Omega}\times\overline{\Omega}) - \{(z,z) : z \in \overline{\Omega}\})$. Also, $L(z,a)$ is non-zero for all (z,a) in $\overline{\Omega}\times\Omega$ with $z \neq a$.

The kernel $S(z,w)$ is holomorphic in z and antiholomorphic in w on $\Omega\times\Omega$, and $L(z,w)$ is holomorphic in both variables for $z,w \in \Omega$, $z \neq w$. We note here that $S(z,z)$ is real and positive for each $z \in \Omega$, and that $S(z,w) = \overline{S(w,z)}$ and $L(z,w) = -L(w,z)$. Also, the Szegő kernel reproduces holomorphic functions in the sense that

$$h(a) = \langle h, S(\cdot,a)\rangle_{b\Omega}$$

for all $h \in H^2(b\Omega)$ and $a \in \Omega$.

Given a point $a \in \Omega$, the Ahlfors map f_a associated to the pair (Ω,a) is a proper holomorphic mapping of Ω onto the unit disc. It is an n-to-one mapping (counting multiplicities), it extends to be in $A^\infty(\Omega)$, and it maps each boundary curve γ_j one-to-one onto the unit circle. Furthermore, $f_a(a) = 0$, and f_a is the unique function mapping Ω into the unit disc maximizing the quantity $|f_a'(a)|$ with $f_a'(a) > 0$. The Ahlfors map is related to the Szegő kernel and Garabedian kernel via

$$f_a(z) = \frac{S(z,a)}{L(z,a)}. \tag{2.2}$$

Note that $f_a'(a) = 2\pi S(a,a) \neq 0$. Because f_a is n-to-one, f_a has n zeroes. The simple pole of $L(z,a)$ at a accounts for the simple zero of f_a at a. The other $n-1$ zeroes of f_a are given by $(n-1)$ zeroes of $S(z,a)$ in Ω. Let $a_1, a_2, \ldots, a_{n-1}$ denote these $n-1$ zeroes (counted with multiplicity). I proved in [2] (see also [1, page 105]) that, if a is close to one of the boundary curves, the zeroes $a_1, \ldots, a_{n-1}$ become distinct simple zeroes. It follows from this result that, for all but at most finitely many points $a \in \Omega$, $S(z,a)$ has $n-1$ distinct simple zeroes in Ω as a function of z.

3. NEARLY ORTHOGONAL POWER SERIES ON MULTIPLY CONNECTED DOMAINS

The Ahlfors function gives rise to a particularly nice basis for the Hardy space of an n-connected domain with C^∞ smooth boundary. We shall use the notation that we set up in the preceding section. We assume that $a \in \Omega$ is a fixed point in Ω that has been chosen so that the $n-1$ zeroes, $a_1, \ldots, a_{n-1}$, of $S(z,a)$ are distinct and simple. We shall let a_0 denote a and we shall use the shorthand notation $f(z)$ for the Ahlfors map $f_a(z)$.

It was proved in [2] that the set of functions

$$h_{ik}(z) = S(z, a_i) f(z)^k,$$

where $0 \le i \le n-1$ and $k \ge 0$, forms a basis for the Hardy space $H^2(b\Omega)$ and that

$$\langle h_{ik}, h_{jm} \rangle_{b\Omega} = \begin{cases} 0, & \text{if } k \neq m, \\ S(a_j, a_i), & \text{if } k = m. \end{cases} \tag{3.1}$$

I shall prove this result here in order to demonstrate an interesting connection between the Ahlfors map and "power series" on multiply connected domains. First, I will show that the set of functions above spans a dense subset of $H^2(b\Omega)$. Indeed, suppose that $g \in H^2(b\Omega)$ is orthogonal to the span. Notice that the reproducing property of the Szegő kernel yields that

$$\langle g, S(\cdot, a_j) \rangle_{b\Omega} = g(a_j),$$

and therefore g vanishes at $a_0, a_1, \ldots, a_{n-1}$. Suppose we have shown that g vanishes to order m at each a_j, $j = 0, 1, \ldots, n-1$. It follows that g/f^m has removable singularities at each a_j and so it can be viewed as an element of $H^2(b\Omega)$. The value of g/f^m at a_j is $\frac{1}{m!} g^{(m)}(a_j)/f'(a_j)^m$. Since $|f(z)| = 1$ when $z \in b\Omega$, it follows that $1/f(z) = \overline{f(z)}$ when $z \in b\Omega$, and we may write

$$\langle g, S(\cdot, a_j) f^m \rangle_{b\Omega} = \langle g/f^m, S(\cdot, a_j) \rangle_{b\Omega} = \frac{1}{m!} g^{(m)}(a_j)/f'(a_j)^m.$$

(The last equality follows from the reproducing property of the Szegő kernel.) We conclude that g vanishes to order $m+1$ at each a_j. By induction, g vanishes to infinite order at each a_j and hence, $g \equiv 0$. This proves the density.

To prove (3.1), let us suppose first that $k > m$. The fact that $\overline{f} = 1/f$ on $b\Omega$ and the reproducing property of the Szegő kernel now yield that

$$\langle h_{ik}, h_{jm} \rangle_{b\Omega} = \int_{z \in b\Omega} S(z, a_i) f(z)^{k-m} \, \overline{S(z, a_j)} \, ds =$$

$$\int_{z \in b\Omega} S(a_j, z) \, \left[S(z, a_i) f(z)^{k-m} \right] \, ds = S(a_j, a_i) f(a_j)^{k-m}.$$

The identity now follows because $f(a_j) = 0$ for all j. If $k = m$, then

$$\langle h_{ik}, h_{jm} \rangle_{b\Omega} = \int_{z \in b\Omega} S(a_j, z) \, S(z, a_i) \, ds = S(a_j, a_i),$$

and identity (3.1) is proved. It is now easy to see that the functions h_{ik} are linearly independent. Indeed, identity (3.1) reveals that we need only check that, for fixed k, the n functions h_{ik}, $i = 0, 1, \ldots, n-1$, are linearly independent, and this is true because a relation of the form

$$\sum_{i=0}^{n-1} C_i S(z, a_i) \equiv 0$$

implies, via the reproducing property of the Szegő kernel, that every function g in the Hardy space satisfies

$$\sum_{i=0}^{n-1} \overline{C_i}\, g(a_i) = 0,$$

and it is easy to construct polynomials g that violate such a condition.

To obtain a formula for the Szegő kernel on Ω, we next orthonormalize the sequence $\{h_{ik}\}$ via the Gram-Schmidt procedure. Identity (3.1) shows that the functions in the sequence are orthogonal for different values of k, and so our task is merely to orthonormalize the n functions h_{ik}, $i = 0, 1, \ldots, n-1$ for each k. We obtain an orthonormal set $\{H_{ik}\}$ given by

$$H_{0k}(z) = b_{00} S(z, a) f(z)^k \qquad \text{and,}$$

$$H_{ik}(z) = \sum_{j=1}^{i} b_{ij} S(z, a_j) f(z)^k, \qquad i = 1, \ldots, n-1,$$

where $b_{ii} \neq 0$ for each $i = 0, 1, \ldots, n-1$. Because $|f| = 1$ on $b\Omega$, it follows that *the coefficients* b_{ij} *do not depend on* k. Notice that H_{ik} does not contain a term involving $S(z, a)$ if $i > 0$ because of (3.1) and the fact that $S(a_i, a) = 0$.

The Szegő kernel can be written in terms of our orthonormal basis as

$$S(z, w) = \sum_{i=0}^{n-1} \sum_{k=0}^{\infty} H_{ik}(z)\, \overline{H_{ik}(w)}.$$

The geometric sum

$$\sum_{k=0}^{\infty} f(z)^k\, \overline{f(w)^k} = \frac{1}{1 - f(z)\overline{f(w)}}$$

can be factored from the expression for $S(z, w)$ to yield the formula,

$$S(z, w) = \frac{1}{1 - f(z)\overline{f(w)}} \left(c_0 S(z, a) \overline{S(w, a)} + \sum_{i,j=1}^{n-1} c_{ij} S(z, a_i)\, \overline{S(w, a_j)} \right). \quad (3.2)$$

We shall now determine the coefficients in this formula. At the moment, we only know that these coefficients exist and that they are given as combinations of the Gram-Schmidt coefficients found above. That $c_0 = 1/S(a,a)$ can be seen by setting $z = a$ and $w = a$ in (3.2). To determine the coefficients c_{ij}, suppose $1 \le k \le n-1$ and set $w = a_k$ in (3.2). Note that $f(a_k) = 0$ and that $S(a, a_k) = 0$. Hence,

$$S(z, a_k) = \sum_{i=1}^{n-1} \left(\sum_{j=1}^{n-1} c_{ij} S(a_j, a_k) \right) S(z, a_i).$$

Such a relation can only be true if

$$\sum_{j=1}^{n-1} c_{ij} S(a_j, a_k) = \begin{cases} 1, & \text{if } i = k, \\ 0, & \text{if } i \neq k. \end{cases}$$

This shows that the $(n-1) \times (n-1)$ matrix $[S(a_j, a_k)]$ is invertible and that $[c_{ij}]$ is its inverse. Let us summarize these results in the following theorem.

Theorem 3.1. *The Szegő kernel of an n-connected domain is related to the Ahlfors map $f_a(z)$ associated to a point a in the domain via the formula*

$$S(z,w) = \frac{1}{1 - f_a(z)\overline{f_a(w)}} \left(c_0 S(z,a)\overline{S(w,a)} + \sum_{i,j=1}^{n-1} c_{ij} S(z, a_i)\, \overline{S(w, a_j)} \right)$$

where $c_0 = 1/S(a,a)$ and the coefficients c_{ij} are given as the coefficients of the inverse matrix to the matrix $[S(a_j, a_k)]$.

Theorem 3.1 generalizes in a routine manner to any finitely connected domain Ω_1 such that no boundary component is a point. Such a domain can be mapped to a finitely connected domain with smooth boundary Ω_2 via a biholomorphic mapping Φ. The function Φ' has a single valued holomorphic square root on Ω_1 (see [1, page 43]) and if we define the Szegő kernel on Ω_1 via the natural transformation formula

$$S_1(z,w) = \sqrt{\Phi'(z)} S_2(\Phi(z), \Phi(w)) \overline{\sqrt{\Phi'(w)}}, \tag{3.3}$$

then it is easy to see that the terms in (3.2) transform in exactly the correct manner in which to make (3.2) valid on Ω_1.

We mention that the nearly orthogonal basis $h_{ik}(z) = S(z, a_i) f(z)^k$ defined above can be used to expand a holomorphic function on Ω in a special power series expansion. Indeed, given a homorphic function $G(z)$ in $H^2(b\Omega)$, we may write

$$G(z) = \sum_{i=0}^{n-1} \sum_{k=0}^{\infty} b_{ik} S(z, a_i) f(z)^k = \sum_{i=0}^{n-1} S(z, a_i) \sum_{k=0}^{\infty} b_{ik} f(z)^k.$$

The coefficients b_{ij} may be computed by means of the inner product on $b\Omega$, or they may also be computed by inductively equating coefficients of Taylor expansions at each of the points a_i, $i = 0, \ldots, n-1$. The expansion for G can also be written in the form

$$G(z) = \sum_{i=0}^{n-1} S(z, a_i) H_i(f(z))$$

where the H_i are holomorphic on the unit disc. The n linear operators given by the mappings that take G to H_i have yet to be studied.

4. COMPLEXITY OF THE KERNEL FUNCTIONS

Formula (3.2) reveals that the Szegő kernel associated to an n-connected domain is composed of the $n+1$ functions, $S(z,a)$, $S(z,a_1)$, $S(z,a_2), \ldots, S(z,a_{n-1})$, and $f_a(z)$. (Note that because $f_a(z) = S(z,a)/L(z,a)$, we may replace $f_a(z)$ in this list of functions by $L(z,a)$ if we like.) We shall now see that the Bergman kernel of an n-connected domain in the plane is composed of the same basic functions that comprise the Szegő kernel.

The Bergman kernel $K(z,w)$ is related to the Szegő kernel via the identity

$$K(z,w) = 4\pi S(z,w)^2 + \sum_{i,j=1}^{n-1} A_{ij} F_i'(z) \overline{F_j'(w)},$$

where the functions $F_i'(z)$ are classical functions of potential theory described as follows. The harmonic function ω_j which solves the Dirichlet problem on Ω with boundary data equal to one on the boundary curve γ_j and zero on γ_k if $k \neq j$ has a multivalued harmonic conjugate. The function $F_j'(z)$ is a globally defined single valued holomorphic function on Ω which is locally defined as the derivative of $\omega_j + iv$ where v is a local harmonic conjugate for ω_j. The Cauchy-Riemann equations reveal that $F_j'(z) = 2(\partial\omega_j/\partial z)$.

Let $\mathcal{F}'$ denote the vector space of functions given by the complex linear span of the set of functions $\{F_j'(z) : j = 1, \ldots, n-1\}$. It is a classical fact that $\mathcal{F}'$ is $n-1$ dimensional. Notice that $S(z,a_i)L(z,a)$ is in $A^\infty(\Omega)$ because the pole of $L(z,a)$ at $z=a$ is cancelled by the zero of $S(z,a_i)$ at $z=a$. A theorem due to Schiffer (see [5,1,2]) states that the $n-1$ functions $S(z,a_i)L(z,a)$, $i = 1, \ldots, n-1$ form a basis for $\mathcal{F}'$. We may now write

$$K(z,w) = 4\pi S(z,w)^2 + \sum_{i,j=1}^{n-1} \lambda_{ij} S(z,a_i) L(z,a) \overline{S(w,a_j) L(w,a)}, \qquad (4.1)$$

which, together with (3.2) allows us to see that the Bergman kernel is composed of exactly the same basic functions that make up the Szegő kernel.

A recipe is given for explicitly computing all the elements appearing in formula (4.1) in [4]. It is interesting that all the elements of the kernel function can be computed by means of one dimensional line integrals and simple linear algebra.

We have proved formula (4.1) on a domain with smooth boundary. If a finitely connected domain Ω_1 does not have smooth boundary, and if none of its boundary components are points, there is a conformal mapping Φ of Ω_1 onto a domain Ω_2 whose boundary is smooth. The transformation formula for the Bergman kernels under biholomorphic mappings,

$$K_1(z,w) = \Phi'(z)K_2(\Phi(z),\Phi(w))\overline{\Phi'(w)},$$

together with the transformation formula for the Szegő kernels (3.3), can then be used to show that (4.1) is valid on Ω_1.

5. THE AHLFORS MAP AND OTHER KERNEL FUNCTIONS

I showed in [3] (see also [1]) how the Szegő projection can be used to solve the Dirichlet problem. The method gives rise to a formula for the Poisson kernel of a bounded n-connected domain Ω with C^∞ smooth boundary in terms of the Szegő kernel (see [2]). The Poisson kernel $p(z,w)$ is given by

$$p(a,w) = \frac{|S(w,a)|^2}{S(a,a)} + \sum_{j=1}^{n-1}(\omega_j(a) - \lambda_j(a))\mu_j(w)$$

where ω_j are the harmonic measure functions defined in §3, $\mu_j(w)$ is a real valued function that is a linear combination of $S(a_k,w)S(w,a)$, $k = 1,\ldots,n-1$, and

$$\lambda_j(a) = \int_{\zeta\in\gamma_j} \frac{|S(\zeta,a)|^2}{S(a,a)}\,ds$$

is a function in $C^\infty(\overline{\Omega})$ that has the same boundary values as $\omega_j(a)$, i.e., equal to one on γ_j and equal to zero on the other boundary components. The Ahlfors map is the principal ingredient of the main term, $|S(w,a)|^2/S(a,a)$ in the Poisson kernel; the other term, $\sum_{j=1}^{n-1}(\omega_j(a) - \lambda_j(a))\mu_j(w)$, is in $C^\infty(\overline{\Omega}\times\overline{\Omega})$.

The Garabedian kernel can also be expressed in terms of the Ahlfors map. Let $z \in \Omega$ and $w \in b\Omega$ and use identity (2.1) and the fact that $\overline{f_a} = 1/f_a$ on $b\Omega$ to rewrite formula (3.2) in the form

$$L(z,w) = \frac{f_a(w)}{f_a(z) - f_a(w)}\left(c_0 S(z,a)L(w,a) + \sum_{i,j=1}^{n-1} \bar{c}_{ij} S(z,a_i)L(w,a_j)\right).$$

Since both sides of this identity are holomorphic in z and w, this identity holds for $z,w \in \Omega$, $z \neq w$. Note that the constants c_0 and c_{ij} are the same as the constants in (3.2).

The complimentary kernel $\Lambda(z,w)$ to the Bergman kernel (see [1,page 134]), may also be expressed in terms of the Ahlfors map via

$$\Lambda(w,z) = 4\pi L(w,z)^2 + \sum_{i,j=1}^{n-1} \lambda_{ij} L(w,a_i)S(w,a)S(z,a_j)L(z,a),$$

$z, w \in \Omega$, $z \neq w$.

Finally, we mention that the gradient of the Green's function on a finitely connected domain Ω with C^∞ smooth boundary is composed of finitely many functions of one variable in $C^\infty(\overline{\Omega})$. It is shown in [2] that

$$\frac{\partial G}{\partial \bar{w}}(z,w) = \pi \left(\frac{S(z,w)\,\overline{L(w,z)}}{S(z,z)} - i \sum_{j=1}^{n-1} (\omega_j(z) - \lambda_j(z))\overline{g_j(w)} \right)$$

for all $z, w \in \Omega$, $z \neq w$, where $g_j(w)$ is a linear combination of the holomorphic functions $S(w, a_k)L(w, a)$, $k = 1, \ldots, n-1$.

Research supported by NSF grant DMS-9623098

References

1. S. Bell, *The Cauchy transform, potential theory, and conformal mapping*, CRC Press, Boca Raton, 1992.
2. ______, *Complexity of the classical kernel functions of potential theory*, Indiana Univ. Math. J. **44** (1995), 1337–1369.
3. ______, *The Szegő projection and the classical objects of potential theory in the plane*, Duke Math. J. **64** (1991), 1–26.
4. ______, *Recipes for classical kernel functions associated to a multiply connected domain in the plane*, Complex Variables, Theory and Applications **29** (1996), 367–378.
5. M. Schiffer, *Various types of orthogonalization*, Duke Math. J. **17** (1950), 329–366.

5 SOME GENERALIZED LAPLACE TRANSFORMATIONS

Erwin A. K. Brüning

Department of Mathematicis & Applied Mathematics
University of Durban-Westville
Private Bag X54001, Durban 4000, ZA

ebruning@pixie.udw.ac.za

Abstract: In the summation theory for divergent power series the Laplace transform with index $k > 0$ (for analytic functions of exponential size at most $k > 0$) and the Borel transform with index $k > 0$ play a prominent role. In the investigation into a more flexible summation theory we encountered two new classes of transformations: The generalized Laplace transformations and the generalized Borel transformations, both are parametrized by certain Radon probability measures on the positive half line. Here we introduce these two classes of transformations and discuss some of their basic properties. The Laplace transform (resp. the Borel transform) with index $k > 0$ appears as the case of a special Radon probability measure.

1.1 INTRODUCTION. MOTIVATION

We are going to specify a class of 'admissible' Radon probability measures on the positive half line which have moments of every order (Section 3). For each element of this class we will define two transformations which we propose to call *generalized Laplace transformation* (GLT) respectively *generalized Borel transformation* (GBT) by reasons which will be explained later.

These transformations emerged from an investigation into the problem of developing a sufficiently flexible reconstruction theory for analytic functions, given their asymptotic expansion at a boundary point of the domain of analyticity. These new transformations are used in this new reconstruction theory in a similar way as the Laplace- and Borel transformation enter the Watson-Nevanlinna reconstruction theory, but allowing a weaker concept of an asymptotic expansion, a greater flexibility in the growth rate of the expansion coefficients with the order n, and a wider class of geometrical shapes of the boundary of the domain of analyticity near the expansion point [2].

The generalized Laplace transformation is also of interest in those cases where the presently available "Laplace transformations with index k" cannot be applied because the required growth restrictions are not given (see next section).

There are formal similarities of our GLT with the integral transforms considered in [3]. If we write our transform in the form (1) of [3] the kernel h however is not a square integrable function but a point measure.

The Laplace transform and its inverse, the Borel transform, have a long and extensive history. In order to have a reference point we recall in Section 2 very briefly the basic facts about a more recent development, the *Laplace and Borel transform with index* $k > 0$. This review is based on [1].

1.2 LAPLACE AND BOREL TRANSFORM WITH POSITIVE INDEX K. A REMINDER

A *sector* $S = S(d, \alpha, \rho)$ *in direction* d *with opening* α *and radius* $\rho > 0$ is the following subset of the complex plane:

$$S(d, \alpha, \rho) = \{z = re^{i\varphi} \mid 0 < r < \rho,\ d - \frac{\alpha}{2} < \varphi < d + \frac{\alpha}{2}\}.$$

For $\rho = +\infty$ we speak about an *infinite sector* and write $S(d, \alpha)$ for it. A *closed sector* is a set of the form

$$\overline{S} = \overline{S}(d, \alpha) = \{z = re^{i\varphi} \mid 0 < r \leq \rho,\ d - \frac{\alpha}{2} \leq \varphi \leq d + \frac{\alpha}{2}\}$$

where ρ is always finite.

An analytic function $f : S(d, \alpha, \rho) \longrightarrow C$ is called *bounded at the origin* if, and only if, it is bounded on every closed sub-sector $\overline{S_1}$ of $S(d, \alpha, \rho)$. An analytic function $f : S(d, \alpha, \rho) \longrightarrow C$ is said to be *continuous at the origin* if, and only if, a complex number, denoted by $f(0)$, exists such that $f(0) = \lim_{S \ni z \to 0} f(z)$, uniformly in every closed sub-sector of $S = S(d, \alpha, \rho)$. Finally, an analytic function $f : S(d, \alpha) \longrightarrow C$ is said to be of *exponential size at most* k, for some $k > 0$, if, and only if, the following condition is satisfied: For every $\varphi \in (d - \frac{\alpha}{2}, d + \frac{\alpha}{2})$ and every $r_0 > 0$ there exist constants ε, c_1, $c_2 > 0$ such that $|f(z)| \leq c_1 e^{c_2 |z|^k}$ holds for every $z = re^{i\tau}$, $r \geq r_0$, and $|\varphi - \tau| < \varepsilon$. With the help of these definitions we can formulate concisely the basic results about the Laplace and Borel transform with index k.

Theorem 1 *Let $f : S(d, \alpha) \longrightarrow C$ be analytic and continuous at the origin. If f is of exponential size at most $k > 0$ in $S = S(d, \alpha)$ then the Laplace transform with index k of f, $\mathcal{L}_k[f]$, is defined by*

$$\mathcal{L}_k[f](z) = z^{-k} \int_0^{\infty(\tau)} f(u) e^{-(\frac{u}{z})^k} d(u^k)$$

with integration along $\arg u = \tau$ *for any* τ *with* $|d - \tau| < \alpha$. $\mathcal{L}_k[f]$ *is defined and analytic in a region which contains a finite sector* $\hat{S} = S(d, \beta, \rho)$ *for some* $\rho > 0$ *and opening* $\beta < \frac{\pi}{k} + \alpha$, *arbitrarily close to* $\frac{\pi}{k} + \alpha$. $\mathcal{L}_k[f]$ *is bounded at the origin.*

Theorem 2 *Let* $S = S(d, \alpha, \rho)$ *be a finite sector in direction* d *with opening* $\alpha > \frac{\pi}{k}$ *for some* $k > 0$ *and let* $f : S \longrightarrow C$ *be analytic, and bounded at the origin. Then the Borel transform with index* k *of* f, $\mathcal{B}_k[f]$, *is defined by*

$$\mathcal{B}_k[f](u) = \frac{1}{2i\pi} \int_{\gamma_k(\tau)} z^k f(z) e^{(\frac{u}{z})^k} d(z^{-k})$$

for a certain closed piecewise smooth contour $\gamma_k(\tau)$ *inside* S *for any real* τ *with* $|\tau - d| < \frac{1}{2}(\alpha - \frac{\pi}{k})$. $\mathcal{B}_k[f]$ *is defined and analytic in the sector* $\tilde{S} = S(d, \alpha - \frac{\pi}{k})$.

For details about the precise choice of the integration contour $\gamma_k(\tau)$ we have to refer to [1]. These results suggest to consider the Laplace and the Borel transform with index k on the following 'natural' domains of definition dom $\mathcal{L}_k$ respectively dom $\mathcal{B}_k$. dom $\mathcal{L}_k$ is the set of all functions f which are analytic on some infinite sector $S = S(d, \alpha)$, continuous at the origin, and of exponential size at most k in S. Similarly, dom $\mathcal{B}_k$ is the set of all functions f, analytic on some finite sector $S = S(d, \alpha, \rho)$ with opening $\alpha > \frac{\pi}{k}$ and bounded at the origin. The above results then ensure that the Laplace transform with index k, $\mathcal{L}_k$, maps its domain dom $\mathcal{L}_k$ into the domain dom $\mathcal{B}_k$ of the Borel transform with index k, and similarly, that the Borel transform with index k maps its domain dom $\mathcal{B}_k$ into the domain dom $\mathcal{L}_k$ of the Laplace transform with index k. These two transformations are inverse to each other according to the following result.

Theorem 3 *For* $f \in$ dom $\mathcal{L}_k$ *one has* $\mathcal{B}_k[\mathcal{L}_k[f]] = f$ *and for* $f \in$ dom $\mathcal{B}_k$ *one has* $\mathcal{L}_k[\mathcal{B}_k[f]] = f$.

1.3 ADMISSIBLE MEASURES AND ADMISSIBLE ENTIRE FUNCTIONS

Obviously, the *exponential function* plays a distinguished role in the theory of the Laplace and Borel transformations with index k. The natural question arises:

> *Can one have similar transformations where the exponential function is replaced by some other functions? Which other functions?*

To answer these questions we first introduce the set $\mathcal{M}$ of all *admissible measures*. By definition, $\mathcal{M}$ is the set of all nonnegative Radon probability measures on the positive half-line $\mathcal{R}_+ = [0, \infty)$ with the following three properties:

1. ν has moments $\nu(n) = \int_0^\infty t^n d\nu(t)$ of every order n and $\nu(n) > 0$ for $n = 0, 1, \ldots$.
2. $\limsup_{n\to\infty} (\frac{1}{\nu(n)})^{\frac{1}{n}} = 0$, i.e.,

$$\Phi_\nu(z) = \sum_{n=0}^{\infty} \frac{z^n}{\nu(n)} \tag{1.1}$$

is an entire function of z.

3. The domain G_ν of all $w \in C$ for which $\lim_{T\to\infty} \int_0^T \Phi_\nu(tz)d\nu(t)$ exists uniformly in $z \in K$ for some compact neighbourhood K of w contains an open complex neighbourhood $U((0, R_\nu))$ of some real interval $(0, R_\nu)$.

Now the set $\mathcal{E}$ of *admissible entire functions* simply is the set of all functions Φ_ν, $\nu \in \mathcal{M}$. Actually, in order to be able to formulate a number of important and more specific results later we will restrict these classes somewhat.

In order to illustrate these concepts we give some elementary examples.

Example 1.1 *a) For the measure $d\nu(t) = e^{-t}dt$ on $\mathcal{R}_+$ we have $\nu(n) = n!$ and thus $\Phi_\nu(z) = e^z$.*
b) Consider the class of measures ν on $\mathcal{R}_+ = [0, \infty)$ defined by $\nu = \nu_{k,\gamma}$, $0 < \gamma < \infty$, $\frac{1}{2} \leq k \leq \infty$ with $d\nu_{k,\gamma}(t) = \gamma k t^{k-1} e^{-\gamma t^k} dt$. The moments of these measures are $\nu(n) = \gamma^{-\frac{n}{k}}\, \Gamma(\frac{n}{k} + 1)$ and thus the associated entire function ϕ_ν is

$$\Phi_\nu(z) = \sum_{n=0}^{\infty} \frac{z^n}{\nu(n)} = \sum_{n=0}^{\infty} \frac{(\gamma^{\frac{1}{k}} z)^n}{\Gamma(\frac{n}{k} + 1)} = E_{\frac{1}{k}}(\gamma^{\frac{1}{k}} z) \tag{1.2}$$

where E_α denotes the Mittag-Leffler function of index α. The asymptotic behaviour of the Mittag-Leffler functions is well known [4]: Thus one has, for $|\arg z| \leq \frac{\pi}{2k}$,

$$\lim_{|z|\to\infty} e^{-\gamma z^k}\, \Phi_\nu(z) = k.$$

Now it follows (see [2]) that all the measures considered in this example also satisfy Condition 3 and thus these measures are admissible in the above sense.

An admissible measure ν and its associated entire function Φ_ν satisfy a simple but important integral equation as the following lemma shows (see [2]).

Lemma 1.1 *For any $\nu \in \mathcal{M}$ and all $w \in G_{\nu,0}$, the connected component of $G_\nu \cap \{w \in C|\ |w| < 1\}$ in $G_\nu \backslash \{1\}$, one has*

$$\int_0^\infty \Phi_\nu(tw)d\nu(t) = \frac{1}{1 - w}. \tag{1.3}$$

The entire functions Φ_ν, $\nu \in \mathcal{M}$, will take the role of the exponential function in our definition of the generalized Borel transformation. In our suggestion for the definition of the generalized Laplace transformation no exponential function enters directly; it is absorbed in the measure $\nu \in \mathcal{M}$.

1.4 GENERALIZED LAPLACE TRANSFORMATION

Fix a measure ν in the class $\mathcal{M}$ and let $g : S(d, \alpha) \longrightarrow C$ be analytic on some infinite sector, and continuous at the origin. Clearly, for any $T > 0$,

$$z \to F_T(g; z) = \int_0^T g(tz)d\nu(t)$$

is well defined and analytic on $S(d,\alpha)$. Denote by $G(\nu, g)$ the set of all those points $w \in C$ for which a compact neighborhood K exists such that $F_T(g;z)$ has a limit as $T \to \infty$, uniformly in $z \in K$. Clearly, it can happen that $G(\nu, g)$ is empty. For $G(\nu, g)$ to be not empty the function g has to satisfy some growth restrictions determined by the measure ν. Now the domain $\mathcal{D}(\mathcal{L}_\nu)$ of the *generalized Laplace transformation* $\mathcal{L}_\nu$ is defined as the set of those functions g which are analytic on some infinite sector $S(d,\alpha)$, continuous at the origin, and for which the set $G(\nu, g)$ is not empty. For $g \in \mathcal{D}(\mathcal{L}_\nu)$ we define, for all $z \in G(\nu, g)$,

$$\mathcal{L}_\nu[g](z) = \int_0^\infty g(tz)d\nu(t) = \lim_{T\to\infty} \int_0^T g(tz)d\nu(t). \tag{1.4}$$

Then $\mathcal{L}_\nu[g]$ is analytic on $G(\nu, g)$. It is not always straightforward to decide whether a point z belongs to the set $G(\nu, g)$. Therefore in practice one might want to work with the set $G_1(\nu, g)$ of those points $w \in C$ for which a compact neighbourhood K and a function $h_K \in L^1(d\nu)$ exist such that for all $z \in K$ the estimate $|g(tz)| \le h_K(t)$ holds for all $t > 0$. Obviously one has $G_1(\nu, g) \subset G(\nu, g)$.

Though for $g \in \mathcal{D}(\mathcal{L}_\nu)$, $g(0) = \lim_{S\ni z\to 0} g(z)$ exists uniformly in every closed sub-sector S of the domain of analyticity $S(d,\alpha)$ of g, the point $z = 0$ does not necessarily belong to the domain of analyticity $G(\nu, g)$ of the generalized Laplace transform $\mathcal{L}_\nu[g]$ of g. To see this it suffices to look at simple examples. However, the point $z = 0$ is a boundary point of this set.

The GLT as defined above is clearly homogenous in the sense that for all $g \in \mathcal{D}(\mathcal{L}_\nu)$ and all $\lambda \in C$ one has $\lambda g \in \mathcal{D}(\mathcal{L}_\nu)$ and $\mathcal{L}_\nu[\lambda g] = \lambda \mathcal{L}_\nu[g]$. However $\mathcal{L}_\nu$ is additive only in the following restricted sense: If $g_1, g_2 \in \mathcal{D}(\mathcal{L}_\nu)$ are such that the intersection $G(\nu, g_1) \cap G(\nu, g_2)$ is not empty then $g_1 + g_2 \in \mathcal{D}(\mathcal{L}_\nu)$ and $\mathcal{L}_\nu[g_1 + g_2] = \mathcal{L}_\nu[g_1] + \mathcal{L}_\nu[g_2]$. But clearly, given $\nu \in \mathcal{M}$, one can specify a subspace of $\mathcal{D}(\mathcal{L}_\nu)$ on which $\mathcal{L}_\nu$ is linear.

The above transformations $\mathcal{L}_\nu$, $\nu \in \mathcal{M}$, have been named generalized Laplace transformations because of the following: For the measure $\nu_k = \nu_{k,1}$ according to Example 1 it agrees with the Laplace transformation $\mathcal{L}_k$ with index $k > 0$. To see this take any $g \in \operatorname{dom} \mathcal{L}_k$; then g is of exponential size at most k in some sector $S = S(d,\alpha)$. One easily finds thus a closed subsector K such that for all $z \in K$ the function $t \to g(tz)$ is bounded by some function h_K in $L^1(d\nu)$. Hence it follows $g \in \mathcal{D}(\mathcal{L}_{\nu_k})$ and thus

$$\mathcal{L}_{\nu_k}[g](z) = \int_0^\infty g(tz)kt^{k-1}e^{-t^k}dt = \int_0^\infty g(tz)e^{-t^k}d(t^k).$$

With the new integration variable $u = tz$ this equals

$$z^{-k}\int_0^{\infty(\tau)} g(u)e^{-(\frac{u}{z})^k}d(u^k) = \mathcal{L}_k[g](z)$$

with the specifications given in Theorem 1, and it shows that the Laplace transformations with index $k > 0$ are a special case of the transformations introduced above.

A systematic investigation of the range of the GLT has not yet been completed. If a specific form of the asymptotic behaviour of the entire function Φ_ν is taken into account, for instance those of Section 4 and/or 5 of [2] then some estimates for the domain of analyticity of $\mathcal{L}_\nu[g]$ and thus of the range of $\mathcal{L}_\nu$ can be obtained. This then amounts to considering a certain subclass $\mathcal{M}_0$ of the admissible measures (see [2]).

For $\nu \in \mathcal{M}_0$ one has, by definition, for $r \to \infty$

$$|\Phi_\nu(re^{i\theta})| \cong \text{const.}e^{r^k h(\theta)} \tag{1.5}$$

with the indicator function h of the entire function Φ_ν of order $k > 0$. Then a point $z = re^{i\theta}$ belongs to the basic set G_ν if, and only if,

$$r^k h(\theta) < \gamma$$

where $\gamma > 0$ is determined by the convergence of the integral $\int_0^\infty e^{at^k}\, d\nu(t)$. For $a < \gamma$ it converges while for $a > \gamma$ it diverges. The indicator function h is assumed to be positive for $|d - \theta| < \alpha_\nu$ for some $d \in [0, 2\pi)$ and some $\alpha_\nu \in (0, \pi)$ and negative for $\alpha_\nu < |d - \theta| \le \pi$. If now g is a function which is analytic on a sector $S = S(d, \alpha)$ and satisfies some growth restrictions there, for instance one has, for some continuous nonnegative functions a and b on some subsector S_1 of S,

$$|g(tz)| \le b(z)e^{a(z)t^k} \tag{1.6}$$

for $z \in S_1$ and all $t > 0$ then one can show that the set $G_1(\nu, g)$ contains at least the set $S_1 \cap \{z : a(z) < \gamma\}$ and thus typically a finite sector. Furthermore the function $\mathcal{L}_\nu[g]$ then is bounded at the origin. If g is of exponential size at most k on S then g satisfies the above bound with $a(z) = c_2|z|^k$ and $b(z) = c_1$.

With the notation introduced above we summarize:

Theorem 4 *a) Let $\nu \in \mathcal{M}$ be an admissible measure. Then the generalized Laplace transformayion is well defined on the domain $\mathcal{D}(\mathcal{L}_\nu)$. The generalized Laplace transform $\mathcal{L}_\nu[g]$ of $g \in \mathcal{D}(\mathcal{L}_\nu)$ is given by Equation (1.4).*

b) If $\nu \in \mathcal{M}_0$, i.e., the associated entire function Φ_ν satisfies (1.5), then $\mathcal{L}_\nu[g]$ is well defined for all functions g which are analytic in some sector $S(d,\alpha)$ and satisfy the growth condition (1.6). $\mathcal{L}_\nu[g]$ is analytic in some finite subsector of $S(d,\alpha)$.

1.5 GENERALIZED BOREL TRANSFORMATION

As we have indicated at the end of the previous section, for measures in the subclass $\mathcal{M}_0$ of admissible measures $\mathcal{M}$ the explicit knowledge of the asymptotic behaviour of the admissible function Φ_ν is available. This allows a fairly explicit discussion of the generalized Borel transformation and is thus preferred. The domain $\mathcal{D}(\mathcal{B}_\nu)$ of the *generalized Borel transformation* $\mathcal{B}_\nu$ consists of all functions f which are analytic on some domain G of the complex plane for which the point $z = 0$ is a point of the boundary ∂G and which contains a closed sector, and f is bounded on this sector (Actually one can allow here more general shapes than sectors, see [2]). Denote by γ a closed piecewise smooth contour in G 'through the point $z = 0$' similar to the path $\gamma_k(\tau)$ in Theorem 2 (actually γ is the 'limit' of integration contours $C_{z_0,z}$ as $z_0 \to 0$ as specified in Section 2 of [2]). For such functions f we define

$$\mathcal{B}_\nu[f](z) = \frac{1}{2i\pi}\int_\gamma \frac{f(u)}{u}\Phi_\nu(\frac{z}{u})du \tag{1.7}$$

In a sufficiently small neighborhood of the point $z = 0$ the integration contour γ can be parametrized by $u = re^{i\theta_1}$, respectively by $u = (r_0 - r)e^{i\theta_2}$, with $0 \le r \le r_0$ for appropriate values of θ_1 and θ_2. Taking into account the asymptotic behaviour of the entire function Φ_ν the contributions of these parts of the integration contour to the defining integral of $\mathcal{B}_\nu[f](z)$ can be estimated by

$$\text{const}\int_0^{r_0} \frac{1}{r}e^{(\frac{|z|}{r})^k h(\theta)}dr$$

with $\theta = \arg z - \theta_1$. Hence this contribution is finite for all points $z = |z|e^{i\arg z}$ for which $h(\theta) < 0$. And similarly for the other part of the integration contour near $z = 0$. Since the remainder of the contour γ has a distance of at least r_0 from the point $z = 0$ its contribution to the above integral is finite for all z. Thus one deduces that $\mathcal{B}_\nu[f]$ is analytic on some infinite sector. Its growth is determined by the asymptotic behaviour of Φ_ν.

The generalized Borel transformation (GBT) is *normalized* according to the following relation: The GBT of $f(u) = u^m$ is $\mathcal{B}_\nu[f](z) = \frac{z^m}{\nu(m)}$.

For a fixed domain G of the above form the GBT $\mathcal{B}_\nu$ is *linear* on the subspace $\mathcal{D}(\mathcal{B}_\nu, G)$ of all functions which are analytic on G and bounded on a common closed subsector.

In Example 1 we have seen that for the admissible measure $d\nu(t) = e^{-t}dt$ the function Φ_ν is just the exponential function. It follows that the GBT $\mathcal{B}_\nu$ for this measure as defined above is equal to the Borel transform with index $k = 1$ considered in Theorem 2. Accordingly the transformations $\mathcal{B}_\nu$ of Equation 1.7 are called generalized Borel transformations.

Theorem 5 *For $\nu \in \mathcal{M}_0$ the generalized Borel transformation $\mathcal{B}_\nu$ is well defined on the domain $\mathcal{D}(\mathcal{B}_\nu)$ by Equation 1.7. For $f \in \mathcal{D}(\mathcal{B}_\nu)$ the generalized Borel transform $\mathcal{B}_\nu[f]$ is analytic in some infinite sector. $\mathcal{B}_\nu$ is normalized according to the relation $\mathcal{B}_\nu[f](z) = \frac{z^m}{\nu(m)}$ for $f(u) = u^m$.*

The characterizations of the ranges of the GLT $\mathcal{L}_\nu$ and of the GBT $\mathcal{B}_\nu$ being incomplete one cannot show that these two transformations are inverse to each other. Nevertheless some important partial results in this direction are known which we indicate now.

Suppose $f : G \longrightarrow C$ belongs to the domain of $\mathcal{B}_\nu$. For all points z in the interior $I(\gamma)$ of the integration contour γ of formula (1.7) one clearly has

$$f(z) = \frac{1}{2i\pi}\int_\gamma \frac{f(u)}{u}\frac{1}{1-\frac{z}{u}}du.$$

If we now consider only those $z \in I(\gamma)$ for which $\frac{z}{u} \in G_{\nu,0}$ for all u along the integration contour γ then Lemma 1.1 implies

$$f(z) = \frac{1}{2i\pi}\int_\gamma \frac{f(u)}{u}(\lim_{T\to\infty}\int_0^T \Phi_\nu(t\frac{z}{u})d\nu(t))du$$

and one shows (see [2]) that this equals

$$\lim_{T\to\infty}\int_0^T \frac{1}{2i\pi}\int_\gamma \frac{f(u)}{u}\Phi_\nu(\frac{tz}{u})dud\nu(t) = \lim_{T\to\infty}\int_0^T \mathcal{B}_\nu[f](tz)d\nu(t)$$

and we deduce in this case

$$f = \mathcal{L}_\nu[\mathcal{B}_\nu[f]], \tag{1.8}$$

i.e., $\mathcal{L}_\nu$ is a left inverse of $\mathcal{B}_\nu$, respectively $\mathcal{B}_\nu$ is a right inverse of $\mathcal{L}_\nu$, on suitable sub-domains.

Next suppose that $f \in \mathcal{D}(\mathcal{L}_\nu)$ is actually analytic in some neighborhood of $z = 0$. According to formulas (1.4) and (1.7) we then have, for all points w in the domain of analyticity of f, sufficiently close to $z = 0$,

$$\mathcal{B}_\nu[\mathcal{L}_\nu[f]](w) = \frac{1}{2i\pi}\int_\gamma \mathcal{L}_\nu[f](u)\frac{1}{u}\Phi_\nu(\frac{w}{u})du.$$

By methods as explained in [2] one can justify that the limit operations involved in this integral can be exchanged in their order to obtain

$$\lim_{T\to\infty}\int_0^T \frac{1}{2i\pi}\int_\gamma f(tu)\frac{1}{u}\Phi_\nu(\frac{w}{u})dud\nu(t)$$

and since Φ_ν is entire we have

$$\frac{1}{2i\pi}\int_\gamma f(tu)\frac{1}{u}\Phi_\nu(\frac{w}{u})du = \sum_{n=0}^\infty \frac{w^n}{\nu(n)}\frac{1}{2i\pi}\int_\gamma \frac{f(tu)}{u^{n+1}}du = \sum_{n=0}^\infty \frac{w^n}{\nu(n)}\frac{t^n}{n!}f^{(n)}(0).$$

Thus the above limit is equal to

$$\lim_{T\to\infty}\int_0^T \sum_{n=0}^\infty \frac{w^n}{\nu(n)}\frac{t^n}{n!}f^{(n)}(0)d\nu(t) =$$

$$= \lim_{T\to\infty}\sum_{n=0}^\infty \frac{w^n}{n!}f^{(n)}(0)\frac{1}{\nu(n)}\int_0^T t^n d\nu(t) = f(w),$$

i.e., on a determining subset one has

$$\mathcal{B}_\nu[\mathcal{L}_\nu[f]] = f. \tag{1.9}$$

We summarize this in the following theorem.

Theorem 6 *On 'suitable' domains the generalized Laplace and the generalized Borel transformations are inverse to each other, more precisely:*
a) For $f \in \mathcal{D}(\mathcal{B}_\nu)$ one has $f = \mathcal{L}_\nu[\mathcal{B}_\nu[f]]$, i.e., the generalized Borel transformation is a right inverse of the generalized Laplace transformation, for the same admissible measure ν.
b) If $f \in \mathcal{D}(\mathcal{L}_\nu)$ is analytic in some neighborhood of $z = 0$ then $\mathcal{B}_\nu[\mathcal{L}_\nu[f]] = f$, i.e., the generalized Borel transformation is also a left inverse of the generalized Laplace transformation.

1.6 CONCLUSION

Two classes of new transformations have been proposed which have proved to be decisive in a recently developed reconstruction theory for analytic functions from their asymptotic expansion. A number of (elementary) properties of these transformations could be derived. And these transformations have been related to the Laplace and Borel transformation with index k. Clearly, many things still have to be done. First of all, a more explicit specification for the domains of both transformations as well as a more concrete characterization of the ranges are needed for a systematic study of these transformations. Then their properties under asymptotic expansions and their potential for summability methods are of great interest (Here one successfull instance with regard to asymptotic expansion in the sense of Poincaré is known, see [2]). Finally we think that these transformation will help in the following type of problems.

Suppose that g is a function analytic in some sector $S(d, \alpha)$, continuous at the origin but not of exponential size at most k, for any $k > 0$. Then clearly, the theory of Laplace transform with index k does not apply. Nevertheless one expects to have measures $\nu \in \mathcal{M}$ such that the set $G(\nu, g)$ introduced earlier is not empty. Then the function g is in the domain of the GLT $\mathcal{L}_\nu$ and thus $\mathcal{L}_\nu[g](z)$ is a well defined analytic function on $G(\nu, g)$. The challenge then is to find a class of growth restrictions on the functions in this case which allow to develop sufficiently many interesting details of the theory of generalized Laplace and Borel transformations in this situation too.

References

[1] Balser, W. (1994). *From Divergent Power Series to Analytic Functions.* Theory and Application of Multisummable Power Series. Berlin: Lecture Notes in Mathematics No. 1582, Springer-Verlag.

[2] Brüning, E. (1996). "How to reconstruct an analytic function from its asymptotic expansion?" *Complex Variables*, Vol. 30, 199-220.

[3] Saitoh, S. (1994). "One Approach to some general integral transforms and its applications" *Integral Transforms and Special Functions*, Vol. 3, No. 1, 49-85.

[4] Sansone, G. and J. Gerretsen (1960). *Lectures on the theory of a complex variable.* Groningen: Walters-Nordhoff.

6 ASYMPTOTIC BEHAVIOUR OF REPRODUCING KERNELS, BEREZIN QUANTIZATION AND MEAN-VALUE THEOREMS

Miroslav Engliš

Mathematical Institute of the Academy of Sciences,
Žitná 25, 11567 Prague 1, Czech Republic
englis@math.cas.cz

Abstract: Let Ω be a domain in $\mathbf{C}^n$, F and G positive measurable functions on Ω such that $1/F$ and $1/G$ are locally bounded, A^2_α the space of all holomorphic functions on Ω square-integrable with respect to the measure $F^\alpha G\,dm$, where dm is the $2n$-dimensional Lebesgue measure, $K_\alpha(x,y)$ the reproducing kernel for A^2_α (if it exists), and $B_\alpha f(y) = K_\alpha(y,y)^{-1}\int_\Omega f(x)|K_\alpha(x,y)|^2 F(x)^\alpha G(x)\,dm(x)$ the Berezin operator on Ω. In this paper we present some results on the asymptotic behavior of K_α and B_α as $\alpha \to +\infty$. For instance, if $-\log F$ is convex then $\lim_{\alpha\to+\infty} K_\alpha(x,x)^{1/\alpha} = 1/F(x)$ for any integrable G, and $K_\alpha(x,y)$ has a zero for all sufficiently large α whenever F is not real-analytic. Applications to mean value theorems and to quantization on curved phase spaces are also discussed.

1. INTRODUCTION

The subject of this talk is best illustrated by the following familiar example. Let $\mathbf{D}$ be the unit disc in the complex plane $\mathbf{C}$ and consider the family of measures

$$\begin{aligned} d\mu_\alpha(x) &= (1-|x|^2)^{\alpha-2}\cdot\pi^{-1}\,dm(x) \\ &=: F(x)^\alpha\cdot G(x)\,dm(x), \qquad F(x) = 1-|x|^2,\; G=\pi^{-1}F^{-2}, \end{aligned}$$

indexed by a parameter $\alpha > 1$ (dm is the Lebesgue measure). The Bergman spaces

$$A^2_\alpha = \{f \in L^2(\mathbf{D}, d\mu_\alpha) : f \text{ holomorphic on } \mathbf{D}\}.$$

The research was supported by grants GA AV ČR A1019701 and GA ČR 201/96/0411.

then have a reproducing kernel given by

$$K_\alpha(x,y) = \frac{\alpha - 1}{(1 - x\overline{y})^\alpha}.$$

In particular, on the diagonal $x = y$

$$K_\alpha(x,x) = \frac{\alpha - 1}{(1 - |x|^2)^\alpha} = \frac{\alpha - 1}{F(x)^\alpha}$$

is, up to the factor $\alpha - 1$, the reciprocal of the weight we have started with, $F(x)^\alpha = (1 - |x|^2)^\alpha$. In a sense, this is true also for $x \neq y$: namely, the weight function $F(x)^\alpha = (1 - |x|^2)^\alpha$ extends to a sesqui-analytic function $F(x,y)^\alpha = (1 - x\overline{y})^\alpha$ on $\mathbf{D} \times \mathbf{D}$, and

$$K_\alpha(x,y) = (\alpha - 1) \cdot F(x,y)^{-\alpha}$$

for all $x, y \in \mathbf{D}, \alpha > 1$.

Similar situation can be seen to prevail also for some other domains Ω and measures $d\mu_\alpha = F^\alpha G\, dm$:

(a) Ω =the unit ball in $\mathbf{C}^n$, $F = 1 - \|z\|^2$, $G = \pi^{-n} F^{-n-1}$ (so $G\, dm$ is the invariant measure on the ball)
(b) $\Omega = \mathbf{C}^n$, $F = e^{-\|z\|^2}$, $G = 1$ (the Segal-Bargmann, or Fock, spaces)
(c) (a generalization of (a)) Ω a bounded symmetric domain in $\mathbf{C}^n$, $F(z) = K(z,z)^{-1/p}$ and $G = \pi^{-n} F^{-p}$, where $K(z,z)$ is the Bergman kernel (with respect to the Lebesgue measure) and p is the genus of Ω. (This means that $G\, dm$ is the invariant measure on Ω.)

In all these cases,

$$(*) \qquad K_\alpha(x,y) = P_\Omega(\alpha)/F(x,y)^\alpha$$

where $F(x,y)$ is the sesqui-analytic extension of $F(x)$ and P_Ω is a monic polynomial in α, $\deg P_\Omega = \dim \Omega$.

(d) Ω a domain in $\mathbf{C}$ whose complement contains at least two points, F is such that $F(z)^{-1}\,|dz|$ is the Poincaré metric on Ω, and $G = \pi^{-1} F^{-2}$ (so $G\, dm$ is the Poincaré measure). Again, $F(x)$ admits a sesqui-analytic extension $F(x,y)$ to a neighborhood of the diagonal $x = y$ and $(*)$ holds up to an exponentially small error term:

$$K_\alpha(x,y) = F(x,y)^{-\alpha} \cdot [\alpha - 1 + O(\gamma^\alpha)], \qquad 0 \le \gamma = \gamma(x,y) < 1.$$

(e) $\Omega = \mathbf{C} \setminus \{0\}$, the punctured plane, $F(z) = |z|$, $G = \pi^{-1} F^{-2}$. The same assertion as in (d) holds, except with $\alpha - 1 + O(\gamma^\alpha)$ replaced by $\alpha + O(\gamma^\alpha)$.
(f) some pseudoconvex domains in $\mathbf{C}^2$, with appropriate F and G:

$$\Omega = \{z \in \mathbf{C}^2 : |z_1|^2 + |z_2|^{2/p} < 1\} \quad (p > 0),$$
$$F(z) = (1 - |z_1|^2)^p - |z_2|^2, \qquad G = (1 - |z_1|^2)^{2p-2} p\pi^{-2} F^{-3};$$
$$\Omega = \{z \in \mathbf{C}^2 : |z_2|^2 e^{|z_1|^2} < 1\},$$
$$F(z) = e^{-|z_1|^2} - |z_2|^2, \qquad G = \pi^{-2} F^{-3} e^{-2|z_1|^2};$$
$$\Omega = \{z \in \mathbf{C}^2 : |z_2|^2 (1 + |z_1|^2)^p < 1\} \quad (p > 0),$$
$$F(z) = (1 + |z_1|^2)^{-p} - |z_2|^2, \qquad G = (1 + |z_1|^2)^{-2p-2} p\pi^{-2} F^{-3}.$$

In each case,

$$K_\alpha(x, y) = F(x, y)^{-\alpha} \cdot (\alpha - 2)[\alpha + q(x, y)]$$

where $q(x, y)$ is bounded on $\Omega \times \Omega$.

It seems now natural to ask whether something can be said about the behaviour of the kernels K_α as $\alpha \to +\infty$ in the general situation of an arbitrary domain Ω and functions F, G. In this talk we present some partial results in this direction, and also briefly mention some applications to mean-value theorems and in mathematical physics.

2. ASYMPTOTIC BEHAVIOUR OF THE KERNELS

Let Ω be a domain in $\mathbf{C}^n$, $F, G > 0$ measurable functions on Ω such that $1/F, 1/G$ are locally bounded, $d\mu_\alpha = F^\alpha G \, dm$. Owing to the local boundedness of $1/F$ and $1/G$, the Bergman space $A^2_\alpha = A^2(\Omega, d\mu_\alpha)$ then admits a reproducing kernel $K_\alpha(x, y)$. We want to know whether

$$K_\alpha \sim F^{-\alpha},$$

in some sense, as $\alpha \to +\infty$. More precisely, one can ask the following questions:

Question 1. *Is it true that*

$$\text{(Q1)} \qquad \lim_{\alpha \to \infty} K_\alpha(x, x)^{1/\alpha} = 1/F(x)?$$

Question 2. *If F admits a sesqui-analytic extension, i.e. $F(x) = F(x, x)$ for a sesqui-analytic function $F(x, y)$ on $\Omega \times \Omega$, is it true that*

$$\text{(Q2)} \qquad \lim_{\alpha \to \infty} K_\alpha(x, y)^{1/\alpha} = 1/F(x, y)?$$

Question 3. *If F admits a sesqui-analytic extension $F(x, y)$, does one have an asymptotic expansion*

$$\text{(Q3)} \qquad K_\alpha(x, y) = F(x, y)^{-\alpha} \cdot \alpha^m B_0(x, y) \Big[1 + \frac{B_1(x, y)}{\alpha} + \frac{B_2(x, y)}{\alpha^2} + \dots \Big]$$

as $\alpha \to +\infty$, with some integer m and functions $B_0, B_1, B_2, \dots$?

Of course, (Q3) $\implies$ (Q2) $\implies$ (Q1). We are going to say something about each of these questions in turn. In brief, the answer to Question 1 probably is "if and only if $-\log F$ is plurisubharmonic and the spaces A^2_α are nonzero for sufficiently large α". In that case, the answer to the second question turns out to be affirmative whenever the roots on the right-hand side of (Q2) make sense, i.e. if $K_\alpha(x,y)$ does not vanish; moreover, the existence of the sesqui-analytic extension $F(x,y)$ of $F(x)$ then need not be hypothesized but follows as a corollary. (However, we do not know of any criterion for telling if $K_\alpha(x,y)$ is zero-free.) For the last question we have no good answer, but are able to tell, at least, what the functions $B_0, B_1, \dots$ and the number m have to be if (Q3) holds.

QUESTION 1. Assume, for simplicity, that F is lower semicontinuous, so that $-\log F$ is upper semicontinuous. Recall that a function $\phi : \Omega \to \mathbf{R}$ is plurisubharmonic (PSH) if it is upper semicontinuous and its restriction to any complex line is a subharmonic function; i.e. if the function of one complex variable

$$z \mapsto \phi(x_0 + zv)$$

is subharmonic on its domain of definition, for any $x_0 \in \Omega$ and $v \in \mathbf{C}^n$. Examples of PSH functions are convex functions (i.e. if Ω is convex as a domain in $\mathbf{R}^{2n} \simeq \mathbf{C}^n$ and ϕ is a convex function on $\Omega \subset \mathbf{R}^{2n}$, then ϕ is PSH) and real parts of holomorphic functions (f holomorphic on $\Omega \implies \operatorname{Re} f$ is PSH). Define functions $F^\#$ and F^* by

$$\begin{aligned} -\log F^\# &= \sup\{\psi : \ \psi \text{ is PSH}, \psi \leq -\log F\} \\ -\log F^* &= \sup\{\operatorname{Re} : \ g \text{ holomorphic}, \operatorname{Re} g \leq -\log F\}. \end{aligned}$$

Thus $-\log F^\#$ is the greatest PSH minorant of $-\log F$; in particular, $F = F^\#$ iff $-\log F$ is PSH. Clearly

$$1/F^* \leq 1/F^\# \leq 1/F.$$

Our first result says that the limit in (Q1) must live between the left and the middle term.

Theorem 1. ([7]) *Assume that F is lower semicontinuous and $1 \in A^2_\alpha$ for some $\alpha \geq 0$. Then*

$$1/F^*(x) \leq \liminf_{\alpha\to\infty} K_\alpha(x,x)^{1/\alpha} \leq \limsup_{\alpha\to\infty} K_\alpha(x,x)^{1/\alpha} \leq 1/F^\#(x) \leq 1/F(x).$$

Corollary 1A. *If $\lim_{\alpha\to\infty} K_\alpha(x,x)^{1/\alpha} = 1/F(x)\ \forall x$, then $-\log F$ is PSH.*

Corollary 1B. *If Ω is convex, $-\log F$ is convex, and $1 \in A^2_\alpha$ for some $\alpha \geq 0$, then $\lim_{\alpha\to\infty} K_\alpha(x,x)^{1/\alpha} = 1/F(x)\ \forall x \in \Omega$.*

Proof. A convex function is the supremum of affine functions ($f(x) = \operatorname{Re}(\langle x, a\rangle + b)$, $a \in \mathbf{C}^n, b \in \mathbf{C}$) lying below it; hence, $F^* = F$. □

Remarks. (1) The statement of Theorem 1 is independent of G. One can think that the main (coarsest) asymptotics of K_α are governed by F only, with G entering only on a finer level (such as the coefficients $B_0, B_1, \dots$ in (Q3)).

(2) The assumption that F be lower semicontinuous has been made only to avoid some technicalities; otherwise it is unnecessary. See [7].

(3) On the other hand, the hypothesis that $1 \in A^2_\alpha$ for some α cannot be omitted: for $\Omega = \mathbf{C}^n$, $F = G = 1$ (i.e. $d\mu_\alpha = dm$ for all α), we have $A^2_\alpha = \{0\}$, so $K_\alpha(x,y) \equiv 0$ $\forall \alpha$, while $F(x,y) \equiv 1$.

(4) Using the methods of Lin and Rochberg [14] (Lemma 2.8) one can show that (Q1) holds for $\Omega = \mathbf{D}$ and $-\log F$ strictly subharmonic whenever the function $\tau = (-\Delta \log F)^{-1/2}$ satisfies $\lim_{|z|\to 1} \tau(z) = 0$ and there exists a constant $C > 0$ such that $|\tau(z) - \tau(\xi)| \le C|z - \xi|$ $\forall z, \xi \in \mathbf{D}$. □

These results suggest the following

Conjecture. *$\lim K_\alpha(z,z)^{1/\alpha} = 1/F(z)$, for all $z \in \Omega$, whenever $1 \in A^2_\alpha$ for some α and $-\log F$ is PSH.*

QUESTION 2. Of course, for the root $K_\alpha(x,y)^{1/\alpha}$ to make sense, $K_\alpha(x,y)$ must be zero-free. Also, the answer to Question 1 must be affirmative, i.e. $K_\alpha(x,x)^{1/\alpha} \to 1/F(x)$ as $\alpha \to \infty$. It turns out that these are the only obstacles:

Theorem 2. ([7]) *Assume that the limit $\lim_{\alpha\to\infty} K_\alpha(x,x)^{1/\alpha}$ exists and equals $1/F(x)$, for all $x \in \Omega$. Suppose further that there exist a sequence $\alpha_j \to +\infty$ and a simply-connected open set $U \subset \Omega$ such that*

$$K_\alpha(x,y) \neq 0 \qquad \text{for all } \alpha = \alpha_1, \alpha_2, \dots \text{ and } x, y \in U.$$

Then:

- *$F(x)$ admits a zero-free sesqui-analytic extension $F(x,y)$ to $U \times U$;*
- *$\lim_{j\to\infty} K_{\alpha_j}(x,y)^{1/\alpha_j} = 1/F(x,y)$ $\forall x, y \in U$, if the branches of the roots are chosen to be positive for $x = y$;*
- *$|F(x,y)|^2 \ge F(x,x)F(y,y)$ $\forall x, y \in U$ (the "reverse Schwarz" inequality).*

Observe that the existence of the sesquianalytic extension $F(x,y)$ is even obtained as a consequence and need not be assumed a priori. The theorem can thus often be used to prove the existence of zeroes of the reproducing kernels:

Corollary 2. *If Ω is convex, $1 \in A^2_\alpha$ for some $\alpha \ge 0$ and $-\log F$ is convex, but F is not real-analytic at a point $w \in \Omega$, then for every neighborhood U of w there exists $\alpha_U > 0$ such that $\forall \alpha \ge \alpha_U$, $K_\alpha(x,y)$ has a zero in $U \times U$.*

Similar conclusion holds if F is real-analytic but does not admit a sesquianalytic extension $F(x,y)$ to $\Omega \times \Omega$ (for instance, for topological reasons; see the next example) or if the sesquianalytic extension fails to satisfy the "reverse Schwarz" inequality: for instance, if

- $\Omega = \mathbf{C}, G = 1, F(z) = (|z|^2 + 1)e^{-|z|^2}$, or
- $\Omega = \mathbf{D}, G = 1, F(z) = (|z|^2 - 1)(|z|^2 + \frac{3}{4})(|z|^2 - \frac{11}{4})$.

Then $K_\alpha(z,z)^{1/\alpha} \to 1/F(z)$ $\forall z \in \Omega$ by Corollary 1B, but weird things are happening off the diagonal (with $K_\alpha(x,y)^{1/\alpha}$ for $x\overline{y} = -1$ or $x\overline{y} = -3/4$, respectively). It can be shown that the functions $|K_\alpha(x,y)|^{1/\alpha}$ are always locally uniformly bounded; it would be interesting to see what is their behaviour as $\alpha \to +\infty$ in this case. The following example shows that it by no means needs to be simple.

Example. Let Ω be a domain in $\mathbf{C}$ whose complement contains at least two points, $\phi : \mathbf{D} \to \Omega$ the uniformization map, and define F, G on Ω by

$$F(\phi(\xi)) := (1 - |\xi|^2) \cdot |\phi'(\xi)| \qquad \text{and} \qquad G := \pi^{-1} F^{-2}.$$

(It is easy to see that the definition of F is correct; in fact $|dz|/F(z)$ is the familiar Poincaré metric on Ω.) Denote further

$$G = \{\omega \in \mathrm{Aut}(\mathbf{D}) : \ \phi \circ \omega = \phi\}$$

the covering group of ϕ,

$$d(\eta, \xi) = \left| \frac{\xi - \eta}{1 - \overline{\eta}\xi} \right|$$

the pseudohyperbolic distance of two points $\eta, \xi \in \mathbf{D}$, and

$$\rho(x,y) = \inf\{d(\xi, \eta) \ : \ x = \phi(\xi), y = \phi(\eta)\}$$

the Kobayashi distance of x and y in Ω (see [13], Theorem 3.3.7). It is easy to see that the infimum is always attained at some (ξ, η) (and, hence, at all $(\omega(\xi), \omega(\eta))$ with $\omega \in G$ as well). Define the integer-valued function $n(x,y)$ on $\Omega \times \Omega$ by

$$n(\phi(\xi), y) = \mathrm{card}\{\eta \in \mathbf{D} : \ \phi(\eta) = y \text{ and } d(\xi, \eta) = \rho(\phi(\xi), y)\}.$$

(It is readily seen that the right-hand side depends only on $\phi(\xi)$, i.e. the definition is correct.) Set

$$\mathcal{L} = \{(x,y) \in \Omega \times \Omega : \ n(x,y) > 1\}.$$

Then $\mathcal{L}$ has zero Lebesgue measure, its complement in $\Omega \times \Omega$ is a dense open subset containing the diagonal, the function $F(x)$ has a (single-valued) sesqui-analytic extension $F(x,y)$ to $(\Omega \times \Omega) \setminus \mathcal{L}$, and the modulus $|F(x,y)|$ of the latter even extends to a continuous function on all of $\Omega \times \Omega$ (even though $F(x,y)$ itself usually does not). In fact,

$$|F(x,y)|^2 = \frac{F(x)F(y)}{1 - \rho(x,y)^2}.$$

Explicitly, $F(x,y)$ is given by

$$F(x,y) = (1 - \xi\overline{\eta})\, \phi'(\xi)^{1/2}\, \overline{\phi'(\eta)^{1/2}}$$

for any ξ, η satisfying $x = \phi(\xi)$, $y = \phi(\eta)$, $d(\xi, \eta) = \rho(x, y)$; the right-hand side is independent of the choice of ξ, η if $(x, y) \notin \mathcal{L}$. Let now $\alpha = 2q$ run through the set of even positive integers. The kernel $K_\alpha(x, y)$ then satisfies (see §1 in [5])

$$K_{2q}(\phi(\xi), \phi(\eta)) = \frac{2q-1}{\pi \phi'(\xi)^q \overline{\phi'(\eta)^q}} \sum_{\omega \in G} \frac{\omega'(\xi)^q}{(1 - \overline{\eta}\omega(\xi))^{2q}}.$$

A short computation reveals that

$$\left| \frac{\omega'(\xi)}{(1 - \overline{\eta}\omega(\xi))^2} \right| = \frac{1 - d(\eta, \omega(\xi))^2}{(1 - |\xi|^2)(1 - |\eta|^2)}.$$

It follows that as $q \to +\infty$, the sum is dominated by the terms for which $d(\eta, \omega(\xi))$ is the smallest, i.e. $d(\eta, \omega(\xi)) = \rho(\phi(\xi), \phi(\eta))$; there are precisely $n(\phi(\xi), \phi(\eta))$ such terms. Thus for $(x, y) \in (\Omega \times \Omega) \setminus \mathcal{L}$,

$$K_{2q}(x, y) \sim \frac{2q-1}{\pi} F(x, y)^{-2q}$$

as $q \to +\infty$ (here $a \sim b$ means that $\lim a/b = 1$); so, in a sense, $K_{2q}(x, y)^{1/2q} \to 1/F(x, y)$. On the other hand, for $(x, y) \in \mathcal{L}$

$$\frac{\pi}{2q-1} |K_{2q}(x, y)| \cdot |F(x, y)|^{2q} \sim \left| \sum_{j=1}^{m} e^{iqt_j} \right|$$

where $t_1, \ldots, t_m \in [0, 2\pi)$ and $m = n(x, y) > 1$. This can have both 0 and m as its cluster values; thus $|K_{2q}(x, y)|^{1/2q}$ can display an oscillatory behaviour.

To make things more specific, consider the annulus $\Omega = \{z \in \mathbf{C} : 1 < |z| < R\}$. Then

$$\mathcal{L} = \{(x, y) \in \Omega \times \Omega : \; x\overline{y} < 0\},$$

i.e. the "singular locus" $\mathcal{L}$ consists of points (x, y) with opposite signs. The sesqui-analytic extension of F to $(\Omega \times \Omega) \setminus \mathcal{L}$ is given by ([10])

$$F(x, y) = \frac{2 \log R}{\pi} \exp\left(\frac{1}{2} \log x\overline{y}\right) \sin\left(\frac{\pi \log x\overline{y}}{2 \log R}\right), \qquad -\pi < \operatorname{Im} \log(x\overline{y}) < +\pi,$$

the modulus of which indeed extends continuously to all of $\Omega \times \Omega$; and the reproducing kernels are given by

$$K_{2q}(x, y) = \left(\frac{\pi}{2 \log R}\right)^{2q} \frac{2q-1}{\pi (x\overline{y})^q} \sum_{k \in \mathbf{Z}} \sin^{-2q} \left(\frac{\pi \log x\overline{y}}{2 \log R} + ki \frac{\pi^2}{\log R}\right),$$

see [10], Section 2. Thus for $(x, y) \notin \mathcal{L}$,

$$K_{2q}(x, y) = \frac{2q-1}{\pi} F(x, y)^{-2q} \cdot [1 + O(\gamma^q)] \qquad \text{as } q \to +\infty,$$

where $0 < \gamma < 1$. On the other hand for $(x,y) \in \mathcal{L}$,

$$|K_{2q}(x,y)|^{1/2q} \sim \frac{1}{|x\overline{y}|^{1/2}} \cdot \frac{\pi}{2\log R}\frac{1}{\sqrt{r}}|\cos 2q\theta|^{1/2q},$$

where $\sqrt{r}e^{\pm i\theta} = \sin\left(\frac{\pi\log|x\overline{y}|}{2\log R} \pm i\frac{\pi^2}{2\log R}\right)$, i.e.

$$r = \frac{1}{2}\left(\cosh\frac{\pi^2}{\log R} - \cos\frac{\pi\log|x\overline{y}|}{\log R}\right) > 0,$$

$$\theta = \operatorname{arctg}\left(\operatorname{tgh}\frac{\pi^2}{2\log R}\cdot\operatorname{cotg}\frac{\pi\log|x\overline{y}|}{2\log R}\right) \in (-\pi/2, +\pi/2).$$

As $x\overline{y}$ ranges through the interval $(-R^2, -1)$, θ assumes all values in the interval $(-\pi/2, +\pi/2)$, so there is e.g. an $x\overline{y}$ for which $\theta = \pi/8$, say; but then $|\cos 2q\theta|$ equals 0 for $q = 4k+2$ and 1 for $q = 4k$, so

$$|K_{8k}(x,y)|^{1/8k} \to 1/|F(x,y)| \qquad \text{while} \qquad |K_{8k+4}(x,y)|^{1/(8k+4)} \to 0. \quad \square$$

QUESTION 3. Unfortunately, in this case we have no general result to offer regarding the validity of the asymptotic expansion (Q3). However, granted the existence of such expansion *a priori*, it turns out that we can already tell what the integer m and the coefficients $B_0, B_1, B_2, \ldots$ must be. Observe that, first of all, (Q3) implies (Q1), so by our Corollary 1A a necessary condition for (Q3) is that $-\log F$ be PSH. Also, $B_k(x,y)$ are sesqui-analytic functions, so they are uniquely determined by their restrictions $B_k(x,x) =: B_k(x)$ to the diagonal $x = y$.

Theorem 3. ([9]) *Assume that $-\log F$ is strictly PSH and C^∞ on Ω and (Q3) holds. Let $g_{j\bar{k}}$ be the Lévi matrix*

$$g_{j\bar{k}}(x) = \frac{-\partial^2 \log F(x)}{\partial x_j \partial \overline{x}_k},$$

$(g^{\bar{j}k})$ its matrix inverse, $g = \det(g_{j\bar{k}})$ its determinant, and denote by $\widetilde{\Delta}$ the differential operator

$$\widetilde{\Delta} = \sum_{j,k} g^{\bar{j}k}\frac{\partial^2}{\partial\overline{x}_j\partial x_k}.$$

Then:

(i) $m = n$, the dimension of Ω,

and the coefficients $B_0, B_1, \ldots$ are given by universal (i.e. not depending on F, G and Ω) formulas involving the quantities $g_{j\bar{k}}, g, g^{\bar{j}k}, G$ and their derivatives: for instance,

(ii) $B_0 = \pi^{-n} g/G$,

(iii) $B_1 = \widetilde{\Delta} \log(\sqrt{g}/G)$,

(iv) the formula for B_2 *is already quite complicated, but for* Ω *a planar domain and* $G = \pi^{-n} g$, *it simplifies to* $B_2(x,x) = -\frac{1}{3}\widetilde{\Delta}^2 \log g$.

Remark. Theorem 3 has a nice geometric interpretation. The strict plurisubharmonicity of $-\log F$ implies that the matrix $(g_{j\bar{k}})$ is positive definite and defines thus an Hermitian metric on Ω. The metrics arising in this way are precisely the Kähler metrics. The measure $g\,dm = d\nu$ is then the corresponding volume element, $\widetilde{\Delta}$ is the Laplace-Beltrami operator and $\widetilde{\Delta} \log g = R$ the scalar curvature of the metric. Thus choosing for $G\,dm$ the normalized volume $\pi^{-n} d\nu$ (i.e. $G = \pi^{-n} g$), Theorem 3 says that

$$B_0 = 1, \qquad B_1 = -\tfrac{1}{2} R, \qquad B_2 = -\tfrac{1}{3} \widetilde{\Delta} R,$$

(the last for a planar domain) and it is conceivable that the higher-order coefficients B_k can be expressed in terms of the curvature tensor $R_{j\bar{k}l\bar{m}}$ (of the metric $g_{j\bar{k}}$) and its covariant derivatives. Compare this to Fefferman's formula for the behaviour of the (unweighted) Bergman kernel at the boundary (see e.g. Chapter 12, §3, Theorem 5 in [3]). □

SOME OPEN PROBLEMS:

1. *For which domains* Ω *with Kähler metrics* $g_{j\bar{k}}$ *as above does* (Q3) *hold?*
2. *When are* $B_j(x,y)$ *constants?*

   This we have seen to be true for all the examples in the Introduction except the last one ($\mathbf{D}, \mathbf{B}^n, \mathbf{C}^n$, a bounded symmetric domain, a planar domain with the Poincaré metric, and the punctured plane). Also, an argument similar to §2 in [5] indicates that this property is inherited when passing to a holomorphic quotient (so, for instance, its validity for the planar domains with the Poincaré metric in fact follows from its being true on $\mathbf{D}$ and $\mathbf{C}$, by uniformization). By Theorem 3, a necessary condition is that $G/g =$ const. and that the scalar curvature $R(x) = -2B_1(x,x)$ be constant; thus, in particular, for planar domains this happens only in the situations from examples (d) and (e) in the Introduction.

3. QUANTIZATION

We would like to mention here that the original motivation for Question 3 came from the work of Berezin [4] on quantization on curved phase spaces (in his paper, domains in $\mathbf{C}^n$ with a Kähler form $\omega = d\bar{d}\Psi$; the generalization to manifolds — involving Bergman kernels on holomorphic line bundles — was done only later by Peetre [15]). The basic ingredient in his construction is the *Berezin transform*, which is the integral operator on Ω given by

$$(\dagger) \qquad B_\alpha f(y) = \int_\Omega f(x) \frac{|K_\alpha(x,y)|^2}{K_\alpha(y,y)} F(x)^\alpha G(x)\, dm(x)$$

where $F = e^{-\Psi}$ and $G = \det(\partial\Psi/\partial x_j\partial\overline{x}_k)$ is the volume density. In particular, $-\log F = \Psi$ is strictly-PSH. The crucial step is to establish the asymptotic expansion for the Berezin transform

$$(\ddagger) \qquad B_\alpha f = f + \frac{Q_1 f}{\alpha} + \frac{Q_2 f}{\alpha^2} + \dots$$

as $\alpha \to \infty$ ($1/\alpha$ plays the role of the Planck constant) where Q_1 is the Laplace-Beltrami operator and $Q_2, Q_3, \dots$ are certain differential operators. This is a direct consequence of (Q3), since the latter reduces the integral (†) to integrals of the form

$$\int_\Omega h(x,y)\left[\frac{F(x,x)F(y,y)}{|F(x,y)|^2}\right]^\alpha dm(x)$$

whose asymptotics can be obtained by the Laplace method. See [6] for details. Also, if we take (Q3) for granted, *a priori* formulas for the operator-coefficients $Q_k, k > 1$, can be obtained along the lines of Theorem 3 as well.

4. MEAN VALUE THEOREMS

Consider again the Berezin transform from the previous paragraph:

$$B_\alpha f(y) = \int_\Omega f(x)\frac{|K_\alpha(x,y)|^2}{K_\alpha(y,y)}F(x)^\alpha G(x)\,dm(x).$$

As the integration kernel $|K_\alpha(x,y)|^2/K_\alpha(y,y)$ is nonnegative and

$$B_\alpha 1 = 1$$

by the reproducing property of K_α, $B_\alpha f(y)$ can be regarded as a certain mean-value of f over Ω. This has a particularly clear interpretation on *bounded symmetric domains.* The latter are domains $\Omega \subset \mathbf{C}^n$ on which the group $\mathrm{Aut}(\Omega)$ of biholomorphic self-maps acts transitively. It can be shown that every such domain is biholomorphic to one which is circular, convex and centered at the origin (its *Harish-Chandra realization*). Let Ω be a bounded symmetric domain in its Harish-Chandra realization, $F(z) = K(z,z)^{-1}$ the reciprocal of the Bergman kernel function with respect to the Lebesgue measure, and $G = 1$. The corresponding Berezin transforms B_α are then $\mathrm{Aut}(\Omega)$-invariant,

$$B_\alpha(f \circ \phi) = (B_\alpha f)\circ\phi, \qquad \forall\phi \in \mathrm{Aut}(\Omega),$$

and satisfy

$$B_\alpha f(y) = \int_\Omega f(\phi(x))\,d\mu_\alpha(x) \qquad \text{if } \phi \in \mathrm{Aut}(\Omega), \phi(0) = y.$$

In other words $B_\alpha f(y)$ can be interpreted as a certain "invariant mean value of f around y". Hence a function satisfying $B_\alpha f = f$ is said to have an *invariant mean-value property*.

On $\mathbf{C}^n$, functions having the ordinary mean-value property are precisely the harmonic functions — the functions annihilated by the Laplace operator Δ. The counterpart on bounded symmetric domains of harmonic functions are the $\mathcal{M}$-*harmonic* functions: f is $\mathcal{M}$-harmonic if

$$Df = 0$$

for all Aut(Ω)-invariant differential operators on Ω which annihilate the constants.

What is the connection between the invariant mean-value property and $\mathcal{M}$-harmonicity?

It can be shown that if f is $\mathcal{M}$-harmonic, then $B_\alpha f = f$ for any α. (This is an easy consequence of an old characterization of $\mathcal{M}$-harmonic functions due to Godement [12].) However, the situation in the converse direction turns out to be more delicate.

If $B_\alpha f = f$ for some α and f is bounded, then f is $\mathcal{M}$-harmonic. This result is due to Fürstenberg [11]. Unfortunately, it breaks down if the boundedness condition is relaxed: Ahern, Flores and Rudin showed that there exist functions $f \in L^1(\mathbf{B}^d, dm)$ on the unit ball $\mathbf{B}^d$ of dimension $d \geq 12$ which satisfy $B_0 f = f$ yet are not $\mathcal{M}$-harmonic [1], and later Arazy and Zhang extended this negative result to all bounded symmetric domains other than balls [2]. Finally, for balls of dimension ≤ 11 the converse also breaks down once we consider B_α with $\alpha > 0$ — it can be shown that on any ball there exist functions $f \in L^1(\mathbf{B}^d, d\mu_\alpha)$ satisfying $B_\alpha f = f$ which are not $\mathcal{M}$-harmonic as soon as $\alpha \geq 7/6$.

Using the asymptotic expansion of the Berezin transform (‡), one can obtain the following positive improvement upon the result of Fürstenberg.

Theorem 4. ([8]) *Let Ω be a bounded symmetric domain of rank r, $\alpha \geq 0$, and $f \in L^1(\Omega, dm)$ a function satisfying*

$$f = B_\alpha f = B_{\alpha+1} f = \cdots = B_{\alpha+r} f.$$

Then f is $\mathcal{M}$-harmonic.

(Recall that the rank of a bounded symmetric domain in the Harish-Chandra realization is the maximum possible dimension of a polydisc which is contained in Ω and whose boundary lies on $\partial\Omega$. For instance, the bounded symmetric domains of rank 1 are (up to biholomorphic equivalence) precisely the unit balls $\mathbf{B}^d$.)

We have also the following local version of this result.

Theorem 5. ([8]) *Let Ω be a bounded symmetric domain, $y \in \Omega$, and assume that $f \in L^1(\Omega, dm)$ is C^∞ in a neighborhood of y. If*

$$B_{\alpha_k} f(y) = f(y)$$

for some sequence $\alpha_k \to \infty$, *then*

$$Df(y) = 0$$

for every $\mathrm{Aut}(\Omega)$*-invariant differential operator* D *which annihilates the constants.*

REFERENCES

[1] P. Ahern, M. Flores, W. Rudin. *An invariant volume-mean-value property.* J. Funct. Anal., 111, pp. 380–397, 1993.

[2] J. Arazy, G. Zhang. *L^q-estimates of spherical functions and an invariant mean-value property.* Integr. Eq. Oper. Theory, 23, pp. 123–144, 1995.

[3] M. Beals, C. Fefferman, R. Grossman. *Strictly pseudoconvex domains in* $\mathbf{C}^n$. Bull. Amer. Math. Soc., 8, pp. 125–326, 1983.

[4] F.A. Berezin. *Quantization.* Math. USSR Izvestiya, 8, pp. 1109–1163, 1974.

[5] M. Engliš. *Asymptotics of the Berezin transform and quantization on planar domains.* Duke Math. J., 79, pp. 57–76, 1995.

[6] M. Engliš. *Berezin quantization and reproducing kernels on complex domains.* Trans. Amer. Math. Soc., 348, pp. 411–479, 1996.

[7] M. Engliš. *Asymptotic behavior of reproducing kernels of weighted Bergman spaces.* Trans. Amer. Math. Soc., 349, pp. 3717–3735, 1997.

[8] M. Engliš. *A mean-value theorem on bounded symmetric domains.* Proc. Amer. Math. Soc., to appear.

[9] M. Engliš. *A calculation of asymptotics of the Laplace integral on a Kähler manifold.* In preparation.

[10] M. Engliš, J. Peetre. *On the correspondence principle for the quantized annulus.* Math. Scand., 78, pp. 183–206, 1996.

[11] H. Fürstenberg. *Poisson formula for semisimple Lie groups.* Ann. Math., 77, pp. 335–386, 1963.

[12] R. Godement. *Une généralisation des représentations de la moyenne pour les fonctions harmoniques.* C. R. Acad. Sci. Paris, 234, pp. 2137–2139, 1952.

[13] M. Jarnicki, P. Pflug. *Invariant distances and metrics in complex analysis.* Walter de Gruyter, 1993.

[14] P. Lin, R. Rochberg. *Hankel operators on the weighted Bergman spaces with exponential type weights.* Integral Equations Operator Theory, 21, pp. 460–483, 1995.

[15] J. Peetre. *The Berezin transform and Ha-plitz operators.* J. Operator Theory, 24, pp. 165–186, 1990.

7 HILBERT SPACES OF EIGEN-FUNCTIONS OF THE LAPLACIAN

Keiko FUJITA

Faculty of Culture and Education,
Saga University, 840-8502, Japan,
keiko@cc.saga-u.ac.jp

Abstract: In this paper, we introduce Hilbert spaces of eigenfunctions of the Laplacian with a reproducing kernel. They appear as an image under the Fourier transformation of a Hilbert space of holomorphic functions on the complex sphere. Related topics are in [2], [3], [4] and [5].

The author would like to thank the referee for his useful comments.

1 EIGENSPACES OF THE LAPLACIAN

Let $\tilde{\mathbf{E}} = \mathbf{C}^{n+1}$, $n \geq 2$, be the complex Euclidean space. We denote by $\mathcal{O}(\tilde{\mathbf{E}})$ the space of entire functions on $\tilde{\mathbf{E}}$ with the topology of uniform convergence on compact sets. We define the complex Laplacian Δ on $\tilde{\mathbf{E}}$ by

$$\Delta_z = \frac{\partial^2}{\partial z_1^2} + \frac{\partial^2}{\partial z_2^2} + \cdots + \frac{\partial^2}{\partial z_{n+1}^2}.$$

We denote the space of entire eigenfunctions of the Laplacian with eigenvalue λ^2 by $\mathcal{O}_{\Delta-\lambda^2}(\tilde{\mathbf{E}})$; that is,

$$\mathcal{O}_{\Delta-\lambda^2}(\tilde{\mathbf{E}}) = \{f \in \mathcal{O}(\tilde{\mathbf{E}}); \Delta_z f(z) = \lambda^2 f(z)\}.$$

$\mathcal{P}^k_\Delta(\tilde{\mathbf{E}})$ denotes the space of k-homogeneous harmonic polynomials on $\tilde{\mathbf{E}}$. The dimension of $\mathcal{P}^k_\Delta(\tilde{\mathbf{E}})$ is given by

$$N(k,n) = \frac{(2k+n-1)(k+n-2)!}{k!(n-1)!} = O(k^{n-1}).$$

Let $P_{k,n}(t)$ be the Legendre polynomial of degree k and of dimension $n+1$. Note that $\overline{P_{k,n}(t)} = P_{k,n}(\bar{t})$. The coefficient of its highest power is given by

$$\gamma_{k,n} \equiv \frac{\Gamma(k+(n+1)/2)2^k}{N(k,n)\Gamma((n+1)/2)k!}. \tag{1}$$

Let $R > 0$ and $\lambda \in \mathbf{C}$. Put

$$L_{k,\lambda,R} \equiv \begin{cases} |\lambda|^{2k} P_{k,n}\left(\frac{1}{2}\left(\frac{R^2}{|\lambda|^2} + \frac{|\lambda|^2}{R^2}\right)\right), & \lambda \neq 0, \\ \frac{\gamma_{k,n}}{2^k} R^{2k}, & \lambda = 0. \end{cases}$$

Note that $L_{k,0,R} = \lim_{\lambda \to 0} L_{k,\lambda,R}$.

The harmonic extension $\tilde{P}_{k,n}(z, w)$ of $P_{k,n}(z \cdot w)$ is defined by

$$\tilde{P}_{k,n}(z, w) = (\sqrt{z^2})^k (\sqrt{w^2})^k P_{k,n}\left(\frac{z}{\sqrt{z^2}} \cdot \frac{w}{\sqrt{w^2}}\right),$$

where $z \cdot w = z_1 w_1 + z_2 w_2 + \cdots + z_{n+1} w_{n+1}$ and $z^2 = z \cdot z$.

$\tilde{P}_{k,n}(z, w)$ is a k-homogeneous polynomial in z and in w, $\tilde{P}_{k,n}(z, w) = \tilde{P}_{k,n}(w, z)$, $\Delta_z \tilde{P}_{k,n}(z, w) = \Delta_w \tilde{P}_{k,n}(z, w) = 0$ and $\tilde{P}_{k,n}(\overline{z}, \overline{w}) = \overline{\tilde{P}_{k,n}(z, w)}$. Note that $\tilde{P}_{k,n}(z, w) = \gamma_{k,n}(z \cdot w)^k$ if $z^2 = 0$ or $w^2 = 0$.

Let $L(z)$ be the Lie norm given by

$$L(z) = \{\|z\|^2 + (\|z\|^4 - |z^2|^2)^{1/2}\}^{1/2},$$

where $\|z\|^2 = z \cdot \overline{z}$. For $|\lambda| \leq R$ put

$$\tilde{S}_{\lambda,R} = \{z \in \tilde{\mathbf{E}}\,;\ z^2 = \lambda^2,\ L(z) = R\}.$$

If $|\lambda| = R$, then $\tilde{S}_{\lambda,R}$ is the real sphere with complex radius λ :

$$\tilde{S}_{\lambda,|\lambda|} = \lambda S_1 = \{\lambda\omega\,;\ \omega \in S_1\},$$

where S_1 is the n-dimensional unit sphere. Since the rotation group acts transitively on $\tilde{S}_{\lambda,R}$, there is a unique normalized invariant measure on $\tilde{S}_{\lambda,R}$ which we denote by $d\tilde{S}_{\lambda,R}$ or $dS_\lambda = d\tilde{S}_{\lambda,|\lambda|}$.

Let $L^2(\tilde{S}_{\lambda,R})$ be the space of square integrable functions on $\tilde{S}_{\lambda,R}$ with the inner product

$$(f, g)_{\tilde{S}_{\lambda,R}} = \int_{\tilde{S}_{\lambda,R}} f(w)\overline{g(w)}\, d\tilde{S}_{\lambda,R}(w). \tag{2}$$

The following lemmas are known:

Lemma 1.1 ([11, Lemma 1.3] and [12, Lemmas 3.2 and 3.3])
Let $f_k \in \mathcal{P}^k_\Delta(\tilde{\mathbf{E}})$, $g_l \in \mathcal{P}^l_\Delta(\tilde{\mathbf{E}})$ and $|\lambda| \leq R$. Then we have the orthogonality;

$$(f_k, g_l)_{\tilde{S}_{\lambda,R}} = \delta_{kl} L_{k,\lambda,R} \int_{S_1} f_k(w)\overline{g_l(w)}\, dS_1(w).$$

Further, for $f_k \in \mathcal{P}^k_\Delta(\tilde{\mathbf{E}})$, we have the reproducing formula;

$$\begin{aligned} f_k(z) &= \frac{N(k,n)}{L_{k,\lambda,R}} \int_{\tilde{S}_{\lambda,R}} f_k(w) \tilde{P}_{k,n}(z, \overline{w})\, d\tilde{S}_{\lambda,R}(w) \\ &= \frac{N(k,n)}{L_{k,\lambda,R}} \int_{\tilde{S}_{\lambda,R}} f_k(\overline{w}) \tilde{P}_{k,n}(z, w)\, d\tilde{S}_{\lambda,R}(w). \end{aligned}$$

Especially, we have

$$\begin{aligned}\tilde{P}_{k,n}(z,w) &= \frac{N(k,n)}{L_{k,\lambda,R}}\int_{\tilde{S}_{\lambda,R}}\tilde{P}_{k,n}(\zeta,w)\tilde{P}_{k,n}(z,\overline{\zeta})d\tilde{S}_{\lambda,R}(\zeta)\\ &= 2^k N(k,n)\gamma_{k,n}\int_{\tilde{S}_{0,1}}(\zeta\cdot w)^k(z\cdot\overline{\zeta})^k d\tilde{S}_{0,1}(\zeta).\end{aligned}$$

LEMMA 1. 2 **([13, Corollary 2.3])** *Let $f\in\mathcal{O}_{\Delta-\lambda^2}(\tilde{\mathbf{E}})$. Define*

$$f_k(z)=2^k N(k,n)\int_{\tilde{S}_{0,1}} f(w)(z\cdot\overline{w})^k d\tilde{S}_{0,1}(w),\quad z\in\tilde{\mathbf{E}}.$$

Then $f_k\in\mathcal{P}^k_\Delta(\tilde{\mathbf{E}})$ and f is expanded as follows:

$$f(z)=\sum_{k=0}^{\infty}\tilde{j}_k(i\lambda\sqrt{z^2})f_k(z),$$

where the convergence is in the topology of $\mathcal{O}_{\Delta-\lambda^2}(\tilde{\mathbf{E}})$ and $\tilde{j}_k(t)=\tilde{J}_{k+(n-1)/2}(t)$ is the entire Bessel function;

$$\tilde{J}_\mu(t)=\sum_{l=0}^{\infty}\frac{(-1)^l\Gamma(\mu+1)}{\Gamma(\mu+l+1)l!}\left(\frac{t}{2}\right)^{2l}=\Gamma(\mu+1)\left(\frac{t}{2}\right)^{-\mu}J_\mu(t).$$

LEMMA 1. 3 *Let $f_k\in\mathcal{P}^k_\Delta(\tilde{\mathbf{E}})$. Then $\tilde{j}_k(i\lambda\sqrt{z^2})f_k(z)\in\mathcal{O}_{\Delta-\lambda^2}(\tilde{\mathbf{E}})$ and for $R\geq|\lambda|$ we have an integral representation:*

$$\tilde{j}_k(i\lambda\sqrt{z^2})f_k(z)=\frac{k!\gamma_{k,n}N(k,n)}{L_{k,\lambda,R}}\int_{\tilde{S}_{\lambda,R}}\exp(z\cdot w)f_k(\overline{w})d\tilde{S}_{\lambda,R}(w),\quad z\in\tilde{\mathbf{E}}.$$

PROOF. By Lemma 7.3 in [8], for $z,w\in\tilde{\mathbf{E}}$, we have

$$\exp(z\cdot w)=\sum_{k=0}^{\infty}\frac{1}{k!\gamma_{k,n}}\tilde{j}_k(i\sqrt{z^2}\sqrt{w^2})\tilde{P}_{k,n}(z,w).\tag{3}$$

Thus by Lemma 1. 1 , we have

$$\begin{aligned}&\frac{k!\gamma_{k,n}N(k,n)}{L_{k,\lambda,R}}\int_{\tilde{S}_{\lambda,R}}\exp(z\cdot w)f_k(\overline{w})d\tilde{S}_{\lambda,R}(w)\\ &=\frac{k!\gamma_{k,n}N(k,n)}{L_{k,\lambda,R}}\int_{\tilde{S}_{\lambda,R}}\sum_{k=0}^{\infty}\frac{1}{k!\gamma_{k,n}}\tilde{j}_k(i\lambda\sqrt{z^2})\tilde{P}_{k,n}(z,w)f_k(\overline{w})d\tilde{S}_{\lambda,R}(w)\\ &=\tilde{j}_k(i\lambda\sqrt{z^2})f_k(z).\end{aligned}$$

q.e.d.

Put $\|f\|^2_{\tilde{S}_{\lambda,R}} = (f,f)_{\tilde{S}_{\lambda,R}}$. Since $\tilde{j}_k(0) = 1$, for $f_k \in \mathcal{P}^k_\Delta(\tilde{\mathbf{E}})$ and $f(z) = \sum_{k=0}^{\infty} \tilde{j}_k(i\lambda\sqrt{z^2}) f_k(z) \in \mathcal{O}_{\Delta-\lambda^2}(\tilde{\mathbf{E}})$, we have $\|f\|^2_{\tilde{S}_{0,1}} = \|\sum_{k=0}^{\infty} \tilde{j}_k(i\lambda\sqrt{z^2}) f_k(z)\|^2_{\tilde{S}_{0,1}} = \|\sum_{k=0}^{\infty} f_k\|^2_{\tilde{S}_{0,1}}$. Further by Lemma 1. 1, $\|f\|^2_{\tilde{S}_{0,1}} = \sum_{k=0}^{\infty} \|f_k\|^2_{\tilde{S}_{0,1}}$.

We consider the following space:

$$\mathcal{E}^2_{\Delta-\lambda^2}(\tilde{\mathbf{E}}; R) = \left\{ f(z) = \sum_{k=0}^{\infty} \tilde{j}_k(i\lambda\sqrt{z^2}) f_k(z) \in \mathcal{O}_{\Delta-\lambda^2}(\tilde{\mathbf{E}}); f_k \in \mathcal{P}^k_\Delta(\tilde{\mathbf{E}}), \right.$$
$$\left. \sum_{k=0}^{\infty} \frac{(N(k,n)k!)^2 2^k \gamma_{k,n}}{L_{k,\lambda,R}} \|f_k\|^2_{\tilde{S}_{0,1}} < \infty \right\}. \tag{4}$$

It is a Hilbert space with the inner product $((f\,,\,g))^{\lambda,R}$;

$$((f\,,\,g))^{\lambda,R} = \sum_{k=0}^{\infty} \frac{(N(k,n)k!)^2 2^k \gamma_{k,n}}{L_{k,\lambda,R}} (f_k, g_k)_{\tilde{S}_{0,1}}, \tag{5}$$

where $f_k,\ g_k \in \mathcal{P}^k_\Delta(\tilde{\mathbf{E}})$ and $f(z) = \sum_{k=0}^{\infty} \tilde{j}_k(i\lambda\sqrt{z^2}) f_k(z)$, $g(z) = \sum_{k=0}^{\infty} \tilde{j}_k(i\lambda\sqrt{z^2}) g_k(z)$.

The following lemma implies that the Hilbert space $\mathcal{E}^2_{\Delta-\lambda^2}(\tilde{\mathbf{E}}; R)$ has a reproducing kernel.

LEMMA **2. 1** *Let $f \in \mathcal{E}^2_{\Delta-\lambda^2}(\tilde{\mathbf{E}}; R)$. Then we have the following estimate:*

$$\begin{cases} |f(z)| \le \exp(RL^*(z))\sqrt{((f,f))^{\lambda,R}}, & |\lambda| \le R, \\ |f(z)| \le \exp\left(\dfrac{|\lambda|^2}{R} L^*(z)\right)\sqrt{((f,f))^{\lambda,R}}, & |\lambda| \ge R, \end{cases}$$

where $L^*(z)$ is the dual Lie norm given by

$$L^*(z) = \sup\{|z \cdot \zeta|\,;\ L(\zeta) \le 1\} = \{(\|z\|^2 + |z^2|)/2\}^{1/2}.$$

PROOF. By Lemmas 1. 2, 1. 3 and 1. 1, for $|\lambda| \le R$, we have

$$\begin{aligned} |f(z)| &= \left| \sum_{k=0}^{\infty} \tilde{j}_k(i\lambda\sqrt{z^2}) f_k(z) \right| \\ &= \left| \sum_{k=0}^{\infty} \frac{k!\gamma_{k,n} N(k,n)}{L_{k,\lambda,R}} \int_{\tilde{S}_{\lambda,R}} \exp(z \cdot w) f_k(\overline{w}) d\tilde{S}_{\lambda,R}(w) \right| \end{aligned}$$

$$\begin{aligned}
&\leq \exp(RL^*(z)) \lim_{N\to\infty} \left\{ \int_{\tilde{S}_{\lambda,R}} \left| \sum_{k=0}^{N} \frac{k!\gamma_{k,n}N(k,n)}{L_{k,\lambda,R}} f_k(w) \right|^2 d\tilde{S}_{\lambda,R}(w) \right\}^{1/2} \\
&= \exp(RL^*(z)) \lim_{N\to\infty} \left\{ \sum_{k=0}^{N} (k!\gamma_{k,n}N(k,n))^2 \frac{2^k}{\gamma_{k,n}L_{k,\lambda,R}} \|f_k\|^2_{\tilde{S}_{0,1}} \right\}^{1/2} \\
&= \exp(RL^*(z)) \lim_{N\to\infty} \left\{ \sum_{k=0}^{N} \frac{(k!N(k,n))^2\gamma_{k,n}2^k}{L_{k,\lambda,R}} \|f_k\|^2_{\tilde{S}_{0,1}} \right\}^{1/2} \\
&= \exp(RL^*(z))\sqrt{((f,f))^{\lambda,R}}.
\end{aligned}$$

Thus we have the first estimate for $|\lambda| \leq R$.

If $R \leq |\lambda|$, then $|\lambda| \leq |\lambda|^2/R$. Therefore the second estimate for $|\lambda| \geq R$ comes from the first one for $|\lambda| \leq R$ because $L_{k,\lambda,R} = L_{k,\lambda,|\lambda|^2/R}$ and $((f,f))^{\lambda,|\lambda|^2/R} = ((f,f))^{\lambda,R}$. q.e.d.

In Section 5, we concretely construct a reproducing kernel.

If $w^2 = \lambda^2$, then $(\Delta_z - \lambda^2)e^{z\cdot w} = 0$. This indicates that an eigenfunction appears as an image under the Fourier transformation on the complex sphere (Section 6).

3 HIBERT SPACES ON THE COMPLEX SPHERE

We define the complex sphere $\tilde{S}_\lambda$ with radius $\lambda \in \mathbf{C}$ by

$$\tilde{S}_\lambda = \{z \in \tilde{\mathbf{E}};\, z^2 = \lambda^2\}.$$

Especially for $\lambda = 0$, sometimes we call $\tilde{S}_0$ the complex light cone. Put

$$\tilde{S}_\lambda(R) = \{z \in \tilde{S}_\lambda;\, L(z) < R\}.$$

Note that $\partial\tilde{S}_\lambda(R) = \tilde{S}_{\lambda,R}$, $\tilde{S}_\lambda(\infty) = \tilde{S}_\lambda$ and $\tilde{S}_\lambda(R) = \emptyset$ for $R \leq |\lambda|$.

Put $\mathcal{H}^k(\tilde{S}_{\lambda,R}) = \mathcal{P}^k_\Delta(\tilde{\mathbf{E}})|_{\tilde{S}_{\lambda,R}}$. By Lemma 1. 1 , endowed with an inner product (2) on $\mathcal{H}^k(\tilde{S}_{\lambda,R})$, $k = 0,1,2,\cdots$, they are mutually orthogonal finite dimensional subspaces of $L^2(\tilde{S}_{\lambda,R})$. Let $L^2\mathcal{O}(\tilde{S}_{\lambda,R})$ be the closed subspace of $L^2(\tilde{S}_{\lambda,R})$ generated by $\mathcal{H}^k(\tilde{S}_{\lambda,R})$, $k = 0,1,\cdots$. Then by the definition, we have the following lemma:

LEMMA **3. 1** *Let $f \in L^2\mathcal{O}(\tilde{S}_{\lambda,R})$ and define $\tilde{f}_k$ by*

$$\tilde{f}_k(z) = \frac{N(k,n)}{L_{k,\lambda,R}} \int_{\tilde{S}_{\lambda,R}} f(w)\tilde{P}_{k,n}(z,\overline{w})d\tilde{S}_{\lambda,R}(w), \quad z \in \tilde{\mathbf{E}}.$$

Then $f_k = \tilde{f}_k|_{\tilde{S}_{\lambda,R}} \in \mathcal{H}^k(\tilde{S}_{\lambda,R})$ *and the expansion* $\sum_{k=0}^{\infty} f_k$ *converges to* f *in the topology of* $L^2\mathcal{O}(\tilde{S}_{\lambda,R})$*; that is, we have the Hilbert direct sum decomposition:*

$$L^2\mathcal{O}(\tilde{S}_{\lambda,R}) = \bigoplus_{k=0}^{\infty} \mathcal{H}^k(\tilde{S}_{\lambda,R}).$$

The mapping $f \mapsto f_k$ *is the orthogonal projection of* $L^2\mathcal{O}(\tilde{S}_{\lambda,R})$ *onto* $\mathcal{H}^k(\tilde{S}_{\lambda,R})$.

Note that $L^2\mathcal{O}(\tilde{S}_{\lambda,|\lambda|}) = L^2(S_\lambda)$. But $L^2\mathcal{O}(\tilde{S}_{\lambda,R}) \underset{\neq}{\subset} L^2(\tilde{S}_{\lambda,R})$ if $|\lambda| < R$.

LEMMA **3. 2** *For* $|\lambda| < R$, $L^2\mathcal{O}(\tilde{S}_{\lambda,R})$ *is isomorphic to the Hardy space*

$$H^2(\tilde{S}_\lambda(R)) = \left\{ f \in \mathcal{O}(\tilde{S}_\lambda(R))\,;\ \sup_{|\lambda| \le R' < R} \int_{\tilde{S}_{\lambda,R'}} |f(z)|^2 d\tilde{S}_{\lambda,R'}(z) < \infty \right\},$$

where $\mathcal{O}(\tilde{S}_\lambda(R))$ *is the space of holomorphic functions on* $\tilde{S}_\lambda(R)$ *with the topology of uniform convergence on compact sets.*

Since $\lim_{k\to\infty} (L_{k,\lambda,R})^{1/k} = R^2$ for $|\lambda| \le R$ and $|\tilde{P}_{k,n}(z,w)| \le L(z)^k L(w)^k$ (see (18) and Lemma 7. 6 in Appendix I),

$$K_{\lambda,R}(z,w) = \sum_{k=0}^{\infty} \frac{N(k,n)}{L_{k,\lambda,R}} \tilde{P}_{k,n}(z,\overline{w}) \tag{6}$$

is well-defined for $L(z)L(w) < R^2$. Then $\overline{K_{\lambda,R}(z,w)} = K_{\lambda,R}(w,z)$ and $K_{R,R}(z,\overline{w})$ is the classical Poisson kernel and $K_{0,R}(z,\overline{w})$ restricted to $\tilde{S}_0 \times \tilde{\mathbf{E}}$ is the Cauchy kernel on $\tilde{S}_0$ (see [7]):

$$K_{R,R}(z,w) = K_{1,1}(z/R, w/R), \quad K_{1,1}(z,\overline{w}) = \frac{1 - z^2 w^2}{(1 + z^2 w^2 - 2z\cdot w)^{(n+1)/2}},$$

$$K_{0,R}(z,w) = K_{0,1}(z/R, w/R), \quad K_{0,1}(z,\overline{w})|_{\tilde{S}_0 \times \tilde{\mathbf{E}}} = \frac{1 + 2z\cdot w}{(1 - 2z\cdot w)^n}.$$

$K_{\lambda,R}(z,w)$ is a reproducing kernel for $H^2(\tilde{S}_\lambda(R))$; that is, we have

PROPOSITION **3. 3** *For* $f \in H^2(\tilde{S}_\lambda(R))$ *we have the following integral representation :*

$$f(z) = (f(w), K_{\lambda,R}(w,z))_{\tilde{S}_{\lambda,R}} = \int_{\tilde{S}_{\lambda,R}} f(w) K_{\lambda,R}(z,w) d\tilde{S}_{\lambda,R}(w), \quad z \in \tilde{S}_\lambda(R).$$

PROOF. A proof is given by a simple calculation with (6), Lemmas 1. 1 , 3. 1 and 3. 2 . q.e.d.

4 HILBERT SPACES OF EIGENFUNCTIONS II

In (4), we defined $\mathcal{E}^2_{\Delta-\lambda^2}(\tilde{\mathbf{E}};R)$ as a weighted Hilbert direct sum of $\mathcal{H}^k(\tilde{S}_{0,1})$. Here, we redefine $\mathcal{E}^2_{\Delta-\lambda^2}(\tilde{\mathbf{E}};R)$ by using an integral over $\tilde{S}_0$.

We define a measure $d\tilde{S}_{0(\lambda,R)}$ on $\tilde{S}_0$ by

$$\int_{\tilde{S}_0} f(z)d\tilde{S}_{0(\lambda,R)}(z) = \int_0^\infty \int_{\tilde{S}_{0,1}} f(rz')d\tilde{S}_{0,1}(z')\rho_{\lambda,R}(r)dr,$$

where $\rho_{\lambda,R}$ is a C^∞ function which satisfies

$$\int_0^\infty r^{2k}\rho_{\lambda,R}(r)dr = \frac{(N(k,n)k!)^2\gamma_{k,n}2^k}{L_{k,\lambda,R}} \equiv \frac{C(k,n)}{L_{k,\lambda,R}}, \quad k=0,1,2,\cdots. \tag{7}$$

By a theorem of Duran ([1]), such a function exists. Especially for $|\lambda| = R$, K.Ii ([6]) and R.Wada ([11]) concretely constructed such a function as a C^∞ function of exponential type $-R$, which we denote by $\rho_R(r)$; that is, we have

$$\int_0^\infty r^{2k}\rho_R(r)dr = C(k,n)/R^{2k}, \quad k=0,1,2,\cdots.$$

Furthermore, using this function, in [2], we defined a measure dE_R on $\mathbf{E}\equiv\mathbf{R}^{n+1}$ by

$$\int_{\mathbf{E}} f(x)dE_R(x) \equiv \int_0^\infty \int_{S_1} f(r\omega)dS_1(\omega)\rho_R(r)dr.$$

Now, $\mathcal{E}^2_{\Delta-\lambda^2}(\tilde{\mathbf{E}};R)$ defined in (4) can be redefined by

$$\mathcal{E}^2_{\Delta-\lambda^2}(\tilde{\mathbf{E}};R) = \left\{f\in\mathcal{O}_{\Delta-\lambda^2}(\tilde{\mathbf{E}})\ ;\ \int_{\tilde{S}_0} |f(z)|^2 d\tilde{S}_{0(\lambda,R)}(z) < \infty\right\}.$$

By Lemma 1. 1 , especially for $\lambda=0$, $\mathcal{E}^2_{\Delta}(\tilde{\mathbf{E}};R)$ can be also defined by

$$\mathcal{E}^2_{\Delta}(\tilde{\mathbf{E}};R) = \left\{f\in\mathcal{O}_{\Delta}(\tilde{\mathbf{E}})\ ;\ \int_{\mathbf{E}} |f(x)|^2 dE_R(x) < \infty\right\}.$$

5 REPRODUCING KERNEL

Now, we consider a reproducing kernel for $\mathcal{E}^2_{\Delta-\lambda^2}(\tilde{\mathbf{E}};R)$.

Since $L_{k,\lambda,R} = L_{k,\lambda,|\lambda|^2/R}$, we have

$$\mathcal{E}^2_{\Delta-\lambda^2}(\tilde{\mathbf{E}};R) = \mathcal{E}^2_{\Delta-\lambda^2}(\tilde{\mathbf{E}};|\lambda|^2/R).$$

Therefore, without loss of generality, we may assume $|\lambda|\le R$ because $|\lambda| < |\lambda|^2/R$ for $|\lambda| > R$. Thus in the sequel, we consider only the case of $|\lambda|\le R$.

For $z, w \in \tilde{\mathbf{E}}$, define

$$E_R^\lambda(z,w) = \sum_{k=0}^{\infty} \frac{2^k N(k,n) L_{k,\lambda,R}}{\gamma_{k,n} C(k,n)} \tilde{j}_k(i\lambda\sqrt{z^2}) \overline{\tilde{j}_k(i\lambda\sqrt{w^2})} \tilde{P}_{k,n}(z,\overline{w}). \tag{8}$$

Then $E_R^\lambda(z,w) = \overline{E_R^\lambda(w,z)}$ and $\Delta_z E_R^\lambda(z,w) = \lambda^2 E_R^\lambda(z,w)$. It is easy to see that

$$E_R^\lambda(\cdot\,, w) \in \mathcal{E}^2_{\Delta-\lambda^2}(\tilde{\mathbf{E}}; R).$$

$E_R^\lambda(z,w)$ has an integral representation:

$$E_R^\lambda(z,w) = \int_{\tilde{S}_{\lambda,R}} \exp(z\cdot x)\overline{\exp(x\cdot w)}\, d\tilde{S}_{\lambda,R}(x), \quad |\lambda| \leq R.$$

PROOF. By (3), (7), (8) and Lemma 1. 1 ,

$$\begin{aligned}
&\int_{\tilde{S}_{\lambda,R}} \exp(z\cdot x)\overline{\exp(x\cdot w)}\, d\tilde{S}_{\lambda,R}(x) \\
&= \int_{\tilde{S}_{\lambda,R}} \sum_{k=0}^{\infty} \frac{1}{k!\gamma_{k,n}} \tilde{j}_k(i\lambda\sqrt{z^2}) \tilde{P}_{k,n}(z,x) \overline{\sum_{l=0}^{\infty} \frac{1}{l!\gamma_{l,n}} \tilde{j}_l(i\lambda\sqrt{w^2}) \tilde{P}_{l,n}(w,x)}\, d\tilde{S}_{\lambda,R}(x) \\
&= \sum_{k=0}^{\infty} \frac{L_{k,\lambda,R}}{N(k,n)(k!\gamma_{k,n})^2} \tilde{j}_k(i\lambda\sqrt{z^2}) \overline{\tilde{j}_k(i\lambda\sqrt{w^2})} \tilde{P}_{k,n}(z,\overline{w}) \\
&= \sum_{k=0}^{\infty} \frac{2^k N(k,n) L_{k,\lambda,R}}{\gamma_{k,n} C(k,n)} \tilde{j}_k(i\lambda\sqrt{z^2}) \overline{\tilde{j}_k(i\lambda\sqrt{w^2})} \tilde{P}_{k,n}(z,\overline{w}) = E_R^\lambda(z,w).
\end{aligned}$$

q.e.d.

Put

$$(f, g)_{\tilde{S}_0,\lambda,R} = \int_{\tilde{S}_0} f(z)\overline{g(z)}\, d\tilde{S}_{0(\lambda,R)}.$$

THEOREM **5. 1** *$E_R^\lambda(z,w)$ is a reproducing kernel for $\mathcal{E}^2_{\Delta-\lambda^2}(\tilde{\mathbf{E}}; R)$; that is, for $f \in \mathcal{E}^2_{\Delta-\lambda^2}(\tilde{\mathbf{E}}; R)$ we have the following integral representation:*

$$f(z) = (f(w), E_R^\lambda(w,z))_{\tilde{S}_0,\lambda,R} = \int_{\tilde{S}_0} f(w) E_R^\lambda(z,w)\, d\tilde{S}_{0(\lambda,R)}(w), \quad z \in \tilde{\mathbf{E}}. \tag{9}$$

PROOF. By (8), (7) and Lemma 1. 2 , we have

$$\begin{aligned}
&\int_{\tilde{S}_0} f(w) E_R^\lambda(z,w)\, d\tilde{S}_{0(\lambda,R)}(w) \\
&= \int_{\tilde{S}_0} f(w) \sum_{k=0}^{\infty} \frac{2^k N(k,n) L_{k,\lambda,R}}{C(k,n)} \tilde{j}_k(i\lambda\sqrt{z^2}) (z\cdot\overline{w})^k\, d\tilde{S}_{0(\lambda,R)}(w)
\end{aligned}$$

$$= \sum_{k=0}^{\infty} \frac{L_{k,\lambda,R}}{C(k,n)} \tilde{j}_k(i\lambda\sqrt{z^2}) f_k(z) \int_0^{\infty} r^{2k} \rho_{\lambda,R}(r)dr$$
$$= \sum_{k=0}^{\infty} \tilde{j}_k(i\lambda\sqrt{z^2}) f_k(z) = f(z).$$

q.e.d.

For $f \in \mathcal{E}^2_{\Delta}(\tilde{\mathbf{E}}; R)$ the integral representation (9) reduces to that of Theorem 1.9 in [2]; that is, $f \in \mathcal{E}^2_{\Delta}(\tilde{\mathbf{E}}; R)$ can also be represented as an integral over $\mathbf{E}$;

$$f(z) = \int_{\mathbf{E}} f(w) E^0_R(z,w) dE_R(w), \quad z \in \tilde{\mathbf{E}}.$$

6 FOURIER TRANSFORMATION

For $f \in H^2(\tilde{S}_\lambda(R))$, we define the Fourier transformation $\mathcal{F}_\lambda$ by

$$\mathcal{F}_\lambda : f \mapsto \mathcal{F}_\lambda f(w) = (\exp(z\cdot w), f(z))_{\tilde{S}_{\lambda,R}} = \int_{\tilde{S}_{\lambda,R}} \exp(z\cdot w)\overline{f(z)} d\tilde{S}_{\lambda,R}(z), \quad w \in \tilde{\mathbf{E}}.$$

Note that it is an antilinear mapping and we have

$$\mathcal{F}_\lambda f(w) = \sum_{k=0}^{\infty} \frac{L_{k,\lambda,R}}{N(k,n)k!\gamma_{k,n}} \tilde{j}_k(i\lambda\sqrt{w^2}) \overline{f_k}(w), \tag{10}$$

where $\overline{f_k}(z) = \overline{f_k(\overline{z})}$ for $f_k \in \mathcal{P}^k_{\Delta}(\tilde{\mathbf{E}})$.

Following K.Ii's idea (see [6]), more generally, we have the following theorem:

THEOREM **6. 1** *The Fourier transformation $\mathcal{F}_\lambda$ is an antilinear unitary isomorphism of $H^2(\tilde{S}_\lambda(R))$ onto $\mathcal{E}^2_{\Delta-\lambda^2}(\tilde{\mathbf{E}}; R)$.*

PROOF. Let $f \in H^2(\tilde{S}_\lambda(R))$. By (5), (10) and Lemma 1. 1 , we have

$$\begin{aligned}
((\mathcal{F}_\lambda f, \mathcal{F}_\lambda f))^{\lambda,R} &= \sum_{k=0}^{\infty} \frac{(N(k,n)k!)^2 2^k \gamma_{k,n}}{L_{k,\lambda,R}} ((\mathcal{F}_\lambda f)_k, (\mathcal{F}_\lambda f)_k)_{\tilde{S}_{0,1}} \\
&= \sum_{k=0}^{\infty} \frac{(N(k,n)k!)^2 2^k \gamma_{k,n}}{L_{k,\lambda,R}} \left(\frac{L_{k,\lambda,R}\,\overline{f_k}}{N(k,n)k!\gamma_{k,n}}, \frac{L_{k,\lambda,R}\,\overline{f_k}}{N(k,n)k!\gamma_{k,n}} \right)_{\tilde{S}_{0,1}} \\
&= \sum_{k=0}^{\infty} \frac{L_{k,\lambda,R}}{L_{k,0,1}} (f_k, f_k)_{\tilde{S}_{0,1}} \\
&= \sum_{k=0}^{\infty} (f_k, f_k)_{\tilde{S}_{\lambda,R}}.
\end{aligned}$$

Therefore $\mathcal{F}_\lambda$ is a unitary isomorphism of $H^2(\tilde{S}_\lambda(R))$ onto $\mathcal{E}^2_{\Delta-\lambda^2}(\tilde{\mathbf{E}};R)$. By the definition, $\mathcal{F}_\lambda$ is an antilinear mapping. q.e.d.

A special case is known as Theorem 4.1 in [2]:

COROLLARY **6.2** *The Fourier transformation $\mathcal{F}_0$ is an antilinear unitary isomorphism of $L^2\mathcal{O}(\tilde{S}_{0,R})$ onto $\mathcal{E}^2_\Delta(\tilde{\mathbf{E}};R)$.*

7 APPENDIX I

In this section, we review some results on Legendre polynomials. Noting that

$$P_{k,1}(t) = \{(t+\sqrt{t^2-1})^k + (t-\sqrt{t^2-1})^k\}/2, \tag{11}$$

all the statements in this section still hold for $n = 1$.

LEMMA **7.1** *Let $t > 1$ and $k \in \mathbf{N}$. Then we have*

$$P_{0,n}(t) \equiv 1, \quad \frac{\Gamma(\frac{n}{2})\Gamma(\frac{k+1}{2})}{2\sqrt{\pi}\Gamma(\frac{k+n}{2})}(t+\sqrt{t^2-1})^k < P_{k,n}(t) < (t+\sqrt{t^2-1})^k. \tag{12}$$

$$1 = P_{k,n}(1) < P_{k,n}(t) < P_{k,n}(t'), \quad 1 < t < t'. \tag{13}$$

$$1 = P_{0,n}(t) < P_{k,n}(t) < P_{k+1,n}(t). \tag{14}$$

PROOF. Proof of (12): By the definition, $P_{0,n}(t) \equiv 1$. For $n = 1$, (12) is clear by (11).

For $n \geq 2$, we have the following Laplace representation:

$$P_{k,n}(t) = \frac{\Gamma(n/2)}{\sqrt{\pi}\Gamma(\frac{n-1}{2})}\int_{-1}^{1}(t+x\sqrt{t^2-1})^k(1-x^2)^{\frac{n-3}{2}}\,dx. \tag{15}$$

Subdividing $[-1,1]$ into $[-1,0]$ and $[0,1]$, we get

$$P_{k,n}(t) = \frac{\Gamma(n/2)}{\sqrt{\pi}\Gamma(\frac{n-1}{2})}\int_{0}^{1}\{(t+x\sqrt{t^2-1})^k + (t-x\sqrt{t^2-1})^k\}(1-x^2)^{\frac{n-3}{2}}\,dx.$$

Since $t \pm x\sqrt{t^2-1} \geq (t \pm \sqrt{t^2-1})x \geq 0$ for $t > 1$ and $x \geq 0$, we have

$$\begin{aligned} P_{k,n}(t) &> \frac{\Gamma(n/2)}{\sqrt{\pi}\Gamma(\frac{n-1}{2})}\{(t+\sqrt{t^2-1})^k + (t-\sqrt{t^2-1})^k\}\int_0^1 x^k(1-x^2)^{\frac{n-3}{2}}\,dx \\ &> \frac{\Gamma(n/2)}{\sqrt{\pi}\Gamma(\frac{n-1}{2})}(t+\sqrt{t^2-1})^k\int_0^1 x^k(1-x^2)^{\frac{n-3}{2}}\,dx \\ &= \frac{\Gamma(n/2)}{\sqrt{\pi}\Gamma(\frac{n-1}{2})}(t+\sqrt{t^2-1})^k\frac{\Gamma(\frac{k+1}{2})\Gamma(\frac{n-1}{2})}{2\Gamma(\frac{k+n}{2})}. \end{aligned}$$

On the other hand, since $0 < t + x\sqrt{t^2-1} < t + \sqrt{t^2-1}$ for $t > 1$ and $-1 \leq x \leq 1$, we have $(t + x\sqrt{t^2-1})^k(1-x^2)^{(n-3)/2} \leq (t+\sqrt{t^2-1})^k(1-x^2)^{(n-3)/2}$. Thus by (15),

$$\begin{aligned} P_{k,n}(t) &< \frac{\Gamma(n/2)}{\sqrt{\pi}\Gamma(\frac{n-1}{2})}(t+\sqrt{t^2-1})^k \int_{-1}^{1} (1-x^2)^{\frac{n-3}{2}} dx \\ &= \frac{\Gamma(n/2)}{\sqrt{\pi}\Gamma(\frac{n-1}{2})}(t+\sqrt{t^2-1})^k \frac{2^{n-2}\Gamma(\frac{n-1}{2})^2}{\Gamma(n-1)} \\ &= (t+\sqrt{t^2-1})^k. \end{aligned}$$

Proof of (13): $P_{k,n}(1) = 1$ is clear. It is known that $\frac{dP_{k,n}(t)}{dt} = \frac{k(n+k-1)}{n} P_{k-1,n+2}(t)$ for $k, n \in \mathbf{N}$. Thus for $t > 1$ and $k \in \mathbf{N}$, $\frac{dP_{k,n}(t)}{dt} > 0$ by (12). This implies (13).

Proof of (14): Since $P_{k+1,1}(t) - P_{k,1}(t) > (t-1)\frac{(t+\sqrt{t^2-1})^k + (t-\sqrt{t^2-1})^k}{2} > 0$, we have (14) for $n = 1$ and $t > 1$.

For $n \geq 2$ and $t > 1$, by (15) we have

$$\begin{aligned} &P_{k+1,n}(t) = tP_{k,n}(t) + \sqrt{t^2-1}\frac{\Gamma(n/2)}{\sqrt{\pi}\Gamma(\frac{n-1}{2})}\int_{-1}^{1} x(t+x\sqrt{t^2-1})^k(1-x^2)^{\frac{n-3}{2}} dx \\ &> P_{k,n}(t) \\ &\quad + \frac{\Gamma(\frac{n}{2})\sqrt{t^2-1}}{\sqrt{\pi}\Gamma(\frac{n-1}{2})}\int_0^1 x\{(t+x\sqrt{t^2-1})^k - (t-x\sqrt{t^2-1})^k\}(1-x^2)^{\frac{n-3}{2}} dx \\ &> P_{k,n}(t). \end{aligned}$$

q.e.d.

LEMMA 7.2 *Let $k \in \mathbf{N}$. The following function is continuous at $t = 0$ and strongly monotone increasing in t and in r;*

$$g_k(t,r) = \begin{cases} t^k P_{k,n}\left(\frac{1}{2}(\frac{t}{r}+\frac{r}{t})\right), & 0 < t \leq r, \\ \frac{\gamma_{k,n}}{2^k} r^k, & t = 0. \end{cases}$$

Note that $g_0(t,r) \equiv 1$.

PROOF. Since $\lim_{t\to 0} t^k P_{k,n}\left(\frac{1}{2}(\frac{t}{r}+\frac{r}{t})\right) = r^k\gamma_{k,n}/2^k$, $g_k(t,r)$ is continuous at $t = 0$.

The function $f(s) = (a/s + s/a)/2, a > 0$ is strongly monotone increasing for $s \geq a$ and $f(a) = 1$. Therefore by (13), $g_k(t,r)$ is strongly monotone increasing in r with $r \geq t$.

Putting

$$h_k(t) \equiv g_k(rt,r)/r^k = \begin{cases} t^k P_{k,n}\left(\frac{1}{2}(t+\frac{1}{t})\right), & 0 < t \leq 1, \\ \frac{\gamma_{k,n}}{2^k}, & t = 0, \end{cases}$$

we have only to show that $h_k(t)$ is strongly monotone increasing in t with $0 \leq t \leq 1$. For $n = 1$, it is clear by (11). For $n \geq 2$, by (15), we have

$$h_k(t) = \frac{\Gamma(n/2)}{\sqrt{\pi}\Gamma(\frac{n-1}{2})} \int_0^1 \left\{ \left(\tfrac{t^2(1-x)+1+x}{2}\right)^k + \left(\tfrac{t^2(1+x)+1-x}{2}\right)^k \right\} (1-x^2)^{\frac{n-3}{2}} dx.$$

Since the domain of integration is $[0,1]$, the integrand is continuous and positive in $x \in [0,1)$ and strongly monotone increasing in t, $h_k(t)$ is strongly monotone increasing in t. q.e.d.

By this lemma, we have $r^k = g_k(r,r) > g_k(1,r) > g_k(0,r) = \frac{\gamma_{k,n}}{2^k} r^k$. Thus taking $r = t_1 + \sqrt{t_1^2 - 1} > 1$, we have

COROLLARY **7. 3** *Let $k \in \mathbf{N}$. For $t > 1$, $P_{k,n}(t)$ is estimated as follows;*

$$\frac{\gamma_{k,n}}{2^k}(t + \sqrt{t^2-1})^k < P_{k,n}(t) < (t + \sqrt{t^2-1})^k. \tag{16}$$

This estimate is the same as that of (12) for $n = 1$ and is better than that of (12) for $n \geq 2$ (see Appendix II).

By Lemma 7. 2 , $\frac{\gamma_{k,n}}{2^k} < 1$ is clear. More precisely, we have

LEMMA **7. 4** *Let $k \in \mathbf{N}$. We have*

$$\frac{\gamma_{0,n}}{2^0} = 1, \frac{\gamma_{k,1}}{2^k} = \frac{1}{2}, \lim_{k\to\infty} \frac{\gamma_{k,n}}{2^k} = 0 < \frac{\gamma_{k+1,n}}{2^{k+1}} < \frac{\gamma_{k,n}}{2^k} < \frac{\gamma_{0,n}}{2^0} = 1, n \geq 2.$$

$$\lim_{k\to\infty} \left(\frac{\gamma_{k,n}}{2^k}\right)^{1/k} = 1, \quad n \geq 1. \tag{17}$$

PROOF. By (1), we have $\gamma_{0,n}/2^0 = 1$, $\gamma_{k,1}/2^k = 1/2$ and

$$\begin{aligned} &\frac{\gamma_{k+1,n}}{2^{k+1}} \times \frac{2^k}{\gamma_{k,n}} \\ &= \frac{(n-1)!\Gamma(k+1+\frac{n+1}{2})}{\Gamma(\frac{n+1}{2})(2k+n+1)(k+n-1)!} \times \frac{\Gamma(\frac{n+1}{2})(2k+n-1)(k+n-2)!}{(n-1)!\Gamma(k+\frac{n+1}{2})} \\ &= \frac{2k+n-1}{2k+2(n-1)} < 1. \end{aligned}$$

Since $\gamma_{k,n}/2^k = O(k^{-(n-1)/2})$, we have (17). q.e.d.

By Corollary 7. 3 and Lemma 7. 4 , as a corollary, we have

COROLLARY **7. 5** *For $t \geq 1$, we have*

$$\lim_{k\to\infty} (P_{k,n}(t))^{1/k} = t + \sqrt{t^2-1}.$$

In fact,

$$\lim_{k\to\infty} |P_{k,n}(t)|^{1/k} = |t+\sqrt{t^2-1}|, \quad t \in \mathbf{C}\setminus[-1,1]$$

is valid, where we take a branch of $t \pm \sqrt{t^2-1}$ as $|t+\sqrt{t^2-1}| > 1$. This fact is classical in the theory on orthogonal polynomials. We can give a proof by using the function $h_k(t)$ considered in the proof of Lemma 7. 2 referencing to [10] (see Appendix II).

If $a, b > 0$, then $t = (a/b + b/a)/2 \geq 1$ and we have

$$P_{k,n}(t) \leq (\max\{a/b, b/a\})^k, \quad \lim_{k\to\infty} (P_{k,n}(t))^{1/k} = \max\{a/b, b/a\}.$$

Therefore, for $L_{k,\lambda,R}$ in the previous sections, we have

$$\begin{cases} L_{k,\lambda,R} \leq R^{2k}, & \lim_{k\to\infty} (L_{k,\lambda,R})^{1/k} = R^2, \quad |\lambda| \leq R, \\ L_{k,\lambda,R} \leq |\lambda|^{4k}/R^{2k}, & \lim_{k\to\infty} (L_{k,\lambda,R})^{1/k} = |\lambda|^4/R^2, \quad |\lambda| > R. \end{cases} \tag{18}$$

Further, noting that

$$\tilde{P}_{k,n}(z,\overline{z}) = |z^2|^k P_{k,n}(\frac{\|z\|^2}{|z^2|}) = |z^2|^k P_{k,n}\left(\frac{1}{2}(\frac{L(z)^2}{|z^2|} + \frac{|z^2|}{L(z)^2})\right) \leq L(z)^{2k}, \tag{19}$$

we have the following lemma:

LEMMA **7. 6** *We have the following estimate:*

$$|\tilde{P}_{k,n}(z,w)| \leq L(z)^k L(w)^k, \quad z, w \in \tilde{\mathbf{E}}.$$

PROOF. For $n = 1$, it is clear by (11).
By Lemma 1. 1 and (19),

$$\begin{aligned} \left|\tilde{P}_{k,n}(z,w)\right| &\leq 2^k N(k,n) \left| \int_{\tilde{S}_{0,1}} (\zeta\cdot z)^k (w\cdot\overline{\zeta})^k d\tilde{S}_{0,1}(\zeta) \right| \\ &\leq 2^k N(k,n) \left\{ \int_{\tilde{S}_{0,1}} |\zeta\cdot z|^{2k} d\tilde{S}_{0,1}(\zeta) \right\}^{\frac{1}{2}} \left\{ \int_{\tilde{S}_{0,1}} |\zeta\cdot w|^{2k} d\tilde{S}_{0,1}(\zeta) \right\}^{\frac{1}{2}} \\ &\leq \left\{\tilde{P}_{k,n}(z,\overline{z})\right\}^{1/2} \left\{\tilde{P}_{k,n}(w,\overline{w})\right\}^{1/2} \\ &\leq L(z)^k L(w)^k. \end{aligned}$$

q.e.d.

For another proof of this lemma, see Lemma 5.5 in [8].

8 APPENDIX II

First we will give a proof of

$$\lim_{k\to\infty} |P_{k,n}(t)|^{1/k} = \max\{|t \pm \sqrt{t^2-1}|\}, \quad t \in \mathbf{C} \setminus [-1,1] \tag{20}$$

referencing to [10] and [9]. To prove (20), we employ the following lemma:

LEMMA **8.1** *Assume that f_k, $k = 1, 2, \cdots$, are holomorphic and have no zero points in the domain $B(R) \equiv \{z;\ |z| < R\}$ and that $|f_k(z)| < 1$ for $z \in B(R)$.*

If there is a point $a \in B(R)$ such that $\lim_{k\to\infty} f_k(a) = 0$, then we have $\lim_{k\to\infty} f_k(z) = 0,\ z \in B(R)$.

A proof of this lemma is given by using the following lemma which is known as an application of maximum modulus principle. For the detail, we refer the reader, for example, to [9].

LEMMA **8.2** *Let $f \in \mathcal{O}(B(R))$. Put*

$$M(r) = \max_{|z|=r} |f(z)|, \ r < R, \quad M(R) = \lim_{r\to R-0} M(r).$$

If $f(z) \neq 0$ on $B(R)$, then we have

$$M(r) \leq M(0)^{(R-r)/(R+r)} M(R)^{2r/(R+r)}, \quad 0 \leq r < R.$$

PROOF of (20). Let $k \in \mathbf{N}$. Put

$$\tilde{h}_k(z) = \begin{cases} \dfrac{\gamma_{k,n}}{2^k}, & z = 0, \\ z^k P_{k,n}\left(\dfrac{1}{2}\left(z + \dfrac{1}{z}\right)\right), & 0 \neq |z| < 1. \end{cases}$$

Since $\tilde{h}_k(z)$ is a polynomial of degree $2k$ and is continuous at 0, $\tilde{h}_k(z)$ is holomorphic on $B(1)$ and continuous on $B[1] \equiv \{z;\ |z| \leq 1\}$. Further it is known that all the zero points of $P_{k,n}(x)$ are on $[-1,1]$ (see [10]). Therefore, $\tilde{h}_k(z)$ has no zero points on $B(1)$ because for $|z| \neq 1$ we have $t = \frac{1}{2}\left(z + \frac{1}{z}\right) \in \mathbf{C} \setminus [-1,1]$. Therefore $\tilde{h}_k(z)$ takes the maximun and the minimum values on $\partial B[1] \equiv \{z;\ |z| = 1\}$. By the definition, $\tilde{h}_k(z)$ takes the maximum value 1 on $|z| = 1$ because for $z = e^{i\theta}$, $\theta \in \mathbf{R}$ we have $t = \cos\theta \in [-1,1]$ and $|P_{k,n}(t)| \leq 1$. Put

$$Q_k(z) = \left(\tilde{h}_k(z)^{1/k} - 1\right)/2.$$

As $\tilde{h}_k(z)$ is holomorphic, has no zero points and $|\tilde{h}_k(z)| < 1$ on $B(1)$, so is $Q_k(z)$. By (17), we have $\lim_{k\to\infty} Q_k(0) = \lim_{k\to\infty}(\tilde{h}_k(0)^{1/k} - 1)/2 = 0$, thus we have

$$\lim_{k\to\infty} Q_k(z) = 0, \ z \in B(1)$$

by Lemma 8. 1 . This implies

$$\lim_{k\to\infty} \tilde{h}_k(z)^{1/k} = 1. \tag{21}$$

For $0 \neq |z| < 1$ and $t = \frac{1}{2}\left(z + \frac{1}{z}\right)$, we have $1/|z| = \max\{|t \pm \sqrt{t^2-1}|\} > 1$, hence (21) implies (20). q.e.d.

Second, we will prove that the estimate (16) is better than that of (12) for $n \geq 2$ by showing

$$a_{k,n} \equiv \frac{\Gamma(\frac{n}{2})\Gamma(\frac{k+1}{2})}{2\sqrt{\pi}\Gamma(\frac{k+n}{2})} < b_{k,n} \equiv \frac{\gamma_{k,n}}{2^k} = \frac{2^{n-2}\Gamma(\frac{n}{2})\Gamma(k+\frac{n-1}{2})}{\sqrt{\pi}\Gamma(k+n-1)}, \quad n \geq 2. \tag{22}$$

PROOF. First we consider the case of $n = 2$. To prove $a_{k,2} < b_{k,2}$ we will show that

$$\frac{b_{k,2}}{a_{k,2}} = \frac{\Gamma(k+\frac{1}{2})\Gamma(\frac{k}{2})}{\Gamma(k)\Gamma(\frac{k+1}{2})} > 1. \tag{23}$$

Since we know that

$$\frac{\Gamma(x)}{\Gamma(x+\frac{1}{2})} = \frac{1}{\sqrt{\pi}} \sum_{j=0}^{\infty} \frac{(2j-1)!!}{(2j)!!} \frac{1}{x+j}, \quad x > 0,$$

we have

$$x > y > 0 \Rightarrow \frac{\Gamma(y)}{\Gamma(y+\frac{1}{2})} > \frac{\Gamma(x)}{\Gamma(x+\frac{1}{2})}.$$

Putting $x = k > y = k/2$, we get (23).

Second, we consider the case of $n \geq 3$. Since

$$a_{k+2,n} = \frac{k+1}{k+n} a_{k,n} < a_{k,n}, \quad b_{k+2,n} = \frac{(k+\frac{n+1}{2})(k+\frac{n-1}{2})}{(k+n)(k+n-1)} b_{k,n} < b_{k,n}$$

and $(k+1)(k+n-1) \leq (k+\frac{n+1}{2})(k+\frac{n-1}{2})$ for $n \geq 3$, we have

$$a_{k,n} < b_{k,n} \Rightarrow a_{k+2,n} < b_{k+2,n} \tag{24}$$

for $n \geq 3$. Because $a_{0,n} = 1/2 < 1 = b_{0,n}$ and

$$\begin{aligned} a_{1,n} = \frac{\Gamma(\frac{n}{2})\Gamma(\frac{1}{2})}{2\sqrt{\pi}\Gamma(\frac{1}{2})\Gamma(\frac{n+1}{2})} = \frac{1}{2\pi} B\left(\frac{n}{2}, \frac{1}{2}\right) &= \frac{1}{2\pi} 2 \int_0^{\pi/2} \cos^{n-1}\theta d\theta \\ &< \frac{1}{\pi} \times \frac{\pi}{2} < \frac{1}{2} = b_{1,n}, \end{aligned}$$

we have $a_{k,n} < b_{k,n}$ for $n \geq 3$ and $k \in \mathbf{N}$ by (24). q.e.d.

Note that $B(p,q)$ is the Beta function.

For $k \in \mathbf{N}$ and $n \geq 2$, (22) is equivalent to the inequalities $2^{n-1}B(k+\frac{n-1}{2}, \frac{n-1}{2}) > B(\frac{k+1}{2}, \frac{n-1}{2})$, $2^{n-1}B(k+\frac{n-1}{2}, \frac{k+n}{2}) > B(\frac{k+1}{2}, k+n-1)$ and $2^{n-1}B(\frac{k+n}{2}, \frac{k+n}{2}-1) > B(\frac{k+1}{2}, \frac{k+n}{2}-1)$.

References

[1] Antonio J. Duran, The Stieltjes moments problem for rapidly decreasing functions, Proc. AMS 107(1989), 731-741.

[2] K.Fujita, Hilbert spaces related to harmonic functions, Tôhoku Math. J. 48(1996), 149-163.

[3] K.Fujita, On some function spaces of eigenfunctions of the Laplacian, Proceedings of the fifth international colloquium on finite or infinite dimensional complex analysis, Peking Univ, 1997, 61-66.

[4] K.Fujita and M.Morimoto, Integral representation for eigenfunctions of the Laplacian, to appear in J. Math. Soc. Japan.

[5] K.Fujita and M.Morimoto, Reproducing kernels related to the complex sphere, in preparation.

[6] K.Ii, On a Bargmann-type transform and a Hilbert space of holomorphic functions, Tôhoku Math. J. 38(1986), 57–69.

[7] M.Morimoto and K.Fujita, Analytic functionals and entire functionals on the complex light cone, Hiroshima Math. J., 25(1995), 493-512.

[8] M.Morimoto, Analytic functionals on the sphere and their Fourier-Borel transformations, Complex Analysis, Banach Center Publications 11 PWN-Polish Scientific Publishers, Warsaw, 1983, 223–250.

[9] G.Pólya and G.Szegö, Problems and Theorems in Analysis I, Springer, 1972.

[10] G.Szegö, Orthogonal Polynomials, AMS, 1939.

[11] R.Wada, On the Fourier-Borel transformations of analytic functionals on the complex sphere, Tôhoku Math. J. 38(1986), 417–432.

[12] R.Wada, Holomorphic functions on the complex sphere, Tokyo J. Math., 11(1988), 205-218.

[13] R.Wada and M.Morimoto, A uniqueness set for the differential operator $\Delta_z + \lambda^2$, Tokyo J. Math., 10(1987), 93-105.

8 AN EXPANSION THEOREM FOR STATE SPACE OF UNITARY LINEAR SYSTEM WHOSE TRANSFER FUNCTION IS A RIEMANN MAPPING FUNCTION

Subhajit Ghosechowdhury

Department of Mathematics
Purdue University, USA
ghose@math.purdue.edu

Abstract: A power series $W(z)$ with complex coefficients which represents a function bounded by one in the unit disk is the transfer function of a canonical unitary linear system whose state space $\mathcal{D}(W)$ is a Hilbert space. If the power series has constant coefficient zero and coefficient of z positive, and if it represents an injective mapping of the unit disk, it appears as a factor mapping in a Löwner family of injective analytic mappings of the disk. The Löwner differential equation supplies a family of Herglotz functions. Each Herglotz function is associated with a Herglotz space of functions analytic in the unit disk. There exists an associated extended Herglotz space. An application of the Löwner differential equation is an expansion theorem for the starting state space in terms of the extended Herglotz spaces of the Löwner family. A generalization of orthogonality called complementation is used in the proof.

A linear system is a matrix

$$\begin{pmatrix} A & B \\ C & D \end{pmatrix}$$

of linear transformations acting on the Cartesian product of two vector spaces realized as a space of column vectors. The upper entry of the column vector is taken in a vector space $\mathcal{H}$ called the state space. The lower entry of the column vector is taken in a vector space $\mathcal{C}$ called the coefficient space. The main transformation A maps the state space into itself. The input transformation B maps the coefficient space into the state space. The output transformation C maps the state space into the coefficient space. The external operator D maps the coefficient space into itself. The transfer function of the linear system is a power series

$$W(z) = \sum W_n z^n$$

whose coefficients are the operators on the coefficient space defined by

$$W_0 = D$$

and by

$$W_{n+1} = CA^n B$$

for every nonnegative integer n. In the present application the state space is a Hilbert space. The coefficient space is the complex numbers considered as a Hilbert space with absolute value as norm. The linear system is unitary in the sense that the matrix is continuous and unitary. The linear system is canonical in the sense that the elements of the state space are pair of power series and that the entries of the matrix have a prescribed action. The main transformation takes $(f(z), g(z))$ into

$$([f(z) - f(0)]/z, zg(z) - W^*(z)f(0))$$

where

$$W^*(z) = \sum W_n^- z^n.$$

The input transformation takes c into

$$([W(z) - W(0)]c/z, [1 - W^*(z)W(0)]c).$$

The output transformation takes $(f(z), g(z))$ into $f(0)$. The external operator is $W(0)$. Such a linear system is determined by its transfer function, which is characterized as a power series representing a function bounded by one in the unit disk. The state space of the linear system is denoted $\mathcal{D}(W)$. A construction of the space by the recently discovered methods of complementation theory was made by Louis de Branges and James Rovnyak [11].

If a Hilbert space $\mathcal{P}$ is contained contractively in a Hilbert space $\mathcal{H}$, a unique Hilbert space $\mathcal{Q}$ exists which is contained contractively in $\mathcal{H}$ and which has these properties: The inequality

$$\|c\|_{\mathcal{H}}^2 \leq \|a\|_{\mathcal{P}}^2 + \|b\|_{\mathcal{Q}}^2$$

holds whenever $c = a + b$ with a in $\mathcal{P}$ and b in $\mathcal{Q}$. Every element c of $\mathcal{H}$ admits a decomposition for which equality holds. The space $\mathcal{Q}$ is called the complementary space to $\mathcal{P}$ in $\mathcal{H}$. The minimal decomposition of an element c of $\mathcal{H}$, which gives equality, is unique. The element a of $\mathcal{P}$ is obtained from c under the adjoint of the inclusion of $\mathcal{P}$ in $\mathcal{H}$. The element b of $\mathcal{Q}$ is obtained from c under the adjoint of the inclusion of $\mathcal{Q}$ in $\mathcal{H}$.

The present notation was introduced by de Branges [9] in a generalization of complementation to Krein spaces. Complementation permits the construction of Hilbert spaces from a given Hilbert space. A starting Hilbert space for constructions is the space $\mathcal{C}(z)$ of square summable power series. The elements of the space are the power series

$$f(z) = \sum a_n z^n$$

with complex coefficients such that the sum

$$\|f(z)\|^2_{\mathcal{C}(z)} = \sum |a_n|^2$$

is finite. If a nontrivial power series $W(z)$ with complex coefficients represents a function bounded by one in the unit disk, multiplication by $W(z)$ is a contractive transformation in $\mathcal{C}(z)$. The range $\mathcal{M}(W)$ of multiplication by $W(z)$ in $\mathcal{C}(z)$ is a Hilbert space which is contained contractively in $\mathcal{C}(z)$ when considered with the unique scalar product such that multiplication by $W(z)$ is an isometry of $\mathcal{C}(z)$ onto $\mathcal{M}(W)$. The complementary space $\mathcal{H}(W)$ to the range of multiplication by $W(z)$ in $\mathcal{C}(z)$ is a Hilbert space whose properties were studied by de Branges and Rovnyak. The elements of the space are convergent power series in the unit disk. An element $f(z)$ of $\mathcal{C}(z)$ belongs to $\mathcal{H}(W)$ if, and only if,

$$\|f(z)\|^2_{\mathcal{H}(W)} = \sup\{\|f(z) + W(z)g(z)\|^2_{\mathcal{C}(z)} - \|g(z)\|^2_{\mathcal{C}(z)}\}$$

is finite where the least upper bound is taken over all elements $g(z)$ of $\mathcal{C}(z)$. A continuous linear functional is defined on the space by taking $f(z)$ into $f(w)$ when w is in the unit disk. The power series

$$\frac{1 - W(z)W(w)^-}{1 - zw^-}$$

belongs to the space $\mathcal{H}(W)$ for every point w of the unit disk. the identity

$$f(w) = \left\langle f(z), \frac{1 - W(z)W(w)^-}{1 - zw^-} \right\rangle_{\mathcal{H}(W)}$$

holds for every element $f(z)$ of the space. The power series $[f(z) - f(0)]/z$ belongs to the space $\mathcal{H}(W)$ whenever $f(z)$ belongs to the space. The inequality for difference-quotient

$$\|[f(z) - f(0)]/z\|^2_{\mathcal{H}(W)} \leq \|f(z)\|^2_{\mathcal{H}(W)} - |f(0)|^2$$

is always satisfied. An example of an element of norm at most one in the space is

$$[W(z) - W(0)]/z.$$

Define the extension space $\mathcal{D}(W)$ of $\mathcal{H}(W)$ to be the space of pairs $(f(z), g(z))$ consisting of an element $f(z)$ of $\mathcal{H}(W)$ and an element

$$g(z) = \sum a_n z^n$$

of $\mathcal{C}(z)$ such that

$$z^{n+1}f(z) - W(z)(a_0 z^n + ... + a_n)$$

belongs to $\mathcal{H}(W)$ for every nonnegative integer n and such that the sequence of numbers

$$\|z^{n+1}f(z) - W(z)(a_0 z^n + ... + a_n)\|^2_{\mathcal{H}(W)} + |a_0|^2 + ... + |a_n|^2$$

is bounded. It follows that the above sequence is nondecreasing. Define

$$\|(f(z), g(z))\|^2_{\mathcal{D}(W)} = \lim_{n\to\infty} \{\|z^{n+1}f(z) - W(z)(a_0 z^n + ... + a_n)\|^2_{\mathcal{H}(W)} + |a_0|^2 + ... + |a_n|^2\}$$

The space $\mathcal{D}(W)$ is a Hilbert space which contains

$$([f(z) - f(0)]/z, zg(z) - W^*(z)f(0))$$

whenever it contains $(f(z), g(z))$, and the identity

$$\|([f(z) - f(0)]/z, zg(z) - W^*(z)f(0))\|^2_{\mathcal{D}(W)} = \|(f(z), g(z))\|^2_{\mathcal{D}(W)} - |f(0)|^2$$

is satisfied. The element

$$(zf(z) - W(z)g(0), [g(z) - g(0)]/z)$$

belongs to the space $\mathcal{D}(W)$ whenever $(f(z), g(z))$ belongs to the space and the identity

$$\|(zf(z) - W(z)g(0), [g(z) - g(0)]/z)\|^2_{\mathcal{D}(W)} = \|(f(z), g(z))\|^2_{\mathcal{D}(W)} - |g(0)|^2$$

is satisfied. The transformation which takes $(f(z), g(z))$ into $f(z)$ is a partial isometry of $\mathcal{D}(W)$ onto $\mathcal{H}(W)$. The transformation which takes $(f(z), g(z))$ into $(g(z), f(z))$ is an isometry of the extension space $\mathcal{D}(W)$ onto the extension space $\mathcal{D}(W^*)$ of $\mathcal{H}(W^*)$. The pair

$$\left(\frac{1 - W(z)W(w)^-}{1 - zw^-}, \frac{W^*(z) - W(w)^-}{z - w^-}\right)$$

belongs to the space $\mathcal{D}(W)$ for every point w of the unit disk. The identity

$$f(w) = \left\langle (f(z), g(z)), \left(\frac{1 - W(z)W(w)^-}{1 - zw^-}, \frac{W^*(z) - W(w)^-}{z - w^-}\right)\right\rangle_{\mathcal{D}(W)}$$

holds for every element $(f(z), g(z))$ of the space. The pair

$$\left(\frac{W(z) - W(w^-)}{z - w^-}, \frac{1 - W^*(z)W(w^-)}{1 - zw^-}\right)$$

belongs to the space $\mathcal{D}(W)$ for every point w of the unit disk. The identity

$$g(w) = \left\langle (f(z), g(z)), \left(\frac{W(z) - W(w^-)}{z - w^-}, \frac{1 - W^*(z)W(w^-)}{1 - zw^-} \right) \right\rangle_{\mathcal{D}(W)}$$

holds for every element $(f(z), g(z))$ of the space.

There exists a contractive transformation of $\mathcal{H}(W)$ into $\mathcal{H}(W^*)$ which takes $f(z)$ into $\tilde{f}(z)$ such that

$$\tilde{f}(w) = \left\langle f(z), \frac{W(z) - W(w^-)}{z - w^-} \right\rangle_{\mathcal{H}(W)}$$

for every point w of the unit disk. If

$$\tilde{f}(z) = \sum a_n z^n$$

then

$$a_n = \langle f(z), g_n(z) \rangle_{\mathcal{H}(W)}$$

where $g_n(z)$ is defined inductively by

$$g_0(z) = \frac{W(z) - W(0)}{z}$$

and

$$g_{n+1}(z) = \frac{g_n(z) - g_n(0)}{z}.$$

It follows that if $W(z)$ has constant coefficient zero and if

$$f(z) = 1$$

then

$$\tilde{f}(z) = W^*(z)/z.$$

There exists an isometry of $\mathcal{H}(W)$ into $\mathcal{D}(W)$ which takes $f(z)$ into $(f(z), \tilde{f}(z))$.

When $W(z)$ represents an injective mapping of the unit disk into itself, the space $\mathcal{H}(W)$ has special properties studied by de Branges [7]. When $W(z)$ has constant coefficient zero, the complex numbers, considered as a Hilbert space with absolute value as norm, are contained isometrically in the space $\mathcal{H}(W)$. The orthogonal complement of the constants is the set of elements of the space which have constant coefficient zero. The Hilbert space $\mathcal{D}$ is defined as the isometric image of the space $\mathcal{D}(W)$ under the transformation which takes $(f(z), g(z))$ into

$$(f(z), -g(z)z/W^*(z)).$$

The reproducing kernel function of the space $\mathcal{D}$ in the first coordinate at the point w is

$$\left(\frac{1-W(z)W(w)^-}{1-zw^-}, -\frac{1-\frac{W(w^-)}{W^*(z)}}{1-\frac{w^-}{z}}\right).$$

The reproducing kernel function of the space $\mathcal{D}$ in the second coordinate at the point w is

$$\left(-\frac{1-\frac{W(z)}{W(w^-)}}{1-\frac{z}{w^-}}, \frac{1-\frac{1}{W^*(z)W(w^-)}}{1-\frac{1}{zw^-}}\right).$$

Note that

$$\|(1,-1)\|_{\mathcal{D}} = \|(1, W^*(z)/z)\|_{\mathcal{D}(W)} = \|1\|_{\mathcal{H}(W)} = 1.$$

Consequently the space

$$\mathcal{P} = \left\{(a,-a) : \text{a is any complex number}\right\}$$

with the norm

$$\|(a,-a)\|_{\mathcal{P}} = |a|$$

is contained isometrically in $\mathcal{D}$. The reproducing kernel function of the space $\mathcal{P}$ in the first coordinate at the point w is $(1,-1)$ and the reproducing kernel function of $\mathcal{P}$ in the second coordinate at the point w is $(-1,1)$. It follows that the reproducing kernel functions of the orthogonal complement of $\mathcal{P}$ in the space $\mathcal{D}$ in the first and second coordinate at the point w are

$$\left(\frac{1-W(z)W(w)^-}{1-zw^-} - 1, -\frac{1-\frac{W(w^-)}{W^*(z)}}{1-\frac{w^-}{z}} + 1\right)$$

and

$$\left(-\frac{1-\frac{W(z)}{W(w^-)}}{1-\frac{z}{w^-}} + 1, \frac{1-\frac{1}{W^*(z)W(w^-)}}{1-\frac{1}{zw^-}} - 1\right)$$

respectively.

Related Hilbert spaces of analytic functions were introduced in 1911 by Gustav Herglotz. A Herglotz space is a Hilbert space whose elements are power series with complex coefficients. A continuous transformation of the space into itself, which has an isometric adjoint, is defined by taking $f(z)$ into $[f(z)-f(0)]/z$. A continuous transformation of the space into the coefficient space is defined by taking $f(z)$ into $f(0)$. The elements of the space are convergent power series in the unit disk. A continuous linear functional is defined on the space by taking $f(z)$ into $f(w)$ when w belongs to the unit disk. The reproducing kernel function for the linear functional is of the form

$$\frac{\phi(z)+\phi(w)^-}{2(1-zw^-)}$$

for a power series $\phi(z)$ called the Herglotz function of the Herglotz space. The function is unique within an added imaginary constant. A Herglotz function is characterized as a power series which represents a function with nonnegative real part in the unit disk. A construction of the Herglotz space $\mathcal{L}(\phi)$ associated with a Herglotz function $\phi(z)$ was made by Lawrence Shulman in his thesis [17]. If a power series $\phi(z)$ represents a function with nonnegative real part in the unit disk, then the power series

$$W(z) = \frac{1-\phi(z)}{1+\phi(z)}$$

represents a function bounded by one in the unit disk. Multiplication by

$$1 + W(z)$$

is an isometric transformation of the space $\mathcal{L}(\phi)$ onto the space $\mathcal{H}(W)$. An example of an element of norm atmost one in the space $\mathcal{L}(\phi)$ is

$$\frac{1}{2}[\phi(z) - \phi(0)]/z$$

when $W(z)$ has constant coefficient zero. The corresponding element of the space $\mathcal{H}(W)$ is

$$-W(z)/z.$$

Define the extension space $\mathcal{E}(\phi)$ of $\mathcal{L}(\phi)$ to be the space of pairs $(f(z), g(z))$ consisting of an element $f(z)$ of $\mathcal{L}(\phi)$ and a power series

$$g(z) = \sum a_n z^n$$

such that

$$z^{n+1} f(z) + a_0 z^n + \ldots + a_n$$

belongs to $\mathcal{L}(\phi)$ for every nonnegative integer n and such that the sequence of numbers

$$\|z^{n+1} f(z) + a_0 z^n + \ldots + a_n\|_{\mathcal{L}(\phi)}$$

is bounded. The sequence is nondecreasing because the difference-quotient transformation in the space $\mathcal{L}(\phi)$ is bounded by one. Define

$$\|(f(z), g(z))\|_{\mathcal{E}(\phi)} = \lim_{n\to\infty} \|z^{n+1} f(z) + a_0 z^n + \ldots + a_n\|_{\mathcal{L}(\phi)}.$$

The space $\mathcal{E}(\phi)$ is a Hilbert space in this norm. The elements

$$([f(z) - f(0)]/z, zg(z) + f(0))$$

and

$$(zf(z) + g(0), [g(z) - g(0)]/z)$$

belong to the space $\mathcal{E}(\phi)$ whenever $(f(z), g(z))$ belongs to the space and have the same norm as $(f(z), g(z))$. The pair

$$\left(\frac{\phi(z)+\phi(w)^-}{2(1-zw^-)}, \frac{\phi^*(z)-\phi(w)^-}{2(z-w^-)}\right)$$

belongs to the space $\mathcal{E}(\phi)$ for every point w of the unit disk. The identity

$$f(w) = \left\langle (f(z), g(z)), \left(\frac{\phi(z)+\phi(w)^-}{2(1-zw^-)}, \frac{\phi^*(z)-\phi(w)^-}{2(z-w^-)}\right)\right\rangle_{\mathcal{E}(\phi)}$$

holds for every element $(f(z), g(z))$ of the space. The pair

$$\left(\frac{\phi(z)-\phi(w^-)}{2(z-w^-)}, \frac{\phi^*(z)+\phi(w^-)}{2(1-zw^-)}\right)$$

belongs to the space $\mathcal{E}(\phi)$ for every point w of the unit disk. The identity

$$g(w) = \left\langle (f(z), g(z)), \left(\frac{\phi(z)-\phi(w^-)}{2(z-w^-)}, \frac{\phi^*(z)+\phi(w^-)}{2(1-zw^-)}\right)\right\rangle_{\mathcal{E}(\phi)}$$

holds for every element $(f(z), g(z))$ of the space.

A parametrization of injective analytic functions was made by Karl Löwner. A power series $f(z)$ with constant coefficient zero is said to be subordinate to a power series $g(z)$ with constant coefficient zero if $f(z) = g(W(z))$ for a power series $W(z)$ with constant coefficient zero which represents a function which is bounded by one in the unit disk. If $f(z)$ and $g(z)$ are power series with constant coefficient zero which represent injective mappings of the unit disk, then $f(z)$ is subordinate to $g(z)$ if, and only if, the region onto which $f(z)$ maps the unit disk is contained in the region onto which $g(z)$ maps the unit disk. The power series $W(z)$ then represents an injective mapping of the unit disk into itself, and the identity

$$f'(0) = g'(0)W'(0)$$

is satisfied.

Subordination is a partial ordering when restricted to power series which have coefficient of z positive. A Löwner family is a maximal totally ordered set of such functions. The existence of Löwner families is an application of the Zorn lemma. A Löwner family has a natural parametrization by the coefficient of z in the power series. An application of Riemann mapping theorem shows that all positive numbers appear as parameters of a Löwner family. If power series $f(z)$ and $g(z)$ with constant coefficient zero and coefficient of z positive represent injective mappings of the unit disk, and if $f(z)$ is subordinate to $g(z)$, then $f(z)$ and $g(z)$ are members of a Löwner family.

A Löwner family of functions $Z(t, z)$ satisfies the Löwner differential equation
:

$$t\frac{\partial Z}{\partial t}(t, z) = \phi(t, z)z\frac{\partial Z}{\partial z}(t, z)$$

The coefficients $\phi(t, z)$ are Herglotz functions with constant coefficient one. The family of Herglotz functions is measurable in the sense that coefficients of z^n in $\phi(t, z)$ is a Lebesgue measurable function of t for every nonnegative integer n. The partial derivative with respect to z is taken in the sense of complex analysis. The partial derivative with respect to t is taken in the sense of absolute continuity with respect to Lebesgue measure. The interpretation of the Löwner equation is that the coefficient of z^n on the left is equal to the coefficient of z^n on the right for every nonnegative integer n.

A converse construction of a Löwner family is made from a measurable family of Herglotz functions. If a measurable family $\phi(t, z)$ of Herglotz functions with constant coefficient one is given, then a unique Löwner family of functions exists which has the given Herglotz functions as coefficient functions in the Löwner equation.

A Löwner family of functions $Z(t, z)$ is propagated according to the Huygens principle. Since $Z(a, z)$ is subordinate to $Z(b, z)$ where $a \leq b$, the Huygens identity

$$Z(a, z) = Z(b, W(b, a, z))$$

holds with $W(b, a, z)$ a power series with constant coefficient zero and coefficient of z equal to a/b which represents an injective mapping of the unit disk into itself. The Huygens identity

$$W(c, a, z) = W(c, b, W(b, a, z))$$

is satisfied when $a \leq b \leq c$. The global Löwner equation

$$t\frac{\partial Z}{\partial t}(t, z) = \phi(t, z)z\frac{\partial Z}{\partial z}(t, z)$$

implies the linear form of the local Löwner equation

$$t\frac{\partial W}{\partial t}(b, t, z) = \phi(t, z)z\frac{\partial W}{\partial z}(b, t, z)$$

in the interval $(0, b]$. A nonlinear form of the local Löwner equation

$$t\frac{\partial W}{\partial t}(t, a, z) = -\phi(t, W(t, a, z))W(t, a, z)$$

in the interval $[a, \infty)$ is obtained from the Huygens identity

$$W(b, a, z) = W(b, t, W(t, a, z))$$

on differentiation of each side with respect to t. The derivative is again taken in the sense of absolute continuity.

Consequences of the nonlinear Löwner equation are the differential equations

$$t\frac{\partial}{\partial t}\frac{1-W(t,a,z)W(t,a,w)^-}{1-zw^-}$$

$$=\frac{\phi(t,W(t,a,z))+\phi(t,W(t,a,w))^-}{1-zw^-}W(t,a,z)W(t,a,w)^-$$

and

$$-t\frac{\partial}{\partial t}\frac{1-\frac{W(t,a,w)^-}{W^*(t,a,z)}}{1-\frac{w^-}{z}}$$

$$=\frac{\phi^*(t,W^*(t,a,z))-\phi(t,W(t,a,w))^-}{z-w^-}W(t,a,w)^-\frac{z}{W^*(t,a,z)}$$

and

$$-t\frac{\partial}{\partial t}\frac{1-\frac{W(t,a,z)}{W(t,a,w^-)}}{1-\frac{z}{w^-}}$$

$$=\frac{\phi(t,W(t,a,z))-\phi(t,W(t,a,w^-))}{z-w^-}W(t,a,z)\frac{w^-}{W(t,a,w^-)}$$

and

$$t\frac{\partial}{\partial t}\frac{1-\frac{1}{W^*(t,a,z)W(t,a,w^-)}}{1-\frac{1}{zw^-}}$$

$$=\frac{\phi^*(t,W^*(t,a,z))+\phi(t,W(t,a,w^-)}{1-zw^-}\frac{z}{W^*(t,a,z)}\frac{w^-}{W(t,a,w^-)}.$$

An expansion theorem for the spaces $\mathcal{D}$ is now formulated in terms of measurable families of elements $(f(t,z),g(t,z))$ of the spaces $\mathcal{E}(\phi(t,W(t,a)))$ with $a\le t\le b$. Measurability means that the coefficient of z^n in $f(t,z)$ and in $g(t,z)$ are Lebesgue measurable functions of t in the half line $[a,\infty)$ for every nonnegative integer n. These conditions imply that the norm

$$\|(f(t,z),g(t,z))\|^2_{\mathcal{E}(\phi(t,W(t,a)))}$$

is a Lebesgue measurable function of t. Such families are considered equivalent if the members are equal elements of $\mathcal{E}(\phi(t,W(t,a)))$ for almost all parameters

t. Hilbert spaces are formed by the set of those equivalence classes such that the integral

$$\int_0^\infty \|(f(t,z), g(t,z))\|^2_{\mathcal{E}(\phi(t,W(t,a)))} t^{-1} dt$$

is finite. Related Hilbert spaces are obtained with integration over a finite interval (a, b).

Theorem. *Assume that a Löwner family of functions $Z(t,z)$ is given with factor mappings $W(b,a,z)$ and coefficient Herglotz functions $\phi(t,z)$. If a and b are positive numbers such that $a < b$, then for every measurable family of elements $(f(t,z), g(t,z))$ of $\mathcal{E}(\phi(t,W(t,a)))$ such that the integral*

$$\int_a^b \|(f(t,z), g(t,z))\|^2_{\mathcal{E}(\phi(t,W(t,a)))} t^{-1} dt$$

is finite, a corresponding element $(F(b,z), G(b,z))$ of the space $\mathcal{D}$, which is orthogonal to the space $\mathcal{P}$, is defined by

$$F(b,z) = 2 \int_a^b f(t,z) W(t,a,z) t^{-1} dt$$

and

$$G(b,z) = 2 \int_a^b g(t,z) \frac{z}{W^*(t,a,z)} t^{-1} dt$$

and the inequality

$$\|(F(b,z), G(b,z))\|^2_{\mathcal{D}} \le 2 \int_a^b \|(f(t,z), g(t,z))\|^2_{\mathcal{E}(\phi(t,W(t,a)))} t^{-1} dt$$

is satisfied. Every element $(F(b,z), G(b,z))$ of the space $\mathcal{D}$ which is orthogonal to $\mathcal{P}$ admits a representation for which equality holds.

Proof of the Theorem. Use is made of a continuous analogue of complementation theory. Define $\mathcal{H}$ as the set of pairs $(F(z), G(z))$ of power series of the form

$$F(z) = 2 \int_a^b f(t,z) W(t,a,z) t^{-1} dt$$

and

$$G(z) = 2 \int_a^b g(t,z) \frac{z}{W^*(t,a,z)} t^{-1} dt$$

for a measurable family of elements $(f(t,z), g(t,z))$ of the spaces $\mathcal{E}(\phi(t,W(t,a)))$ such that the integral

$$\int_a^b \|(f(t,z), g(t,z))\|^2_{\mathcal{E}(\phi(t,W(t,a)))} t^{-1} dt$$

is finite. A Hilbert space $\mathcal{H}$ is obtained in the norm

$$\|(F(z),G(z))\|_{\mathcal{H}}^2 = inf\ \ 2\int_a^b \|(f(t,z),g(t,z))\|_{\mathcal{E}(\phi(t,W(t,a)))}^2 t^{-1}dt$$

with the greatest lower bound taken over all such representations of $(F(z),G(z))$. The elements of the space are pairs of convergent power series in the unit disk. The inequality

$$\|(F(z),G(z))\|_{\mathcal{H}}^2 \le 2\int_a^b \|(f(t,z),g(t,z))\|_{\mathcal{E}(\phi(t,W(t,a)))}^2 t^{-1}dt$$

holds for every element (F(z),G(z)) of the space such that

$$F(z) = 2\int_a^b f(t,z)W(t,a,z)t^{-1}dt$$

and

$$G(z) = 2\int_a^b g(t,z)\frac{z}{W^*(t,a,z)}t^{-1}dt\ .$$

Every element $(F(z),G(z))$ of the space admits a minimal representation for which equality holds. The space $\mathcal{H}$ will be shown isometrically equal to the orthogonal complement of the space $\mathcal{P}$ in the space $\mathcal{D}$ by showing that it has the same reproducing kernel function in the first and second coordinate at every point w of the unit disk.

If w is a point of the disk, a measurable family of elements $(f(t,z),g(t,z))$ of the spaces $\mathcal{E}(\phi(t,W(t,a)))$ is given by

$$f(t,z) = \frac{\phi(t,W(t,a,z)) + \phi(t,W(t,a,w))^-}{2(1-zw^-)}W(t,a,w)^-$$

and

$$g(t,z) = \frac{\phi^*(t,W^*(t,a,z)) - \phi(t,W(t,a,w))^-}{2(z-w^-)}W(t,a,w)^-.$$

The corresponding element of $\mathcal{H}$ is

$$\left(\frac{1-W(b,a,z)W(b,a,w)^-}{1-zw^-} - 1\ ,\ -\frac{1-\frac{W(b,a,w)^-}{W^*(b,a,z)}}{1-\frac{w^-}{z}} + 1\right).$$

This reproducing kernel function of the orthogonal complement of $\mathcal{P}$ in the space $\mathcal{D}$ will be shown equal to the reproducing kernel function in the first coordinate of the space $\mathcal{H}$ at the point w. If an element $(F(z),G(z))$ of $\mathcal{H}$ has the minimal representation

$$F(z) = 2\int_a^b f(t,z)W(t,a,z)t^{-1}dt$$

and

$$G(z) = 2\int_a^b g(t,z)\frac{z}{W^*(t,a,z)}t^{-1}dt$$

with a finite integral

$$\int_a^b \|(f(t,z),g(t,z))\|^2_{\mathcal{E}(\phi(t,W(t,a)))}t^{-1}dt,$$

then the identity

$$\left\langle (F(z),G(z)), \left(\frac{1-W(b,a,z)\,W(b,a,w)^-}{1-zw^-} - 1, -\frac{1-\frac{W(b,a,w)^-}{W^*(b,a,z)}}{1-\frac{w^-}{z}} + 1\right)\right\rangle_{\mathcal{H}}$$

$$= 2\int_a^b \left\langle (f(t,z),g(t,z)), \left(\frac{\phi(t,W(t,a,z)) + \phi(t,W(t,a,w))^-}{2(1-zw^-)}W(t,a,w)^-\right.\right.,$$

$$\left.\left.\frac{\phi^*(t,W^*(t,a,z)) - \phi(t,W(t,a,w))^-}{2(z-w^-)}W(t,a,w)^-\right)\right\rangle_{\mathcal{E}(\phi(t,W(t,a)))} t^{-1}dt$$

$$= 2\int_a^b f(t,w)W(t,a,w)t^{-1}dt = F(w)$$

is satisfied.

When w is a point of the disk, a measurable family of elements $(f(t,z),g(t,z))$ of the spaces $\mathcal{E}(\phi(t,W(t,a)))$ is given by

$$f(t,z) = \frac{\phi(t,W(t,a,z)) - \phi(t,W(t,a,w^-))}{2(z-w^-)}\,\frac{w^-}{W(t,a,w^-)}$$

and

$$g(t,z) = \frac{\phi^*(t,W^*(t,a,z)) + \phi(t,W(t,a,w^-))}{1-zw^-}\,\frac{w^-}{W(t,a,w^-)}.$$

The corresponding element of $\mathcal{H}$ is

$$\left(-\frac{1-\frac{W(b,a,z)}{W(b,a,w^-)}}{1-\frac{z}{w^-}} + 1, \frac{1-\frac{1}{W^*(b,a,z)W(b,a,w^-)}}{1-\frac{1}{zw^-}} - 1\right).$$

This reproducing kernel function of the orthogonal complement of $\mathcal{P}$ in the space $\mathcal{D}$ will be shown equal to the reproducing kernel function in the second coordinate of the space $\mathcal{H}$ at the point w. If an element $(F(z),G(z))$ of $\mathcal{H}$ has the minimal representation

$$F(z) = 2\int_a^b f(t,z)W(t,a,z)t^{-1}dt$$

and

$$G(z) = 2\int_a^b g(t,z)\frac{z}{W^*(t,a,z)}t^{-1}dt$$

with a finite integral

$$\int_a^b \|(f(t,z),g(t,z))\|^2_{\mathcal{E}(\phi(t,W(t,a)))}t^{-1}dt,$$

then the identity

$$\left\langle (F(z),G(z)), \left(-\frac{1-\frac{W(b,a,z)}{W(b,a,w^-)}}{1-\frac{z}{w^-}} + 1\, , \, \frac{1-\frac{1}{W^*(b,a,z)W(b,a,w^-)}}{1-\frac{1}{zw^-}} - 1 \right)\right\rangle_{\mathcal{H}}$$

$$= 2\int_a^b \left\langle (f(t,z),g(t,z)), \left(\frac{\phi(t,W(t,a,z)) - \phi(t,W(t,a,w^-))}{2(z-w^-)} \frac{w^-}{W(t,a,w^-)}, \right.\right.$$

$$\left.\left. \frac{\phi^*(t,W^*(t,a,z)) + \phi(t,W(t,a,w^-))}{2(1-zw^-)} \frac{w^-}{W(t,a,w^-)} \right)\right\rangle_{\mathcal{E}(\phi(t,W(t,a)))} t^{-1}dt$$

$$= 2\int_a^b g(t,w)\frac{w}{W^*(t,a,w)}t^{-1}dt = G(w)$$

is satisfied. Since $\mathcal{H}$ is a Hilbert space, it is isometrically equal to the orthogonal complement of the space $\mathcal{P}$ in the space $\mathcal{D}$.

This completes the proof of the theorem.

Acknowledgement: *I would like to thank Professor Louis de Branges for some helpful discussion during the prepareation of this paper. I would also like to thank Professor Saburou Saitoh for his help and support during the ISAAC'97 Congress at the University of Delaware.*

References

1. N. Aronszajn, *Theory of Reproducing Kernels*, Trans. Amer. Math. Soc. **68** (1950), 337–404.
2. L. de Branges, *Coefficient estimates*, J. Math. Anal. Appl. **82** (1981), 420–450.
3. L. de Branges, *Grunsky spaces of analytic functions*, Bull. Sci. Math. **105** (1981), 401–416.
4. L. de Branges, *Löwner expansions*, J. Math. Anal. Appl. **100** (1984), 323–337.
5. L. de Branges, *A proof of the Bieberbach conjecture*, Acta Math. **154** (1985), 137–152.
6. L. de Branges, *Powers of Riemann mapping functions*, Mathematical Surveys, vol. **21**, Amer. Math. Soc., Providence, 1986, 51–67
7. L. de Branges, *Unitary linear systems whose transfer functions are Riemann mapping functions*, Operator Theory: Advances and Applications, vol. **19**, Birkhauser Verlag, Basel, 1986, 105–124
8. L. de Branges, *Underlying concepts in the proof of the Bieberbach conjecture*, Proceedings of the International Congress of Mathematicians, Amer. Math. Soc., Providence, 1987, 25–42
9. L. de Branges, *Complementation in Krein spaces*, Trans. Amer. Math. Soc. **305** (1988), 277–291.
10. L. de Branges, *Square Summable Power Series*, Bieberbach Conjecture Edition, Springer-Verlag, Heidelberg (to appear).
11. L. de Branges and J. Rovnyak, *Canonical models in quantum scattering theory*, Perturbation Theory and its Applications in Quantum Mechanics, Wiley, New York, 1966, 295–392
12. L. de Branges and J. Rovnyak, *Square Summable Power Series*, Holt, Rinehart and Winston, New York, 1966.
13. S. Ghosechowdhury, *Löwner expansions*, Dissertation, Purdue University, 1997.
14. S. Saitoh, *One approach to Some General Integral Transforms and its applications*, Integral Transforms and Special Functions **3, No. 1** (1995), 49–84.
15. S. Saitoh, *Theory of reproducing kernels and its applications*, Pitman Research Notes in Math. Series 189, Longman Scientific and Technical, Essex, England, 1988.
16. S. Saitoh, *Integral transforms, reproducing kernels and their applications*, Pitman Research Notes in Math. Series 369, Longman Scientific and Technical, Essex, England, 1997.
17. L. Shulman, *Perturbations of unitary transformations*, Amer. J. Math. **91** (1969), 267–288.

9 THE BERGMAN KERNEL AND A GENERALIZED FOURIER–BOREL TRANSFORM

Friedrich Haslinger[1]

Institut für Mathematik
Universität Wien, Austria
has@pap.univie.ac.at

Abstract: In this paper we represent the dual space of a Fréchet space of entire functions again as a space of entire functions. For this purpose we use the Bergman kernel of a certain Hilbert space. In the classical setting the exponential functions provide the isomorphism via Fourier–Borel transform. In our case we use the Bergman kernel instead of the exponential functions in order to establish the isomorphism.

[1] Partially supported by a FWF–grant P11390–MAT of the Austrian Ministry of Sciences.
Key words and phrases: Hilbert spaces of entire functions, Bergman kernel, Fourier–Borel transform.
Mathematics Subject Classification: Primary 32H10; Secondary 32A15, 42B10.

INTRODUCTION

Let $P = (p_m)_{m\geq 1}$ be a decreasing sequence of continuous functions $p_m : \mathbb{C}^n \longrightarrow \mathbb{R}$. We consider spaces $\mathcal{F}_P$ of entire functions the topology of which is given by a sequence of Hilbert norms

$$||f||_m = \left[\int_{\mathbb{C}^n} |f(z)|^2 \exp(-2p_m(z))\, d\lambda(z)\right]^{1/2},$$

where λ denotes the Lebesgue measure on $\mathbb{C}^n$. Suppose that the function $\exp(2p_1)$ is locally integrable. In this way $\mathcal{F}_P$ becomes a Fréchet space. The topology of $\mathcal{F}_P$ is in general stronger than the topology of uniform convergence on compact subsets of $\mathbb{C}^n$. $\mathcal{F}_P$ is in fact the intersection of the Hilbert spaces H_{p_m} of entire functions f such that $||f||_m < \infty$. Each of the Hilbert spaces H_{p_m} possesses a reproducing kernel. These kernels are used to describe the adjoint operators of the inclusions $H_{p_l} \hookrightarrow H_{p_m}$ $(m > l)$ as certain Toeplitz operators mapping H_{p_m} to H_{p_l}. Now it is easy to derive a characterization of nuclearity of the space $\mathcal{F}_P$ in terms of the Bergman kernel (see [W]).
The main topic of this paper is it to describe the dual space $\mathcal{F}'_P$ as a space of entire functions. For this purpose it is essential that the Hilbert space H_{p_0} is large enough, where $p_0(z) = \lim_{m\to\infty} p_m(z)$. The Bergman kernel of the space H_{p_0} will be used to obtain the weight functions p^*_m of the dual Fréchet space of entire functions which will turn out to be isomorphic to the dual space $\mathcal{F}'_P$. In the classical setting the exponential functions $z \mapsto \exp < z, w >$ are used to establish the isomorphism by the Fourier–Borel transform, but for this purpose the exponential functions $z \mapsto \exp < z, w >$ have to belong to the original space $\mathcal{F}_P$. In our case we use the Bergman kernel $K_{p_0}(z, w)$ of the space H_{p_0} instead of the exponential functions in order to establish the isomorphism. The weight functions p^*_m defined by

$$\exp(2p^*_m(w)) = \int_{\mathbb{C}^n} |K_{p_0}(z, w)|^2 \exp(-2p_m(z))\, d\lambda(z),$$

correspond to the Young conjugates of the weights p_m (see [H1]).
We use several properties of the Bergman kernel and have to impose a certain growth condition on the Bergman kernel $K_{p_0}(z, w)$ which makes the calculations work. This growth condtion is always satisfied in the case of radial weight functions $p_m(z) = r_m|z|^\alpha$, where $\alpha > 0$ and $(r_m)_m$ is a strictly decreasing sequence of positive real numbers, such that $\lim_{m\to\infty} r_m = r_0 > 0$. A different approach to similar results can be found in [LG].

Weighted spaces of entire functions appear in a natural way in the context of analytically uniform spaces ([BD], [BT], [E], [M1], [M2]), where the problem of finding a distributional solution of a partial differential equation, via Fourier–Borel transform, can be viewed as a problem of mappings between various weighted spaces of entire functions.

Reproducing kernels in Hilbert spaces of entire functions, for example the Bargmann–Fock space, are also investigated in [S]. They also appear in the context of Hardy and Bergman spaces on model domains (see [FH], [GS], [H2], [H3], [H4]).

BERGMAN KERNELS AND TOEPLITZ OPERATORS

The inner product of the Hilbert space H_{p_m} is given by

$$(f,g)_m = \int_{\mathbb{C}^n} f(z)\overline{g(z)} \exp(-2p_m(z))\, d\lambda(z),$$

for $f, g \in H_{p_m}$. The Fréchet space $\mathcal{F}_P$ is called nuclear if for each $m \in \mathbb{N}$ there is a number $l \in \mathbb{N}$ such that the canonical embedding

$$I_{l,m} : H_{p_l} \hookrightarrow H_{p_m}$$

is a nuclear operator (see [P]).
First we compute the adjoint operator of the canonical embedding $I_{l,m}$, which turns out to be a certain Toeplitz operator.

Theorem 1. *The adjoint $I^*_{l,m}$ of $I_{l,m}$ is a Toeplitz operator of the form*

$$I^*_{l,m}(g) = \mathcal{P}_l(g \exp(-2p_m + 2p_l)),$$

where $g \in H_{p_m}$ and $\mathcal{P}_l$ denotes the orthogonal projection of the Hilbert space

$$L^2_{p_l} = \{f : \mathbb{C}^n \longrightarrow \mathbb{C} \text{ measurable } : \int_{\mathbb{C}^n} |f(z)|^2 \exp(-2p_l(z)\, d\lambda(z) < \infty\}$$

onto the subspace H_{p_l} of all entire functions with the corresponding growth condition.

Proof. Since $I_{l,m}$ is the canonical embedding we have

$$(I_{l,m}(f), g)_m = (f,g)_m,$$

for $f \in H_{p_l}$ and $g \in H_{p_m}$. Hence

$$(f,g)_m = (f, I^*_{l,m}(g))_l, \quad (1)$$

Now we consider the canonical embedding

$$\tilde{I}_{l,m} : L^2_{p_l} \longrightarrow L^2_{p_m}$$

of the corresponding weighted L^2–spaces. From (1) we obtain

$$\int_{\mathbb{C}^n} f(z)\overline{g(z)} \exp(-2p_m(z))\, d\lambda(z) = \int_{\mathbb{C}^n} f(z)[\tilde{I}^*_{l,m}(g)(z)]^- \exp(-2p_l(z))\, d\lambda(z), \quad (2)$$

for $f \in L^2_{p_l}$ and $g \in L^2_{p_m}$. Since $f \mapsto f\exp(-p_l)$ is an isometry between $L^2_{p_l}$ and L^2 and since it follows from (2) that

$$\int_{\mathbb{C}^n} f(z) \exp(-p_l(z))\overline{g(z)} \exp(-2p_m(z) + p_l(z))\, d\lambda(z)$$

$$= \int_{\mathbb{C}^n} f(z)\exp(-p_l(z))[\tilde{I}^*_{l,m}(g)(z)]^- \exp(-p_l(z))\, d\lambda(z), \quad (3)$$

for each $f \in L^2_{p_l}$, we obtain

$$\tilde{I}^*_{l,m}(g) = g\exp(-2p_m + 2p_l). \quad (4)$$

Hence for $f \in H_{p_l}$ and $g \in H_{p_m}$ we have by (4) and the fact that $\mathcal{P}_l(f) = f$ for $f \in H_{p_l}$:

$$\begin{aligned}(I_{l,m}(f), g)_m &= (f, g)_m = (f, g\exp(-2p_m + 2p_l))_l \\ &= (\mathcal{P}_l(f), g\exp(-2p_m + 2p_l))_l \\ &= (f, \mathcal{P}_l(g\exp(-2p_m + 2p_l)))_l, \quad (5)\end{aligned}$$

which proves Proposition 1. □

We already mentioned that the topology of the Hilbert spaces H_{p_m} is in general stronger than the topology of uniform convergence on the compact subsets of $\mathbb{C}^n$. Especially, the point evaluations $f \mapsto f(z)$, $z \in \mathbb{C}^n$ fixed, are continuous linear functional on H_{p_m}. Hence there exists a reproducing kernel $K_{p_m}(z,w)$ for H_{p_m} i.e.

$$f(z) = \int_{\mathbb{C}^n} f(w)K_{p_m}(z,w)\exp(-2p_m(w))\, d\lambda(w),$$

for each $f \in H_{p_m}$.
$K_{p_m}(z,w)$ is called the Bergman kernel of the Hilbert space H_{p_m} ; $K_{p_m}(z,w)$ is conjugate symmetric, holomorphic in z and conjugate holomorphic in w.

Theorem 2. *The operator $I^*_{l,m}$ can be written in the form*

$$I^*_{l,m}(g)(z) = \int_{\mathbb{C}^n} g(w)K_{p_l}(z,w)\exp(-2p_m(w))\, d\lambda(w),$$

where $g \in H_{p_m}$ and $K_{p_l}(z,w)$ denotes the Bergman kernel of the Hilbert space H_{p_l}.

Proof. The orthogonal projection

$$\mathcal{P}_l : L^2_{p_l} \longrightarrow H_{p_l}$$

is of the form

$$\mathcal{P}_l(f)(z) = \int_{\mathbb{C}^n} f(w)K_{p_l}(z,w)\exp(-2p_l(w))\, d\lambda(w),$$

for $f \in L^2_{p_l}$. Now the result follows directly from Theorem 1. □

Using the fact that for an orthonormal basis $(\varphi_k)_k$ of H_{p_l} and for an arbitrary compact subset $C \subset \mathbb{C}^n$ one has

$$\sum_{k=1}^{\infty} \varphi_k(z)\overline{\varphi_k(w)} = K_{p_l}(z,w),$$

uniformly on $C \times C$ (see [K], [R]), one gets

Theorem 3. *The Fréchet space $\mathcal{F}_P$ is nuclear if and only if for each $m \in \mathbb{N}$ there exists a number $l \in \mathbb{N}$ such that*

$$\int_{\mathbb{C}^n} K_{p_l}(w,w) \exp(-2p_m(w))\, d\lambda(w) < \infty. \quad (6)$$

See [W]

Theorem 4. *If the weight functions p_m satisfy the following two conditions:*
(i) for each $m \in \mathbb{N}$ there exists a number $l \in \mathbb{N}$ such that

$$\int_{\mathbb{C}^n} \exp(-2p_m(z) + 2p_l(z))\, d\lambda(z) < \infty,$$

(ii)for each $m \in \mathbb{N}$ there exists a number $l \in \mathbb{N}$ and a constant $C = C(m,l) > 0$ such that

$$\sup\{p_l(z+w) \ : \ |w| \leq 1\} - p_m(z) \leq C,$$

for each $z \in \mathbb{C}^n$, then the space $\mathcal{F}_P$ is nuclear.

This result can be derived from Proposition 3 (see [H1]).

BERGMAN KERNELS AND DUALITY

By construction, we have

$$H_{p_0} \subseteq \bigcap_{m=1}^{\infty} H_{p_l} = \mathcal{F}_P$$

and

$$||f||_0 = \left[\int_{\mathbb{C}^n} |f(z)|^2 \exp(-2p_0(z))\, d\lambda(z)\right]^{1/2} \geq ||f||_m,$$

for each $f \in H_{p_0}$, $m \in \mathbb{N}$.
We now take a continuous linear functional $L \in \mathcal{F}'_P$ and use the Bergman kernel $K_{p_0}(z,w)$ of H_{p_0} in order to define an entire function $\hat{L}$ related to L.

Lemma 1. *Let $L \in \mathcal{F}'_P$. Then*

$$\hat{L} : w \mapsto [L(K_{p_0}(.,w))]^-,$$

for $w \in \mathbb{C}^n$, is an entire function.

Proof. Since $L \in \mathcal{F}'_P$, there exists $m \in \mathbb{N}$ and a constant $C_m > 0$ such that

$$|L(f)| \leq C_m ||f||_m,$$

for each $f \in \mathcal{F}_P$. Hence

$$|L(f)| \leq C_m ||f||_0,$$

for each $f \in H_{p_0}$, which means that $L \in H'_{p_0}$. Now let $(\varphi_k)_k$ be an orthonormal basis of H_{p_0}, then

$$K_{p_0}(.,w) = \sum_{k=1}^{\infty} [\varphi_k(w)]^- \varphi_k,$$

with convergence in H_{p_0}. Therefore

$$[L(K_{p_0}(.,w))]^- = \sum_{k=1}^{\infty} [L(\varphi_k)]^- \varphi_k(w).$$

All we have to show now is that the last series converges uniformly on compact subsets of $\mathbb{C}^n$. Since the sequence $(L(\varphi_k))_k$ belongs to l^2 this follows at once from

$$\left| \sum_{k=1}^{\infty} [L(\varphi_k)]^- \varphi_k(w) \right|^2 \leq \sum_{k=1}^{\infty} |L(\varphi_k)|^2 \sum_{k=1}^{\infty} |\varphi_k(w)|^2 = \text{Const. } K_{p_0}(w,w)$$

and from the fact that $w \mapsto K_{p_0}(w,w)$ is continuous (see [W]). □

We now consider the space of all entire functions $\hat{L}$ and define appropriate weight functions such that the dual sapce $\mathcal{F}'_P$ can be realized as a weighted dual Fréchet space of entire functions. We set

$$\exp(2p^*_m(w)) = \int_{\mathbb{C}^n} |K_{p_0}(z,w)|^2 \exp(-2p_m(z))\, d\lambda(z),$$

for $m \in \mathbb{N}$. Note that

$$\exp(2p^*_0(w)) = \int_{\mathbb{C}^n} |K_{p_0}(z,w)|^2 \exp(-2p_0(z))\, d\lambda(z) = K_{p_0}(w,w).$$

Theorem 5. *Assume that H_{p_0} is dense in $\mathcal{F}_P$ and that for each $m \in \mathbb{N}$ there exists a number $l \in \mathbb{N}$ such that*

$$\int_{\mathbb{C}^n} \int_{\mathbb{C}^n} |K_{p_0}(z,w)|^2 \exp(-2p_m(z) - 4p_0(w) + 2p_l(w))\, d\lambda(z)\, d\lambda(w) < \infty. \quad (7)$$

Then the dual space $\mathcal{F}'_P$, endowed with the strong topology, is topologically isomorphic to the dual Fréchet space $\mathcal{G}_{P^*}$ of entire functions, where

$$\mathcal{G}_{P^*} = \bigcup_{m=1}^{\infty} G_{p_m},$$

where

$$G_{p_m} = \{F : \mathbb{C}^n \longrightarrow \mathbb{C} \text{ entire} : \|F\|^*_m = \sup\{|F(w)| \exp(-p^*_m(w)) : w \in \mathbb{C}^n\} < \infty\}$$

the isomorphism being given by $L \mapsto \hat{L}$, $L \in \mathcal{F}'_P$. In addition the following inequality holds : for each $m \in \mathbb{N}$ there exists a number $l \in \mathbb{N}$ and a constant $C = C(l,m) > 0$ such that

$$C\|L\|'_l \leq \|\hat{L}\|^*_m \leq \|L\|'_m,$$

for $L \in \mathcal{F}'_P$, where

$$\|L\|'_m = \sup\{|L(f)| \ : \ f \in \mathcal{F}_P \ , \ \|f\|_m \leq 1\}.$$

Remarks.

(a) Condition (7) is in a certain sense a sharpened form of the nuclearity of the natural imbedding

$$I_{0,m} : H_{p_0} \hookrightarrow H_{p_m}.$$

The operator $I_{0,m}$ is nuclear if and only if

$$\int_{\mathbb{C}^n} K_{p_0}(z,z) \exp(-2p_m(z))\, d\lambda(z) < \infty$$

(compare Proposition 3). Let $l \to \infty$ in (7), then

$$\begin{aligned} &\int_{\mathbb{C}^n} \int_{\mathbb{C}^n} |K_{p_0}(z,w)|^2 \exp(-2p_m(z) - 2p_0(w))\, d\lambda(z)\, d\lambda(w) \\ &= \int_{\mathbb{C}^n} K_{p_0}(z,z) \exp(-2p_m(z))\, d\lambda(z). \end{aligned}$$

(b) Integration with respect to z in (7) does not cause any troubles, because

$$K_{p_0}(.,w) \in H_{p_m} \quad \text{for any} \quad w \in \mathbb{C}^n,$$

but integration with respect to w does, because

$$\exp(-4p_0(w) + 2p_l(w)) \geq \exp(-2p_0(w)).$$

In this connection it is reasonable to restrict on weight functions p_m such that $p_m(z) \leq 2p_0(z)$, for each $z \in \mathbb{C}^n$ and $m \in \mathbb{N}$, because then $-4p_0(z) + 2p_m(z) \leq 0$. We mention that the estimate

$$|K_{p_0}(z,w)|^2 \leq K_{p_0}(z,z) K_{p_0}(w,w)$$

is too coarse to verify condition (7).

(c) Let $H_{2p_0-p_l}$ denote the Hilbert space of all entire functions f such that

$$||f||_{-l} = \left[\int_{\mathbb{C}^n} |f(z)|^2 \exp(-4p_0(z) + 2p_l(z))\, d\lambda(z)\right]^{1/2} < \infty.$$

If $\mathcal{F}_P$ is nuclear, it is easily seen that the natural embedding

$$H_{2p_0-p_l} \hookrightarrow H_{p_0}$$

is nuclear too. Using the spectral decompositon of this embedding, one can show that there exists on orthogonal basis $(\psi_k)_k$ in $H_{2p_0-p_l}$ such that condition (7) can be expressed in the form

$$\sum_{k=1}^{\infty} ||\psi_k||_m^2 ||\psi_k||_{-l}^2 ||\psi_k||_0^{-4} < \infty. \quad (7')$$

Proof. By Lemma 1, $\hat{L}$ is an entire function. First we show that the mapping $L \mapsto \hat{L}$ is injective. For this aim take a continuous linear functional $L \in \mathcal{F}_P'$ such that

$$L(K_{p_0}(.,w)) = 0$$

for each $w \in \mathbb{C}^n$. Now let $(\varphi_k)_k$ be an orthonormal basis of H_{p_0}. Since L is also an element of H'_{p_0}, we have

$$L(K_{p_0}(.,w)) = \sum_{k=1}^{\infty} L(\varphi_k)[\varphi_k(w)]^- = 0,$$

for each $w \in \mathbb{C}^n$ and since L corresponds to an element $\varphi \in H_{p_0}$, for which

$$\varphi = \sum_{k=1}^{\infty} (\varphi, \varphi_k)_0 \varphi_k = \sum_{k=1}^{\infty} [L(\varphi_k)]^- \varphi_k$$

holds in H_{p_0}, we conclude that $L = 0$ on H_{p_0} and, as H_{p_0} is dense in $\mathcal{F}_P$, $L = 0$ on $\mathcal{F}_P$. Hence, $L \mapsto \hat{L}$ is injective.
In order to show that this mapping is surjective, we take an arbitrary entire function $F \in \mathcal{G}_{P^*}$ and show that there exists a continuous linear functional $L \in \mathcal{F}'_P$ such that $\hat{L} = F$. For $F \in \mathcal{G}_{P^*}$ there exists a number $m \in \mathbb{N}$ such that

$$\|F\|_m^* < \infty.$$

Now set

$$\psi(w) = F(w) \exp(-2p_0(w)).$$

We claim that there exists a number $l \in \mathbb{N}$ such that

$$\int_{\mathbb{C}^n} |\psi(w)|^2 \exp(2p_l(w))\, d\lambda(w) < \infty.$$

To prove this we use condition (7) :

$$\begin{aligned} \textstyle\int_{\mathbb{C}^n} |\psi(w)|^2 \exp(2p_l(w))\, d\lambda(w) &= \textstyle\int_{\mathbb{C}^n} |F(w)|^2 \exp(2p_l(w) - 4p_0(w))\, d\lambda(w) \\ &= \textstyle\int_{\mathbb{C}^n} |F(w)|^2 \exp(-2p_m^*(w)) \exp(2p_m^*(w) + 2p_l(w) - 4p_0(w))\, d\lambda(w) \\ &\leq \|F\|_m^{*\,2} \textstyle\int_{\mathbb{C}^n} \int_{\mathbb{C}^n} |K_{p_0}(z,w)|^2 \exp(-2p_m(z) - 4p_0(w) + 2p_l(w))\, d\lambda(z)\, d\lambda(w) < \infty \end{aligned}$$

Notice that F is also an element of H_{p_0}, because

$$\begin{aligned} \int_{\mathbb{C}^n} |F(w)|^2 \exp(-2p_0(w))\, d\lambda(w) &= \int_{\mathbb{C}^n} |\psi(w)|^2 \exp(2p_0(w))\, d\lambda(w) \\ &\leq \int_{\mathbb{C}^n} |\psi(w)|^2 \exp(2p_l(w))\, d\lambda(w) < \infty. \end{aligned}$$

Now we define

$$L(f) = \int_{\mathbb{C}^n} f(z)\overline{\psi(z)}\, d\lambda(z),$$

for $f \in \mathcal{F}_P$. Then $L \in \mathcal{F}'_P$, because

$$\begin{aligned} |L(f)| &= \left| \int_{\mathbb{C}^n} f(z) \exp(-p_l(z)) \overline{\psi(z)} \exp(p_l(z))\, d\lambda(z) \right| \\ &\leq \|f\|_l \left[\int_{\mathbb{C}^n} |\psi(z)|^2 \exp(2p_l(z))\, d\lambda(z) \right]^{1/2}; \end{aligned}$$

and

$$\begin{aligned}[\hat{L}(w)]^- &= L(K_{p_0}(.,w)) = \int_{\mathbb{C}^n} K_{p_0}(z,w)\overline{\psi(z)}\,d\lambda(z) \\ &= \int_{\mathbb{C}^n} K_{p_0}(z,w)\overline{F(z)}\exp(-2p_0(z))\,d\lambda(z) \\ &= \left[\int_{\mathbb{C}^n} F(z)K_{p_0}(w,z)\exp(-2p_0(z))\,d\lambda(z)\right]^- \\ &= \overline{F(w)},\end{aligned}$$

hence $\hat{L} = F$.

It remains to show that the mapping $L \mapsto \hat{L}$ establishes a topological isomorphism between $\mathcal{F}'_P$ with the strong topology and $\mathcal{G}_{P^*}$ with the inductive limit topology of the Banach spaces $G_{p^*_m}$, $m \in \mathbb{N}$. This follows from the inequalities

$$\begin{aligned}||\hat{L}||^*_m &= \sup\{|\hat{L}(w)|\exp(-p^*_m(w)) : w \in \mathbb{C}^n\} \\ &= \sup\{|L(K_{p_0}(.,w))|\exp(-p^*_m(w)) : w \in \mathbb{C}^n\} \\ &\leq ||L||'_m \sup\{||K_{p_0}(.,w)||_m \exp(-p^*_m(w)) : w \in \mathbb{C}^n\} \\ &= ||L||'_m\end{aligned}$$

and

$$\begin{aligned}||L||'_l &= \sup\{|L(f)| : ||f||_l \leq 1\} \\ &= \sup\left\{\left|\int_{\mathbb{C}^n} f(w)[\hat{L}(w)]^- \exp(-2p_0(w))d\lambda(w)\right| \ : \ ||f||_l \leq 1\right\}\end{aligned}$$

$$= \sup\left\{\left|\int_{\mathbb{C}^n} f(w)\exp(-p_l(w))[\hat{L}(w)]^- \exp(-2p_0(w)+p_l(w))\,d\lambda(w)\right| : ||f||_l \leq 1\right\}$$

$$\leq \left[\int_{\mathbb{C}^n} |\hat{L}(w)|^2 \exp(-4p_0(w)+2p_l(w))\,d\lambda(w)\right]^{1/2}$$

$$= \left[\int_{\mathbb{C}^n} |\hat{L}(w)|^2 \exp(-2p^*_m(w))\exp(2p^*_m(w)-4p_0(w)+2p_l(w))\,d\lambda(w)\right]^{1/2}$$

$$\leq ||\hat{L}||^*_m \left[\int_{\mathbb{C}^n}\int_{\mathbb{C}^n} |K_{p_0}(z,w)|^2 \exp(-2p_m(z)-4p_0(w)+2p_l(w))d\lambda(z)d\lambda(w)\right]^{1/2}.$$

This proves the theorem. □

In sake of simplicity we now restrict our considerations to the case $n = 1$, for $n > 1$ analogous results hold.

Theorem 6. *Let $p_m(z) = r_m|z|^\alpha$, where $\alpha > 0$ and $r_m \downarrow r_0 > 0$. Then the dual space $\mathcal{F}'_P$ can be identified with the dual Fréchet space $\mathcal{G}_{P^*}$ of entire functions F such that*

$$||F||^*_m = \sup\left\{|F(w)|\exp\left(-\frac{r_0^2}{r_m}|w|^\alpha\right) \ : \ w \in \mathbb{C}\right\} < \infty,$$

for some $m \in \mathbb{N}$.

Remark. We once more point out that the case $0 < \alpha \leq 1$ is of special interest, because in this case the Fourier–Borel transform cannot be applied.

Proof. In order to show that all assumptions of Theorem 1 are satisfied we use the fact that the linear span of the polynomials is dense in each of the Hilbert spaces H_{p_m}, $m \in \mathbb{N}$ and $m = 0$. For $\alpha \geq 1$ this follows easily from [T] and for $0 < \alpha < 1$ it follows from the fact that the Taylor series expansion of each function f_θ, where $f_\theta(z) = f(\theta z)$, $0 < \theta < 1$, $f \in H_{p_m}$, is convergent in H_{p_m} and f_θ tends to f in H_{p_m} as θ tends to 1. Especially, H_{p_0} is dense in $\mathcal{F}_P$. Since the weight functions p_m are radial, the monomials $\{1, z, z^2, \dots\}$ constitue an orthogonal basis in each of the Hilbert spaces H_{p_m}, $m \in \mathbb{N}$ and $m = 0$.
Using the integral formula for the Γ-function

$$\Gamma(u) = \int_0^\infty x^{u-1}\, e^{-x}\, dx,$$

$\Re(u) > 0$, we obtain

$$\|z\|_m^2 = \frac{2\pi}{\alpha}\, (2r_m)^{-2(n+1)/\alpha}\, \Gamma(2(n+1)/\alpha),$$

$m \in \mathbb{N}$ and $m = 0$. Since

$$K_{p_m}(z, w) = \sum_{k=0}^{\infty} \|z^k\|_m^{-2}\, z^k \overline{w}^k,$$

we get

$$K_{p_m}(z, w) = \frac{\alpha}{2\pi} \sum_{k=0}^{\infty} (2r_m)^{2(k+1)/\alpha}\, [\Gamma(2(k+1)/\alpha)]^{-1}\, z^k \overline{w}^k,$$

$m \in \mathbb{N}$ and $m = 0$. Now it is easy to compute the weights p_m^* :

$$\exp(2p_m^*(w)) = \int_{\mathbb{C}} |K_{p_0}(z, w)|^2\, \exp(-2p_m(z))\, d\lambda(z)$$

$$= \frac{\alpha^2}{(2\pi)^2} \sum_{k=0}^{\infty} \left\{ (2r_0)^{4(k+1)/\alpha}\, [\Gamma(2(k+1)/\alpha)]^{-2}\, 2\pi \int_0^\infty r^{2k+1}\, \exp(-2r_m r^\alpha)\, dr\, |w|^{2k} \right.$$

$$= \frac{\alpha}{2\pi} \left(\frac{2r_0^2}{r_m}\right)^{2/\alpha} \sum_{k=0}^{\infty} [\Gamma(2(k+1)/\alpha)]^{-1} \left[\left(\frac{2r_0^2}{r_m}\right)^{1/\alpha} |w|^\alpha \right]^{2k}.$$

It is not difficult to determine the asymptotic behavior of $\exp(2p_m^*)$ (see [Bo]) :

$$\exp(2p_m^*(w)) \asymp \exp\left(\frac{2r_0^2}{r_m}\, |w|^\alpha\right),$$

as $|w| \to \infty$.

From this we can see that condition (7) is satisfied:

$$\begin{aligned}\int_{\mathbb{C}}\int_{\mathbb{C}} |K_{p_0}(z,w)|^2 \exp(-2p_m(z) - 4p_0(w) + 2p_l(w))\, d\lambda(z)\, d\lambda(w)\\ = \int_{\mathbb{C}} \exp(2p_m^*(w) - 4p_0(w) + 2p_l(w))\, d\lambda(w)\\ \leq \text{Const.} \int_{\mathbb{C}} \exp\left[\left(\tfrac{2r_0^2}{r_m} - 4r_0 + 2r_l\right) |w|^\alpha\right] d\lambda(w).\end{aligned}$$

Choose l so large that

$$r_l < (2r_m - r_0)r_0/r_m,$$

this is possible, since $r_l \downarrow r_0$ and

$$r_0 < (2r_m - r_0)r_0/r_m.$$

Now, with this choice of l, we have

$$\frac{2r_0^2}{r_m} - 4r_0 + 2r_l < 0,$$

and hence condition (7) is fulfilled. □

Remark. For a fixed weight function p let $K_\tau(z,w)$ denote the Bergman kernel of the Hilbert space of entire functions $f : \mathbb{C}^n \longrightarrow \mathbb{C}$ such that

$$\int_{\mathbb{C}^n} |f(z)|^2 \exp(-2\tau p(z))\, d\lambda(z) < \infty,$$

for $\tau > 0$. Then condition (7) corresponds to the following property: for each $\sigma > \tau$ there exists a number $\mu > 0$ with $0 < \mu < \tau$ such that

$$\int_{\mathbb{C}^n}\int_{\mathbb{C}^n} |K_\tau(z,w)|^2 \exp(-2\sigma p(z) - 2\mu p(w)))\, d\lambda(z)\, d\lambda(w) < \infty, \quad (7'')$$

which is an interesting property in itself.

REFERENCES

[BD] C.A. Berenstein and M.A. Dostal *Analytically uniform spaces and their applications to convolution equations* Lecture Notes in Mathematics, Vol. 256, Springer Verlag, Berlin, 1972.

[BT] C.A. Berenstein and B.A. Taylor *A new look at interpolation theory for entire functions of one variable* Adv. in Math. **33** 1979, pp. 109–143.

[Bo] R.P. Boas *Entire functions* Academic Press, New York, 1954.

[E] L. Ehrenpreis *Fourier analysis in several complex variables* Wiley–Interscience, New York, 1970.

[FH] G. Francsics and N. Hanges *Explicit formulas for the Szegõ kernel on certain weakly pseudoconvex domains* Proc.Amer.Math.Soc.**304** 1995, pp. 3161–3168.

[GS] P. C. Greiner and E. M. Stein *On the solvability of some differential operators of type* $\Box_b$ Proc. Internat. Conf., (Cortona, Italy, 1976–1977), Scuola Norm. Sup. Pisa, 1978, pp. 106–165.

[H1] F. Haslinger *Weighted spaces of entire functions* Indiana Univ. Math. J. **35** 1986, pp. 193–207.

[H2] F. Haslinger *Szegõ kernels of certain unbounded domains in* $\mathbb{C}^2$ Révue Roum. Math. Pures et Appl. **39** 1994, pp. 914–926.

[H3] F. Haslinger *Singularities of the Szegõ kernel for certain weakly pseudoconvex domains in* $\mathbb{C}^2$ J. Functional Analysis **129** 1995, pp. 406–427.

[H4] F. Haslinger *Bergman and Hardy spaces on model domains* Illinois J. of Math. (to appear)

[K] S.G. Krantz *Function theory of several complex variables* Wadsworth & Brooks/Cole, Pacific Grove, CA, 1992.

[LG] P. Lelong and L. Gruman *Entire functions of several complex variables* Grundlehren der mathematischen Wissenschaften Vol. 282, Springer–Verlag, Berlin–New York, 1986.

[M1] A. Martineau *Sur les fonctionelles analytiques et la transformation de Fourier-Borel* J.d'Analyse Math. **9** (1963), pp. 1–144.

[M2] A. Martineau *Equations différentielles d'ordre infini* Bull. Soc. math. France **95** (1967), pp. 109–154.

[P] A. Pietsch *Nukleare lokalkonvexe Räume* Akademie–Verlag, Berlin, 1965.

[R] M. Range *Holomorphic functions and integral representations in several complex variables* Springer–Verlag, Berlin, 1986.

[S] S. Saitoh *Integral transforms, reproducing kernels and their applications* Addison, Pitman Research Notes in Mathematics, Vol. 369, 1997.

[T] B.A. Taylor *On weighted polynomial approximation of entire functions* Pacific J. Math. **36** (1971), pp. 523–536.

[W] J. Wloka *Reproduzierende Kerne und nukleare Räume* Math. Ann. **163** (1966), pp. 167–188.

10 THE BERGMAN KERNEL ON CERTAIN DECOUPLED DOMAINS

Joe Kamimoto

Department of Mathematics, Kumamoto University,
Kumamoto 860-8555, Japan.
joe@math.sci.kumamoto-u.ac.jp

Abstract: In this paper we study the singularities of the Bergman kernel of a decoupled tube domain of finite type: $\Omega_m = \{z \in \mathbb{C}^{n+1}; \Im z_{n+1} > \sum_{j=1}^{n} a_j[\Im z_j]^{2m_j}\}$, where $a_j > 0$, $m_j \in \mathbb{N}$ and $m_n \neq 1$. Suppose k_0 is the cardinality of the set $\{j; m_j \neq 1\}$. Note that $O = (0, \ldots, 0)$ is a weakly pseudoconvex point. First the singularities of the Bergman kernel at O on the diagonal is essentially expressed by using $(k_0 + 1)$-variables; moreover the structure of singularities can be understood completely by using at most k_0-times real blowing-ups. Next in order to investigate the singularities of the Bergman kernel off the diagonal, we give an integral representation by using countably many functions whose singularities are understood directly. We also give an analogous result in the case of the Szegö kernel.

1 INTRODUCTION

In this paper, we study the singularities of the Bergman kernel (and the Szegö kernel) of a decoupled tube domain of finite type:

$$\Omega_m = \left\{ z \in \mathbb{C}^{n+1}; \Im z_{n+1} > \sum_{j=1}^{n} a_j[\Im z_j]^{2m_j} \right\},$$

where $a_j > 0$, $m_j \in \mathbb{N}$ and $m_n \neq 1$.

Let Ω be a domain in $\mathbb{C}^n$. The Bergman space $B(\Omega)$ is the subspace of $L^2(\Omega)$ consisting of holomorphic L^2-functions on Ω. The Bergman projection is the orthogonal projection $\mathbb{B} : L^2(\Omega) \to B(\Omega)$. We can write $\mathbb{B}$ as an integral operator

$$\mathbb{B}f(z) = \int_{\Omega} K(z,w)f(w)dV(w) \quad \text{for } f \in L^2(\Omega),$$

where $K : \Omega \times \Omega \to \mathbb{C}$ is the *Bergman kernel* of the domain Ω and dV is the Lebesgue measure on Ω.

First let us consider the singularities of the Bergman kernel *on* the diagonal. There are many studies about its singularities. In the history of this study, the following asymptotic expansion due to C. Fefferman [10] and Boutet de Monvel and Sjöstrand [6] may be said to be one of the most important results. The Bergman kernel K of a C^∞-smoothly bounded strictly pseudoconvex domain $\Omega \subset \mathbb{C}^n$ is expressed as follows:

$$K(z,z) = \frac{\varphi(z)}{r(z)^{n+1}} + \psi(z) \log r(z), \tag{1.1}$$

where $r \in C^\infty(\bar{\Omega})$ is a defining function of Ω (i.e. $\Omega = \{z \in \mathbb{C}^n; r(z) > 0\}$ and $|dr| > 0$ on $\partial\Omega$) and φ, $\psi \in C^\infty(\bar{\Omega})$ can be expanded asymptotically with respect to r. The above formula (1.1) perfectly shows the form of the singularities in the strictly pseudoconvex case.

Now let us remove the strict pseudoconvexity in the condition of domains. Until now there are many studies about the estimate of the size of the singularities and the boundary limit in some sense in weakly pseudoconvex domains of finite type (see the reference in [4], [13]). But the asymptotic formulas of the Bergman kernel are not well-understood. The author [13] recently gave an asymptotic formula in the case of weakly pseudoconvex tube domains in $\mathbb{C}^2$. In this study a kind of real blowing-up plays a key role on the study of the singularities of the Bergman kernel. In this article we study the singularities of the Bergman kernel for a special class of weakly pseudoconvex decoupled tube domains in $\mathbb{C}^n$ ($n \geq 2$) by using a similar idea. Last we remark that the singularities of the Bergman kernel of decoupled domains have already been studied and the perfect estimate and some boundary limit are obtained in [20], [4]. However, our analysis is strong and new from the viewpoint of the asymptotic formula (see Remark in §3.1).

Next let us consider the regularity of the Bergman kernel *off* the diagonal. In the case of C^∞-smoothly bounded strictly pseudoconvex domains, Kerzman [18] showed the Bergman kernel can be C^∞-smoothly extended to the boundary off the diagonal. Moreover Bell [1] and Boas [3] improved his result in the case of domains of finite type. From the viewpoint of real analytic category, many positive results have been obtained until now. In particular the real analyticity is known in the case of C^ω-smoothly bounded strictly pseudoconvex domains ([2], etc.). But Christ and Geller [9] gave counterexamples to real analyticities of the Bergman kernel and the Szegö kernel in the weakly pseudoconvex, of finite type and two-dimensional case. Christ [8] also gave non-analytic examples of the Szegö kernel in the higher dimensional case, which contain that of Ω_m. In this paper, we investigate the singularities of the Bergman kernel of Ω_m off the diagonal in more detail. We give an integral representation by using countably many functions whose singularities are understood directly.

This paper is organized as follows. We give an integral representation of the Bergman kernel of Ω_m in Section 2. Our analysis is based on this representation. In Section 3 we investigate the singularities of the Bergman kernel K of Ω_m on the diagonal by using finitely many times real blowing-ups. First blowing-up shows the situation of the singularities of K in the non-tangential direction (§3.1). Moreover a recursive formula of blowing-ups perfectly reveals the singularities in tangential directions (§3.2). In Section 4 we investigate the singularities of the Bergman kernel off the diagonal. In Section 5 we give analogous results of the Szegö kernel of Ω_m.

Throughout this article, we use the following notation and symbol. (1) $z_j = x_j+iy_j$ $(j = 1,\dots,n+1)$. (2) For $X = (X_1,\dots,X_{n+1})$, $Y = (Y_1,\dots,Y_{n+1})$, we define $\langle X,Y\rangle = \sum_{j=1}^{n+1} X_jY_j$ and $dX = dX_1\cdots dX_{n+1}$. (3) For $R \subset N$ and $m = (m_j)_{j\in N}$ $(N := \{1,\dots,n\})$, $|R|$ is the cardinality of R and $|1/m|_R = \sum_{j\in R} 1/m_j$. (4) $A \approx B$ means that there exits a positive constant C such that $C^{-1}A \leq B \leq CA$.

I would like to express my sincere thanks to Professors Saburou Saitoh and Takeo Ohsawa for giving me a chance to attend the congress ISAAC '97 at Delaware.

2 INTEGRAL REPRESENTATION

In this section we give an integral representation of the Bergman kernel, which is a clue in our analysis. Korányi [19], Nagel [21], Haslinger [11] and Saitoh [22] obtain similar representations of Bergman kernels or Szegö kernels for certain tube domains. By their studies, the results below are obtained by a simple computation. Let $f_j \in C^\infty(\mathbb{R})$ $(j = 1, \ldots, n)$ be functions such that $f_j(0) = 0$ and $f_j(x) \geq 0$. The tube domain Ω_f is defined by

$$\Omega_f = \mathbb{R}^{n+1} + i\omega_f \subset \mathbb{C}^{n+1}, \quad \text{where}$$
$$\omega_f = \left\{ y \in \mathbb{R}^{n+1}; y_{n+1} > \sum_{j=1}^{n} f_j(y_j) \right\}.$$

Let $\Lambda, \Lambda^* \subset \mathbb{R}^{n+1}$ be the cones defined by

$$\Lambda = \{y \in \mathbb{R}^{n+1}; (ty_1, \ldots, ty_{n+1}) \in \omega_f \text{ for any } t > 0\},$$
$$\Lambda^* = \{\lambda \in \mathbb{R}^{n+1}; \langle y, \lambda \rangle > 0 \text{ for any } y \in \Lambda\},$$

respectively. We call Λ^* the dual cone of ω_f. Actually Λ^* can be computed explicitly:

$$\Lambda^* = \{(\lambda, \mu); -R_j^- \lambda_j < \mu < R_j^+ \lambda_j \text{ for } 1 \leq j \leq n\},$$

where $(R_j^\pm)^{-1} = \lim_{x \to \mp\infty} f_j(x)|x|^{-1} > 0$, respectively. We allow that $R_j^\pm = \infty$. If $\lim_{|x| \to \infty} f_j(x)|x|^{-1-\varepsilon} > 0$ with some $\varepsilon > 0$, then $R_j^\pm = \infty$.

By [22], the Bergman kernel K of Ω_f is expressed as follows:

$$K(z, w) = \frac{1}{(2\pi)^{n+1}} \int_{\Lambda^*} e^{i\langle z - \bar{w}, \lambda \rangle} \frac{d\lambda}{D(\lambda)},$$

where $\langle z - \bar{w}, \lambda \rangle = \sum_{j=1}^{n+1} (z_j - \bar{w}_j)\lambda_j$ and

$$D(\lambda) = \int_{\xi_{n+1} > \sum_{j=1}^{n} f_j(\xi_j)} e^{-2\langle \lambda, \xi \rangle} d\xi.$$

By changing variables and simple computation, we obtain the following proposition.

Proposition 2.1. *The Bergman kernel of Ω_f is expressed as follows:*

$$K(z,w) = \frac{1}{(4\pi)^{n+1}} \int_0^\infty e^{(i/2)[z_{n+1}-\bar{w}_{n+1}]\zeta} \prod_{j=1}^{n} F_j((-i/2)[z_j-\bar{w}_j],\zeta)\zeta d\zeta,$$

where

$$F_j(X_j,\zeta) = \int_{-R_j^-\zeta}^{R_j^+\zeta} e^{X_j\lambda} \frac{d\lambda}{\psi_j(\zeta,\lambda)},$$

$$\psi_j(\zeta,\lambda) = \int_{-\infty}^{\infty} e^{-f_j(\xi)\zeta+\xi\lambda} d\xi.$$

Remark. The geometrical character of decoupled domain reflects the *product* of the functions F_j $(j = 1,\dots,n)$ in the integral representation. In [5], Bonami and Lohoué give an integral representation of the Bergman kernel of the domain $\{z \in \mathbb{C}^n; \sum_{j=1}^n |z_j|^{2m_j} < 1\}$ $(m_j \in \mathbb{N})$:

$$K(z,w) = \frac{1}{\pi^n} \int_0^\infty e^{-\tau} \prod_{j=1}^{n} E_{m_j}(z_j\bar{w}_j\tau^{1/m_j})\tau^{\sum_{j=1}^n 1/m_j} d\tau,$$

where

$$E_m(u) = m \sum_{\nu=0}^{\infty} \frac{u^\nu}{\Gamma(\nu/m + 1/m)} \qquad m \in \mathbb{N}.$$

Note that E_m is the derivative of Mittag-Leffler's function.

3 THE SINGULARITIES ON THE DIAGONAL

In this section, we investigate the situation of the singularities of the Bergman kernel of the following domain on the diagonal:

$$\Omega_m = \mathbb{R}^{n+1} + i\omega_m \subset \mathbb{C}^{n+1}, \quad \text{with}$$

$$\omega_m = \left\{ y \in \mathbb{R}^{n+1}; y_{n+1} > \sum_{j=1}^{n} a_j y_j^{2m_j} \right\},$$

where $a_j > 0$, $m_j \in \mathbb{N}$ and $m_n \neq 1$.

3.1 REAL BLOWING-UP

Let $\pi : \mathbb{C}^{n+1} \to \mathbb{R}^{n+1}$ be the projection defined by $\pi(z_1, \ldots, z_{n+1}) = (\Im z_1, \ldots, \Im z_{n+1})$. Set $O = (0, \ldots, 0)$. It is easy to check that Ω_m is a pseudoconvex domain; moreover $z^0 \in \partial\Omega_m$, with $\pi(z^0) = O$, is a weakly pseudoconvex point of the Catlin multi-type $(2m_1, \ldots, 2m_n, 1)$. Set $P = \{j; m_j \neq 1\}$ and $Q = \{j; m_j = 1\}$.

Now we introduce the transformation σ, which plays a key role in our analysis. Set

$$\Delta = \left\{ \tau = (\tau_j)_{j \in P} \in \mathbb{R}^{|P|}; \sum_{j \in P} \tau_j^{2m_j} < 1 \right\}$$

and $\partial\Delta = \overline{\Delta} \setminus \Delta$. The transformation $\sigma : \overline{\omega_m} \to \overline{\Delta \times \mathbb{R}_+}$ is defined by

$$\sigma : \begin{cases} \tau_j = a_j^{1/(2m_j)} y_j (y_{n+1} - \sum_{k \in Q} a_k y_k^{2m_k})^{-1/(2m_j)} & (j \in P), \\ \varrho = y_{n+1} - \sum_{k \in Q} a_k y_k^{2m_k}. \end{cases} \tag{3.1}$$

Then $\sigma \circ \pi$ is the transformation from $\overline{\Omega_m}$ to $\overline{\Delta \times \mathbb{R}_+}$.

The boundary of ω_m is transfered by σ in the following: $\sigma((\partial\omega_f) \setminus \{O\}) = \partial\Delta \times \mathbb{R}_+$ and $\sigma^{-1}(\Delta) = \{O\}$. This indicates that σ is the real blowing-up of $\partial\omega_m$ at O, so we may say that $\sigma \circ \pi$ is the real blowing-up at the weakly pseudoconvex point z^0. Moreover σ patches the coordinates $(\tau, \varrho) = ((\tau_j)_{j \in P}, \varrho)$ on ω_m, which can be considered as the polar coordinates around O. We call $\tau = (\tau_j)_{j \in P}$ the angular variables and ϱ the radial variable, respectively. If the angular variables τ tend to $\partial\Delta$, then we approach z^0 in tangential directions.

The following theorem asserts that the singularities of the Bergman kernel of Ω_m at z^0, $\pi(z^0) = O$, can be essentially expressed in terms of the polar coordinates (τ, ϱ):

Theorem 3.1. *The Bergman kernel K of Ω_m has the form:*

$$K(z, z) = \Phi(\tau) \varrho^{-2 - \sum_{j=1}^{n} 1/m_j}, \tag{3.2}$$

where $\Phi \in C^\omega(\Delta)$ and is positive on Δ. Moreover $\Phi(\tau)$ is unbounded as τ approaches to $\partial\Delta$.

Proof. In case $f_j(x_j) = a_j x_j^{2m_j}$ in Proposition 2.1, we obtain the following representation:

$$K(z,z) = \frac{\prod_{j=1}^n a_j^{1/m_j}}{(4\pi)^{n+1}} \times$$
$$\int_0^\infty e^{-y_{n+1}\zeta} \prod_{j=1}^n G_j(a_j^{1/(2m_j)} y_j \zeta^{1/(2m_j)}) \zeta^{1+\sum_{j=1}^n 1/m_j} d\zeta, \quad (3.3)$$

where $G_j(u) = \int_{-\infty}^\infty e^{uv} \frac{dv}{\varphi_j(v)}$ and $\varphi_j(v) = \int_{-\infty}^\infty e^{-w^{2m_j}+vw} dw$.

Here if $j \in Q$ (i.e. $m_j = 1$), then $\varphi_j(v) = \sqrt{\pi} e^{v^2/4}$ and $G_j(u) = 2e^{u^2}$. Thus substituting $G_j(u) = 2e^{u^2}$ $(j \in Q)$ into (3.3),

$$K(z,z) = \frac{2^{|Q|} \prod_{j=1}^n a_j^{1/m_j}}{(4\pi)^{n+1}} \times$$
$$\int_0^\infty e^{-[y_{n+1} - \sum_{j \in Q} a_j y_j^2]\zeta} \prod_{j \in P} G_j(a_j^{1/(2m_j)} y_j \zeta^{1/(2m_j)}) \zeta^{1+\sum_{j=1}^n 1/m_j} d\zeta.$$

Introducing the variables $\tau = (\tau_j)_{j \in P}$ and ϱ in (3.1) to the above equation, we can obtain (3.2), where

$$\Phi(\tau) = \frac{2^{|Q|} \prod_{j=1}^n a_j^{1/m_j}}{(4\pi)^{n+1}} \int_0^\infty e^{-\eta} \prod_{j \in P} G_j(\tau_j \eta^{1/(2m_j)}) \eta^{1+\sum_{j=1}^n 1/m_j} d\eta. \quad (3.4)$$

The following lemma about the function G_j is necessary to analyze the properties of Φ.

Lemma 3.1. *Let $\tilde{G}_j$ be the function defined by*

$$G_j(u) = u^{2m_j - 2} e^{u^{2m_j}} \cdot \tilde{G}_j(u), \quad (3.5)$$

then $\tilde{G}_j$ is expanded asymptotically in the form $\tilde{G}_j(u) \sim \sum_{k=0}^\infty c_k u^{-2m_j k}$ as $u \to \infty$, where c_k are constants ($c_0 > 0$).

Proof. See Lemma 6.2 in [13]. □

From the positivity of G_j on $[0, \infty)$ and the behavior at infinity (3.5) in the above lemma, we have

$$G_j(u) \approx u^{2m_j - 2} e^{u^{2m_j}} + 1 \quad \text{for } u \geq 0. \quad (3.6)$$

Substituting (3.6) into (3.4), we have

$$\begin{aligned}\Phi(\tau) &\approx \sum_{K\subset P}\prod_{j\in K}\tau_j^{2m_j-2}\int_0^\infty e^{-\eta[1-\sum_{j\in K}\tau_j^{2m_j}]}\eta^{1+|Q|+|K|+|1/m|_{P\setminus K}}d\eta \\ &\approx \sum_{K\subset P}\prod_{j\in K}\tau_j^{2m_j-2}\left[1-\sum_{j\in K}\tau_j^{2m_j}\right]^{-2-|Q|-|K|-|1/m|_{P\setminus K}}. \qquad (3.7)\end{aligned}$$

The above inequalities imply that $\Phi \in C^\omega(\Delta)$ and Φ is unbounded as τ approaches to $\partial\Delta$. □

Remark. From the above argument we can obtain the precise estimate and the boundary limit of K at z^0 ([20],[4]). In fact the inequalities (3.7) yield the precise estimate of K from above and below. The boundary limit of $K(z,z)\cdot\varrho^{2+\sum_{j=1}^n 1/m_j}$ on non-tangential cone equals

$$\Phi(0) = \frac{2^{|Q|}\prod_{j=1}^n a_j^{1/m_j}}{(4\pi)^{n+1}}\cdot\Gamma\left(2+\sum_{j=1}^n\frac{1}{m_j}\right)\cdot\prod_{j\in P}G_j(0).$$

Moreover (3.7) implies that an approach region of K at z^0 is $\mathcal{U}_\alpha = \{z\in\Omega_m; y_{n+1} > \sum_{j\in Q}a_jy_j^2 + \alpha\sum_{j\in P}a_jy_j^{2m_j}\}$ $(\alpha > 1)$, that is, $K(z,z)\cdot\varrho^{2+\sum_{j=1}^n 1/m_j}$ is bounded near z^0 on $\mathcal{U}_\alpha$.

3.2 RECURSIVE FORMULA

We investigate the structure of the singularities of the function $\Phi(\tau)$ on $\partial\Delta$, which appears in Theorem 3.1, in more detail.

First we remark that $\Phi(\tau)$ defined by (3.4) takes a similar form as $K(z,z)$ in (3.3). Thus the argument in the previous subsection will be applied to $\Phi(\tau)$ in place of $K(z,z)$, and the form of the singularities of $\Phi(\tau)$ can be written in the same fashion as in Theorem 3.1. Moreover we can completely understand the singularities of $\Phi(\tau)$ by finitely many recursive processes of this kind.

We precisely explain this process. We inductively define the sets $P_{[k]}$, $Q_{[k]}\subset N$, the variables $\tau_{[k]} = (\tau_{[k],j})_{j\in P_{[k]}}$, $\varrho_{[k]}$, the set $\Delta_{[k]}\subset\mathbb{R}^{|P_{[k]}|}$ and the function $\Phi_{[k]}(\tau_{[1]};\cdots;\tau_{[k]})$ on $\prod_{j=1}^k\Delta_{[j]}$ in the following way.

First we set $P_{[1]} = P(\neq\emptyset)$, $Q_{[1]} = Q$, $\tau_{[1]} = (\tau_{[1],j})_{j\in P_{[1]}} = (\tau_j)_{j\in P}$, $\varrho_{[1]} = r$, $\Delta_{[1]} = \Delta$ and $\Phi_{[1]}(\tau_{[1]}) = \Phi(\tau)$. Suppose that the sets $P_{[k-1]}(\neq \emptyset)$, $Q_{[k-1]} \subset P$ are settled, then the sets $\Delta_{[k-1]}$, $\partial\Delta_{[k-1]} \subset \mathbb{R}^{|P_{[k-1]}|}$ are defined by

$$\Delta_{[k-1]} = \left\{ \tau_{[k-1]} = (\tau_{[k-1],j})_{j\in P_{[k-1]}};\ \sum_{j\in P_{[k-1]}} \tau_{[k-1],j}^{2m_j} < 1 \right\}$$

and $\partial\Delta_{[k-1]} = \overline{\Delta_{[k-1]}} \setminus \Delta_{[k-1]}$. When we select a point $\tau_{[k-1]}^0 = (\tau_{[k-1],j}^0)_{j\in P_{[k-1]}} \in \partial\Delta_{[k-1]}$, the sets $P_{[k]}$, $Q_{[k]} \subset P_{[k-1]}$ are determined by

$$\begin{cases} P_{[k]} &= \{j \in P_{[k-1]}; \tau_{[k-1],j}^0 = 0\}, \\ Q_{[k]} &= \{j \in P_{[k-1]}; \tau_{[k-1],j}^0 \neq 0\}. \end{cases}$$

If $P_{[k]} \neq \emptyset$, then the variables $\tau_{[k]} = (\tau_{[k],j})_{j\in P_{[k]}}$ and $\varrho_{[k]}$ are defined by

$$\begin{cases} \tau_{[k],j}^{2m_j} &= \tau_{[k-1],j}^{2m_j} \cdot \left(1 - \sum_{j\in Q_{[k]}} \tau_{[k-1],j}^{2m_j}\right)^{-1} \quad (j \in P_{[k]}), \\ \varrho_{[k]} &= 1 - \sum_{j\in Q_{[k]}} \tau_{[k-1],j}^{2m_j}. \end{cases}$$

Then we define the function $\Phi_{[k]}$ on $\prod_{j=1}^k \Delta_{[j]}$ in the following:

$$\Phi_{[k]}(\tau_{[1]}; \cdots; \tau_{[k]}) = \int_0^\infty e^{-\eta} \prod_{j\in P_{[k]}} G_j(\tau_{[k],j}\eta^{1/(2m_j)}) \times$$

$$\prod_{l=2}^k \prod_{j\in Q_{[l]}} \tilde{G}_j(\tau_{[l-1],j} \cdot \prod_{p=l}^k \varrho_{[p]}^{-1} \cdot \eta^{1/(2m_j)})\eta^{a_{[k]}-1} d\eta, \tag{3.8}$$

where the constant $a_{[k]}$ is defined by $a_{[k]} = 2 + \sum_{j=1}^k |Q_{[j]}| + |1/m|_P - \sum_{j=2}^k |1/m|_{Q_{[j]}}$ and $\tilde{G}_j$ is as in Lemma 3.1. In the above inductive process, we have

$$P = P_{[1]} \supsetneqq P_{[2]} \supsetneqq \cdots \supsetneqq P_{[k-1]} \supsetneqq P_{[k]}.$$

So there exists a positive integer $k^0 \leq |P|$ such that $P_{[k^0+1]} = \emptyset$. Thus we have defined $P_{[k]}, Q_{[k]}, \tau_{[k]}, \varrho_{[k]}, \Delta_{[k]}, \Phi_{[k]}, a_{[k]}$ for $k = 1, 2, \ldots, k^0$ recursively. We set $Q_{[k_0+1]} = P_{[k_0]}$ and $\varrho_{[k_0+1]} = 1 - \sum_{j\in P_{[k_0]}} \tau_{[k_0],j}^{2m_j}$.

Theorem 3.2. *The function $\Phi_{[k]}(\tau_{[1]};\cdots;\tau_{[k]})$ is a positive and C^∞-function on $\prod_{j=1}^{k-1}\overline{\Delta_{[j]}}\times\Delta_{[k]}$ and is unbounded as $\tau_{[k]}\in\Delta_{[k]}$ approaches $\tau_{[k]}^0\in\partial\Delta_{[k]}$. Moreover the following recursive formula satisfies:*

$$\Phi_{[k-1]}(\tau_{[1]};\cdots;\tau_{[k-1]})=\prod_{j\in Q_{[k]}}\tau_{[k],j}^{2m_j-2}\frac{\Phi_{[k]}(\tau_{[1]};\cdots;\tau_{[k]})}{\varrho_{[k]}^{a_{[k]}}} \tag{3.9}$$

for $2\le k\le k_0$ and

$$\Phi_{[k_0]}(\tau_{[1]};\cdots;\tau_{[k_0]})=\frac{\varphi(\tau_{[1]};\cdots;\tau_{[k_0]})}{\varrho_{[k_0+1]}^{n+2}}+\psi(\tau_{[1]};\cdots;\tau_{[k_0]})\log\varrho_{[k_0+1]}, \tag{3.10}$$

where $\varphi,\ \psi\in C^\infty(\prod_{k=1}^{k_0}\overline{\Delta_{[k]}})$.

Note that the form of the singularities of $\Phi_{[k^0]}$ in (3.10) is that of the asymptotic expansion (1.1) of Fefferman in the strictly pseudoconvex case. Therefore the above theorem shows that at most $|P|$-times blowing-ups resolve the degeneration of the Levi form in the study of the singularities of K.

Proof of Theorem 3.2. First the properties of G_j and $\tilde G_j$ in Lemma 3.1 imply that $\Phi_{[k]}$ is positive and is a C^∞-function on $\prod_{j=1}^{k-1}\overline{\Delta_{[j]}}\times\Delta_{[k]}$.

Now we consider the case $k\le k_0-1$. Substituting (3.5) into (3.8) in the case $j\in Q_{[k]}$,

$$\begin{aligned}
&\Phi_{[k-1]}(\tau_{[1]};\cdots;\tau_{[k-1]})=\\
&\prod_{j\in Q_{[k]}}\tau_{[k-1],j}^{2m_j-2}\int_0^\infty e^{-\eta[1-\sum_{j\in Q_{[k]}}\tau_{[k-1],j}^{2m_j}]}\prod_{j\in P_{[k]}}G_j(\tau_{[k-1],j}\eta^{1/(2m_j)})\times\\
&\prod_{j\in Q_{[k]}}\tilde G_j(\tau_{[k-1],j}\eta^{1/(2m_j)})\prod_{l=2}^{k-1}\prod_{j\in Q_{[l]}}\tilde G_j(\tau_{[l-1],j}\cdot\prod_{p=l}^{k-1}\varrho_{[p]}^{-1}\eta^{1/(2m_j)})\times\\
&\eta^{a_{[k]}-1}d\eta=\varrho_{[k]}^{-a_{[k]}}\prod_{j\in Q_{[k]}}\tau_{[k-1],j}^{2m_j-2}\int_0^\infty e^{-\eta}\times\\
&\prod_{j\in P_{[k]}}G_j(\tau_{[k],j}\eta^{1/(2m_j)})\prod_{l=2}^{k}\prod_{j\in Q_{[l]}}\tilde G_j(\tau_{[l-1],j}\cdot\prod_{p=l}^{k}\varrho_{[p]}^{-1}\eta^{1/(2m_j)})\times
\end{aligned}$$

$$\eta^{a_{[k]}-1}d\eta = \prod_{j\in Q_{[k]}} \tau_{[k-1],j}^{2m_j-2}\frac{\Phi_{[k]}(\tau_{[1]};\cdots;\tau_{[k]})}{\varrho_{[k]}^{a_{[k]}}}.$$

Next we consider the case of $\Phi_{[k_0]}$. Note that $P_{[k_0]} = Q_{[k_0+1]}$. By Lemma 3.1, we have

$$\Phi_{[k_0]}(\tau_{[1]};\cdots;\tau_{[k_0]}) = \int_0^\infty e^{-\eta[1-\sum_{j\in Q_{[k_0+1]},j}\tau_{[k_0],j}^{2m_j}]}\times$$

$$\prod_{j\in Q_{[k_0+1]}} \tilde{G}_j(\tau_{[k_0-1],j}\eta^{1/(2m_j)}) \prod_{l=2}^{k_0-1}\prod_{j\in Q_{[l]}} \tilde{G}_j(\tau_{[l-1],j}\prod_{p=l}^{k_0-1}\varrho_{[p]}^{-1}\eta^{1/(2m_j)})\times$$

$$\eta^{n+1}d\eta = \frac{\varphi(\tau_{[1]};\cdots;\tau_{[k_0]})}{\varrho_{[k_0+1]}^{n+2}} + \psi(\tau_{[1]};\cdots;\tau_{[k_0]})\log\varrho_{[k_0+1]}.$$

Here φ and ψ are C^∞-smooth on $\prod_{k=1}^{k_0}\overline{\Delta_{[k]}}$. We remark that the second equality and the regularity of φ and ψ are induced by putting together the asymptotic expansion of $\tilde{G}_j$ in Lemma 3.1 and the following lemma. (The proof of the lemma is easy, so it is omited.)

Lemma 3.2. *Let $a \in C^\infty([0,\infty))$ be a function satisfying $a(t) \sim \sum_{j=0}^\infty c_j t^{-j}$ as $t \to \infty$, where c_j are constants ($c_0 \neq 0$). Then*

$$\int_0^\infty e^{-rt}a(t)t^{N-1}dt = \frac{A(r)}{r^N} + B(r)\log r,$$

where $A, B \in C^\infty([0,\infty))$ and $A(0) \neq 0$.

□

4 THE SINGULARITIES OFF THE DIAGONAL

In this section, we investigate the singularities of the Bergman kernel of Ω_m off the diagonal. Christ and Geller [9] showed that the Bergman kernel of the domain $\{(z_1,z_2)\in\mathbb{C}^2;\Im z_2 > [\Re z_1]^{2m}\}$ ($m = 2,3,\ldots$) fails to be real analytic on the boundary off the diagonal. The author [15] gave an integral representation of its Bergman kernel by using countably many functions whose singularities can be understood directly. Here we give an analogous representation in the case of the domain Ω_m.

It is known in [23],[17] that all zeros of the entire function:

$$\tilde{\varphi}_j(v) = \int_{-\infty}^{\infty} e^{-2(a_j w^{2m_j} - vw)} dw \quad (j \in P = \{j; m_j \neq 1\})$$

exist on the imaginary axis and are simple. We denote the set of the zeros of $\tilde{\varphi}_j$ by $\{\pm i p_k^{(j)}; 0 < p_k^{(j)} < p_{k+1}^{(j)} \ (k \in \mathbb{N})\}$. Let $S_K^v : \mathbb{C}^n \times \mathbb{R} \setminus \{O\} \to \mathbb{C}$ be a function defined by

$$S_K^v(z,t) = C_K \int_0^{\infty} \tau^{v+\sum_{j=1}^n 1/m_j} \times$$

$$\exp\left\{ it\tau - \sum_{j=1}^n a_j y_j^{2m_j} \tau - \frac{1}{2} \sum_{j \in Q} a_j z_j^2 \tau - \sum_{j \in P} p_{k_j}^{(j)} \sigma(x_j) z_j \tau^{1/(2m_j)} \right\} d\tau,$$

where $K = (k_j)_{j \in P} \in \mathbb{N}^{|P|}$, $C_K = (2\pi i) \prod_{j \in K} [\tilde{\varphi}_j'(i p_{k_j}^{(j)})]^{-1}$, $v \geq 0$ and $\sigma(x)$ is the sign of x. For $H (\neq \emptyset) \subset P$, the set $\Xi_H \subset \mathbb{C}^n \times \mathbb{R}$ is defined by

$$\Xi_H = \{(z,t); y_1 = \cdots = y_n = t = 0 \text{ and } x_j \neq 0 \text{ if and only if } j \in H\}.$$

Set $\Xi = \bigcup \{\Xi_H; \emptyset \neq H \subset P\}$. It is easy to check that S_K^v belongs to s-th order Gevrey class G^s for all $s \geq \min\{2m_j; j \in H\}$ on Ξ_H, but no better, where $G^s := \{u; \exists C > 0 \text{ s.t. } |\partial^\alpha u| \leq C^{|\alpha|} \Gamma(s|\alpha|) \ \forall \alpha\}$.

The following theorem shows that the Bergman kernel $B(z,t) = K((z, t + i\sum_{j=1}^n a_j y_j^{2m_j}), O)$ can be expressed by the superposition of $\{S_K^1\}_K$.

Theorem 4.1. *If* $|\arg z_j - \pi/2 \pm \pi/2| < \pi/(4m_j - 2)$ $(j \in P)$, *then*

$$B(z,t) = c \int_0^{\infty} e^{-p} H(z,t;p) dp,$$

where

$$H(z,t;p) = \sum_{K \in \mathbb{N}^{|P|}} \frac{S_K^1(z,t) p^{\sum_{j \in P} f_{k_j}^{(j)}}}{\prod_{j \in P} \Gamma(f_{k_j}^{(j)} + 1)} \tag{4.1}$$

for some sequence $f_k^{(j)} = k + O(k^{-1}) > 0$ *as* $k \to \infty$ *and* c *is a constant. Here the series (4.1) absolutely converges with respect to* $p \geq 0$ *for fixed* (z,t)*; moreover there exist positive constants* A *and* $C(z)$ *depending on* $z \in \mathbb{C}^n$ *such that,*

$$|H(z,t;p)| \leq C(z) p^A \quad \text{for } p \geq 1. \tag{4.2}$$

Proof. Although the proof is long, the theorem can be essentially shown by using a similar idea in [15]. □

Remark. In the estimate (4.2), we can take $\sum_{j\in P} m_j/(4m_j - 2)$ as the value of A. But we do not know the best possible value for A.

The singularities of $B(z,t)$ is approximated by S_I^1, where $I = (1,\dots,1)$. Actually if $(z,t)\in\Xi$, then

$$\frac{\partial^k}{\partial t^k}B(z,t) = \frac{\partial^k}{\partial t^k}S_I^1(z,t)\{1+O(a^{-k})\},$$

as $k\to\infty$, where $a>1$ is a constant. This equality can be obtained in a similar fashion as in [15], Section 6. Considering the singularities of S_I^1, the above equality implies the following theorem.

Theorem 4.2. *B fails to be real analytic on the set Ξ. Moreover B belongs to s-the Gevrey class for $s \geq \min\{2m_j; j\in H\}$, but no better on Ξ_H.*

Remark. B is known to be C^∞-smooth away from $\{O\}$ ([1],[3]).

5 THE SZEGÖ KERNEL

5.1 INTEGRAL REPRESENTATION

Let Ω_f be a tube domain satisfying the condition in Section 2. Let $H^2(\Omega_f)$ be the subspace of $L^2(\Omega_f)$ consisting of holomorphic functions F on Ω_f such that

$$\sup_{\epsilon>0}\int_{\partial\Omega_f}|F(z_1,z_2+i\epsilon)|^2 d\sigma(z)<\infty,$$

where $d\sigma$ is the measure on $\partial\Omega_f$ given by Lebesgue measure on $\mathbb{C}^n\times\mathbb{R}$ when we identify $\partial\Omega_f$ with $\mathbb{C}^n\times\mathbb{R}$ (by the map $(z,t+if(\Im z_1,\dots,\Im z_n))\mapsto(z,t)$). The Szegö projection is the orthogonal projection $\mathbb{S}: L^2(\partial\Omega_f)\to H^2(\Omega_f)$ and we can write

$$\mathbb{S}F(z)=\int_{\partial\Omega_f}K^S(z,w)F(w)d\sigma(w),$$

where $K^S:\Omega_f\times\Omega_f\to\mathbb{C}$ is the *Szegö kernel* of the domain Ω_f.

Now we can obtain the following proposition by using an integral representation in [22], Chapter 3.

Proposition 5.1. *The Szegö kernel of Ω_f is expressed as follows:*

$$K^S(z,z) = \frac{1}{(4\pi)^n}\int_0^\infty e^{-y_{n+1}\zeta}\prod_{j=1}^n F_j(y_j,\zeta)d\zeta,$$

where the F_j's are as in Proposition 2.1.

5.2 ON THE DIAGONAL

We also give an analogous result of the Szegö kernel of Ω_m to Theorem 3.1. The following theorem can be obtained in a similar fashion.

Theorem 5.1. *The Szegö kernel K^S of Ω_m has the form:*

$$K^S(z,z) = \Phi^S(\tau)\varrho^{-1-\sum_{j=1}^n 1/m_j},$$

where $\Phi^S \in C^\omega(\Delta)$ and is positive on Δ. Moreover $\Phi^S(\tau)$ is unbounded as τ approaches to $\partial\Delta$.

The singularities of $\Phi^S(\tau)$ can be perfectly understood by the same way as in §3.2.

5.3 OFF THE DIAGONAL

We also give an analogous result of the Szegö kernel of Ω_m to Theorem 3.1. The following theorem can be obtained in a similar fashion. Set $S(z,t) = K^S((z, t + i\sum_{j=1}^n a_j y_j^{2m_j}), O)$.

Theorem 5.2. *If $|\arg z_j - \pi/2 \pm \pi/2| < \pi/(4m_j - 2)$ $(j \in P)$, then*

$$S(z,t) = c^S \int_0^\infty e^{-p} H^S(z,t;p)dp,$$

with

$$H^S(z,t;p) = \sum_{K\in\mathbf{N}^{|P|}} \frac{S_K^0(z,t)p^{\sum_{j\in P} f_{k_j}^{(j)}}}{\prod_{j\in P}\Gamma(f_{k_j}^{(j)}+1)}, \tag{5.1}$$

where $f_k^{(j)}$'s are as in Theorem 4.1 and c^S is a constant. Here the series (5.1) absolutely converges with respect to $p \geq 0$ for fixed (z,t); moreover there exist positive constants A and $C^S(z)$ depending on $z \in \mathbf{C}^n$ such that,

$$\left|H^S(z,t;p)\right| \leq C^S(z)p^A \quad \text{for } p \geq 1.$$

Remark. As for the regularity of the Szegö kernel S, the same statement in Theorem 4.2 satisfies.

References

[1] S. R. Bell: Differentiability of the Bergman kernel and pseudolocal estimates, Math. Z. **192** (1986), 467–472.

[2] ______: Extendibility of the Bergman kernel function, Complex analysis, II (College Park, Md., 1985–86), 33–41, Lecture Notes in Math., **1276**, Springer, Berlin-New York, 1987.

[3] H. P. Boas: Extension of Kerzman's theorem on differentiability of the Bergman kernel function, Indiana Univ. Math. J. **36** (1987), 495–499.

[4] H. P. Boas, E. J. Straube and J. Yu: Boundary limits of the Bergman kernel and metric, Michigan Math. J. **42** (1995), 449–461.

[5] A. Bomami and N. Lohoué: Projecteures de Bergman et Szegö pour une classe de domaines faiblement pseudo-convexes et estimation L^p, Compositio Math. **46** Fasc 2, (1982), 159–226.

[6] L. Boutet de Monvel and J. Sjöstrand: Sur la singularité des noyaux de Bergman et de Szegö, Soc. Math. de France Astérisque **34-35** (1976), 123–164.

[7] D. Catlin: Estimates of invariant metrics on pseudoconvex domains of dimension two, Math. Z. **200** (1989), 429–466.

[8] M. Christ: Remarks on the breakdown of analyticity for $\bar{\partial}_b$ and Szegö kernels, Proceedings of 1990 Sendai conference on harmonic analysis (S. Igari, ed.), Lecture Notes in Math. Springer, 61–78.

[9] M. Christ and D. Geller: Counterexamples to analytic hypoellipticity for domains of finite type, Ann. of Math. **235** (1992), 551–566.

[10] C. Fefferman: The Bergman kernel and biholomorphic mappings of pseudoconvex domains, Invent. Math. **26** (1974), 1–65.

[11] F. Haslinger: Szegö kernels of certain unbounded domains in $\mathbf{C}^2$, Rev. Roumain Math. Pures Appl., **39** (1994), 939–950.

[12] J. Kamimoto: Singularities of the Bergman kernel for certain weakly pseudoconvex domains, J. Math. Sci. Univ. of Tokyo **5** (1998), 99–117.

[13] ______: Asymptotic analysis of the Bergman kernel on weakly pseudoconvex domains, Ph. D. thesis, The University of Tokyo, 1997.

[14] ———: Asymptotic expansion of the Bergman kernel for weakly pseudoconvex tube domains in $\mathbb{C}^n$, to appear in Annales Fac. Sci. Toulouse (1998).

[15] ———: On the singularities of non-analytic Szegö kernels, preprint.

[16] ———: On an integral of Hardy and Littlewood, Kyushu J. of Math. **52** (1998), 249–263.

[17] J. Kamimoto, H. Ki and Y-O. Kim: On the multiplicities of the zeros of Laguerre-Pólya functions, to appear in Proc. of Amer. Math. Soc.

[18] N. Kerzman: The Bergman kernel function. Differentiability at the boundary, Math. Ann. **195** (1972), 149–158.

[19] A. Korányi: The Bergman kernel function for tubes over convex cones, Pacific J. Math. **12** (1962), 1355–1359.

[20] J. D. McNeal: Local geometry of decoupled pseudoconvex domains, Proceedings in honor of Hans Grauert, Aspekte de Mathematik, Vieweg, Berlin (1990), 223–230.

[21] A. Nagel: Vector fields and nonisotropic metrics, Beijing Lectures in Harmonic Analysis, (E. M. Stein, ed.), Princeton University Press, Princeton, NXJ, (1986), 241–306.

[22] S. Saitoh: *Integral transforms, reproducing kernels and their applications*, Pitman Reseach Notes in Mathematics Series **369**, Addison Wesley Longman, UK (1997).

[23] G. Pólya: Über trigonometrische Integrale mit nur reelen Nullstellen, J. Reine Angew. Math. **58** (1927), 6–18.

11 A SAMPLING THEOREM FOR SOLUTIONS OF THE DIRICHLET PROBLEM FOR THE SCHRÖDINGER OPERATOR

Alexander Kheyfits

Department of Mathematics and Computer Science
Bronx Community College of the City University of New York
alexander.kheyfits@bcc.cuny.edu

Abstract: A sampling series is derived for solutions of the Dirichlet problem in the half-space for the stationary Schrödinger operator

INTRODUCTION

The sampling theorem of Whitteker - Kotel'nikov - Shannon (WKS-theorem) and many of its extensions and analogs are developed in detail in the comprehensive monograph by Ahmed I. Zayed [8]. According to this theorem, under some natural assumptions a function f can be reconstructed from its sample values at the points of a given sequence $\{t_l\}$:

$$f(t)=\sum_{l=-\infty}^{\infty} f(t_l)\, s_l(t)$$

The sampling functions s_l are explicitly given later on in the article.

We shall prove a similar representation for solutions of the boundary value problem (where Δ is the Laplace operator):

$$\begin{cases} \Delta u(X) - c(X)u(X) = 0, & X \in R_+^{n+1} \\ u(x) = f(x), & x \in R^n = \partial R_+^{n+1} \end{cases} \tag{1}$$

in the half-space $R_+^{n+1} = \{X = (x, x_{n+1}) \,|\, x \in R^n = \partial R_+^{n+1},\ x_{n+1} > 0\}$, $n \geq 1$.

Sampling representations for many problems can be derived by making use of the Hilbert spaces with reproducing kernel. This method was developed by Saburou Saitoh – see, for example, [7]. However, we apply here the known formula for the solutions of the boundary value problem (1), which gives an explicit representation of the sampling functions.

We suppose that a potential $c(X)$ in (1) satisfies the following conditions:

c1) $0 \le c(X) \in L^p_{loc}(R^{n+1}_+)$ with some $p > \dfrac{n+1}{2}$

c2) Every boundary point $x_0 \in \partial R^{n+1}_+$ has a half-ball vicinity

$B^+(x_0, r_0) = \{X \mid |X - x_0| < r_0,\ x_{n+1} > 0\}$, such that $c \in L^s(B^+(x_0, r_0))$ with some $s > n+1$.

It should be noted that these conditions can be relaxed, but can not be omitted at all. Namely, Fumi-Yuki Maeda [4] has shown that if a potential grows too fast at the boundary, then the problem (1) may have no solution.

Under the conditions **c1), c2)** the problem (1) has the Green's function $G(X,Y)$, whose derivative with respect to the inner normal $\partial G(X,t)/\partial n(t)$ (the c-Poisson kernel) is a positive continuous function on $R^{n+1}_+ \times R^n$ -- see, for example, Marko Bramanti's survey [1].

The boundary function f is supposed to be a real-valued continuous function on R^n such that

cf) The integral $\int_{R^n} |f(t)\partial G(X,t)/\partial n(t)|\, dt$ converges for some (whence, for every) $X \in R^{n+1}_+$

The condition **cf)** can also be weakened [2], but this would result in more tedious formulas, and we omit those statements.

Provided with these conditions, the problem (1) has a solution [2]

$$u(X) = \int_{R^n} f(t) \frac{\partial G(X,t)}{\partial n(t)} dt \tag{2}$$

Next, we assume that the function f belongs to the class of K-band limited functions in Peter McCoy's sense [5]. That is, there exist a bounded domain $\Omega \in R^m$, $m \ge 1$, a sequence $\{t_l\}_{-\infty}^{\infty} \subset R^n$, and a measurable function $K(z,t)$, $z \in \Omega$, $t \in R^n$, such that

K) $K(\cdot,t) \in L^2(\Omega)$ for every $t \in R^n$, and the functions $K(\cdot, t_l)$ form a complete orthogonal sequence in $L^2(\Omega)$.

Moreover, there exists a function $F \in L^2(\Omega)$ such that f has a representation

fK)
$$f(t) = \int_\Omega K(z,t)F(z)d\omega, \tag{3}$$

where $d\omega$ is the Lebesgue measure in Ω.

Note, that (3) is an extension of the Fourier representation of band - limited functions (the Paley - Wiener class) $f(t) = \int_{-a}^{a} e^{itz} F(z)dz,\ F \in L^2(-a,a)$.

The conditions **K)** and **fK)** imply (see, for example, arguments in [5] and in [8, p.p. 46 - 47, 206 - 207]) the validity of an expansion, converging in L^2,

$$f(t) = \sum_{-\infty}^{\infty} f(t_l)\, s_l(K,t) \tag{4}$$

with the sampling functions

$$s_l(K,t) = \frac{\int_\Omega K(z,t)\,\overline{K(z,t_l)}\, d\omega}{\int_\Omega |K(z,t_l)|^2\, d\omega}. \tag{5}$$

Under some additional conditions, the series (4) converges absolutely and uniformly on compact sets in R^n. For instance, it is the case if f is an entire function of exponential type σ of the Paley - Wiener class and the sequence $\{t_l\}$ satisfies the strip inequality $\sup_l |t_l - l\pi/\sigma| < \pi/(4\sigma)$. In particular, if we set $\Omega = (-\pi,\pi)$, $t_l = l$ for $-\infty < l < \infty$, and $K(z,t) = e^{itz}$, then $s_l(K,t) = s_l(t) = \dfrac{(-1)^l \sin \pi t}{\pi(t-l)}$, and the expansion (4) reduces to the WKS-theorem, or in other words, to Lagrange's interpolation series (see, for example, [3, Chap. 4, Sect. 4.4]).

Next, we have to adjust growth properties of the Green's function G to those of the kernel K. We consider the following sampling functions, where $s_l(K,t)$ are given by (5):

$$S_l(X) = \int_{R^n} \frac{\partial G(X,t)}{\partial n(t)} s_l(K,t)dt,\ X \in R_+^{n+1} \tag{6}$$

and use the next condition:

GK) The series $\sum_l |S_l(X)|^q$ with some $1 \le q \le 2$ converges uniformly on compact sets in R_+^{n+1}.

Peter McCoy [5] has shown that under appropriate assumptions it is possible to derive a sampling series for a solution of a boundary value problem from a sampling representation of the boundary data. Here, we consider a boundary value problem in an unbounded domain. To get a sampling series for the boundary data, we use the condition as follows:

ft) The function f in (1) is a restriction on R^n of an entire function $f(z_1,...,z_n)$, $(z_1,...,z_n) \in C^n$, of exponential type at most σ for each $z_j \in C$, j = 1, ..., n, with fixed other variables. Moreover, we assume that $f(t_1,...,t_n) \in L^2(R^n)$ and the sequence $\{t_l\}$, $t_l = (t_{l1}, t_{l2}, ..., t_{ln})$, satisfies the Paley-Wiener inequality $\sup_l |t_{lk} - l\pi/\sigma| < \pi/(4\sigma)$, $\forall k = 1,...,n$.

THE MAIN RESULT

Now we can state our result. A similar assertion is valid for any cone with the smooth boundary.

Theorem *Let the conditions* **c1), c2), cf), K), fK), GK),** *and* **ft)** *are valid. Then the solution* (2) *can be expanded into the sampling series*

$$u(X) = \sum_l f(t_l) S_l(X), \tag{7}$$

where S_l *are given by* (6) *and the series* (7) *converges absolutely and uniformly on compact sets in* R_+^{n+1}.

Remark It is worth emphasizing, that the sampling functions $S_l(X)$ in (7) depend on the kernel K, that is, on the class, which f belongs to, but not upon an individual function in the class.

Proof is straightforward. Substituting the series (4) for f into (2), we get a representation

$$u(X) = \int_{R^n} \frac{\partial G(X,t)}{\partial n(t)} \{\sum_{-\infty}^{\infty} f(t_l) s_l(K,t)\} dt. \tag{8}$$

It is known that under the condition **ft)** the sequence $f(t_l)$ is bounded; moreover, the series $\sum_l |f(t_l)|^r$ converges for every $r \ge 2$. We select r specifically to be conjugate to the exponent q from **GK)**: $1/q + 1/r = 1$. If q = 1, that means the boundedness of $f(t_l), \forall l$. Next, using the condition **GK)** and the Hölder inequality, we immediately get the uniform convergence of the series

$$\left|\sum_l f(t_l)\int_{R^n}\frac{\partial G(X,t)}{\partial n(t)}s_l(K,t)\,dt\right|\le\left\{\sum_l|f(t_l)|^r\right\}^{1/r}\left\{\sum_l|S_l(X)|^q\right\}^{1/q}<\infty.$$

Now, the Fubini theorem justifies interchanging the summation and integration in (8), proving the theorem. ∎

EXAMPLES

1. Let $c(X)=c=const$ be a constant potential in the three-dimensional half-space R^3_+. Direct calculation yields the c-Poisson kernel

$$\frac{\partial G(X,t)}{\partial n(t)}=\frac{x_3(1+\sqrt{c}\,|X-t|)}{2\pi|X-t|^3}e^{-\sqrt{c}|X-t|},$$

where $X=(x_1,\ x_2,\ x_3)$, $x_3>0$, and $t=(t_1,\ t_2)$. We choose $\Omega=(-\pi,\pi)\times(-\pi,\pi)$, $t_{l,m}=(l,m)$ for $-\infty<l,m<\infty$, and $K(z,t)=e^{i\langle z,t\rangle}=e^{i(z_1t_1+z_2t_2)}$. Then

$s_{l,m}(K,t)=\dfrac{(-1)^{l+m}\sin\pi t_1\sin\pi t_2}{\pi^2(t_1-l)(t_2-m)}$, and for the solution (2) to the problem (1)

with an appropriate function f we obtain the expansion $u(X)=$

$$\frac{x_3}{2\pi^3}\sum_{l,m=-\infty}^{\infty}(-1)^{l+m}f(l,m)\{\int_{R^2}\frac{1+\sqrt{c}|X-t|}{|X-t|^3}e^{-\sqrt{c}|X-t|}\frac{\sin\pi t_1\sin\pi t_2}{(t_1-l)(t_2-m)}dt_1dt_2\}.$$

In particular, when $c=0$, we get the representation for the usual harmonic functions in the half-space:

$$u(X)=\frac{x_3}{2\pi^3}\sum_{l,m=-\infty}^{\infty}(-1)^{l+m}f(l,m)\{\int_{R^2}\frac{1}{|X-t|^3}\frac{\sin\pi t_1\sin\pi t_2}{(t_1-l)(t_2-m)}dt_1dt_2\}.\ ∎$$

2. Consider a constant potential $c(X)=c=const$ in the upper half-plane R^2_+. A fundamental solution of the operator $-\Delta+cI$ in the whole plane can be taken as $\frac{1}{2\pi}K_0(\sqrt{c}|X-Y|)$, where $X=(x_1,x_2)$, $Y=(y_1,y_2)$, and K_0 is the Macdonald function (the modified Bessel function of the third kind) [6, Chap. 7, Sect. 8]. Thus, the fundamental solution is normed as $\frac{1}{2\pi}\ln\frac{1}{r}$, $r\to 0$. Using the reflection method, we get the Green's function for the problem (1) with the constant potential c in the half-plane

$$G(X,Y)=\frac{1}{2\pi}\{K_0(\sqrt{c}|X-Y|)-K_0(\sqrt{c}|X-\overline{Y}|)\},\text{ where } x_2>0,$$

$y_2>0$, and $\overline{Y}=(y_1,-y_2)$. Now we can calculate the c-Poisson kernel:

$$\frac{\partial G(X,t)}{\partial n(t)}=-\frac{\sqrt{c}x_2}{\pi\sqrt{(x_1-t)^2+x_2^2}}K_0'(\sqrt{c}\sqrt{(x_1-t)^2+x_2^2})=$$

$$=\frac{\sqrt{c}x_2}{\pi\sqrt{(x_1-t)^2+x_2^2}}K_1(\sqrt{c}\sqrt{(x_1-t)^2+x_2^2}).$$

Therefore, if we set $\Omega=(-\pi,\pi)$, $t_l=l$, $-\infty<l<\infty$, and $K(z,t)=e^{izt}$,

then the functions S_l in (6) -- (7) take the form $S_l(X)=$

$$=\frac{(-1)^l\sqrt{c}x_2}{\pi^2}\int_{-\infty}^{\infty}\frac{\sin\pi t}{(t-l)\sqrt{(x_1-t)^2+x_2^2}}K_1(\sqrt{c}\sqrt{(x_1-t)^2+x_2^2})dt.$$

Thus, in this case $u(x_1,x_2)=$

$$\frac{\sqrt{c}x_2}{\pi^2}\sum_{l=-\infty}^{\infty}(-1)^l f(l)\{\int_{-\infty}^{\infty}\frac{\sin\pi t}{(t-l)\sqrt{(x_1-t)^2+x_2^2}}K_1(\sqrt{c}\sqrt{(x_1-t)^2+x_2^2})d$$

When $c\to 0$, $K_1(A\sqrt{c})\sim\frac{1}{A\sqrt{c}}$, and from the preceding series we get a sampling expansion for the harmonic functions satisfying the boundary value problem (1) with $c=0$ in the upper half-plane $\{z=(x,y)|\ x\in R, y>0\}$:

$$u(x,y)=\frac{1}{\pi}\sum_{l=-\infty}^{\infty}\frac{y+(-1)^l e^{\pi y}[(x-l)\sin\pi x-y\cos\pi x]}{(x-l)^2+y^2}f(l).\quad\blacksquare$$

Acknowledgment I would like to thank the referee for numerous useful remarks.

References

[1] M. Bramanti. *Potential theory for stationary Schrödinger operators: A Survey of results obtained with non-probabilistic methods*. LE MATEMATICHE, **47**, pp. 25-61, 1992.

[2] A. Kheyfits. *Dirichlet problem for the Schrödinger operator in a half-space with boundary data of arbitrary growth at infinity*. Differential and Integral Equations, **10**, pp. 153-164, 1997.

[3] B. Ya. Levin. *Distribution of Zeros of Entire Functions*, AMS Translations of Mathematical Monographs, Vol. 5, Providence, RI, 1964.

[4] F.-Y. Maeda. *On regularity of boundary points for Dirichlet problems of the equation* $\Delta u = qu(q \geq 0)$. Hiroshima Math. J., **1**, pp. 373-404, 1971.

[5] P. McCoy. *Sampling theorems associated with boundary value problems for elliptic partial differential equations in* R^n, Complex Variables. **23**, pp. 269-281, 1993.

[6] F. W. J. Olver. *Asymptotics and Special Functions*, Academic Press, New York - London, 1974.

[7] S. Saitoh. *Theory of Reproducing Kernels and its Applications*, Pitman Res. Notes in Math. Series, Vol. 189, Longman Scientific & Technical, England, 1988.

[8] A. I. Zayed. *Advances in Shannon's Sampling Theory*, CRC Press, Boca Raton, 1993.

12 MULTI-POWER LEGENDRE SERIES IN C^m

Peter A. McCoy

Mathematics Department, U.S.Naval Academy
Annapolis, MD 21402-5002, USA

ABSTRACT. Nehari's theorem links the singularities of Legendre series in C_z with those of associated Taylor's series in C_t. Recently, the theory was generalized to products of m Legendre polynomials in C^m. This paper continues the development to series of products of powers of Legendre polynomials in C^m.

1. INTRODUCTION

There is a classical theorem of Z. Nehari [10] that relates the singularities of Legendre series with those of associated analytic functions of one complex variable. Nehari's theorem was the extension of a theorem of Szego [11] that connected the singularities of real zonal harmonic functions with those of associated analytic functions. Those theorems relied on the Hadamard argument in the multiplication of singularities theorem for the analysis. R.P. Gilbert [2,3][1] extended Hadamard's argument via a procedure which is known as the "envelope method." The envelope method has shown great utility in a broader sense where its function theoretic application has produced analogous theories for the solutions of a numerous classes of partial differential equations [see [1,4,5] and the references therein]. Recently, P.A. McCoy [see [7-9]] used function theoretic methods to develop a multivariable extension of Nehari's theorem that links the singularities of the Legendre expansions on C^m with those of associated functions analytic on C_t .

The "multipower-multivariable" Legendre series under consideration here are of the form

$$F_q(z_1, ..., z_m) = \sum_{k=0}^{\infty} \omega_k a_k P_k^{q_1}(z_1)...P_k^{q_m}(z_m) \tag{1}$$

where $q := (q_1, ..., q_m)$, q_j are positive integers and $(z_1, ..., z_m) \in C^m$. Note that ref [9] concerns the case $q_j = 1$ for $j = 1, ..., m$. The parameters ω_n are the familiar constants in $\int_{-1}^{+1} P_n(\varsigma)P_k(\varsigma)d\zeta = \omega_n^{-1}\delta_{nk}$ given by $\omega_n = n + 1/2$. We write $F(q; Z) := \sum_{k=0}^{\infty} \omega_k a_k F_k(q; Z)$ with $Z := (z_1, z_2, ... z_m)$ where $F_k(q; Z) := P_k^{q_1}(z_1) ... P_k^{q_m}(z_m)$. The focus is on the relationship between the singularities of the multivariable analytic function $F(q; Z)$ and those of the associated analytic function

$$f(t) = \sum_{k=0}^{\infty} a_k t^k \tag{2}$$

with $t \in C_t$. The analysis is based on reciprocal integral transforms $F = T[f]$ and $f = T^{-1}[F]$ with closed form kernels. Function theoretic methods are applied to

[1] Dedicated to R.P. Gilbert on the occasion of his 65th Birthday.

the transform pair to identify the sets of possible singularities of F and f from the ascending and descending operators. The true singularities of the function pair $\{F, f\}$ are identified as those singularities that are retained as singularities under the inverse map. This association puts the singularity theory for the higher dimensional problems into the framework of analytic functions of a single complex variable where a rich and comprehensive theory exits.

2. The Integral Operators

The key to an effective study of singularities is closed form representations for the kernels of T and T^{-1}. These kernels will be derived from the basic kernel [see 9]

$$K(t,j,Z) := \sum_{k=0}^{\infty} \omega_k^{j-1} t^k F_k(Z), \; j = 1,2,3,..., \tag{3}$$

where $F_k(1;Z) = F_k(Z) := P_k(z_1)...P_k(z_m)$, $k = 0,1,2,3,...$. The closed form expression for (3) is identified here as $\Gamma(t;j;Z)$ and is specified in ref [9] where the ascending operator $F := T[f]$ is given by

$$F(Z) := (1/2\pi i) \int_{L_t} \Gamma(1/t;2;Z) \; f(t) \, dt/t. \tag{4}$$

The simple closed contour $L_t \subset C_t$ is homologous to the unit circle $|t| = 1$ modulo the singularities of the integrand. The inverse operator $f := T^{-1}[F]$ is given by

$$f(t) = \int_{L_{z_1}} ... \int_{L_{z_m}} \Gamma(t;m;Z) \, F(Z) \, dZ. \tag{5}$$

Each simple open contour $L_{z_k} \subset C_{z_k}$ has fixed endpoints at $\{-1,+1\}$ and is homologous to the segment $[-1,+1]$ modulo the singularities of the integrand.

The basic transforms for the general theory are constructed from "power kernels" $\Gamma_q(t;j;Z)$. The construction begins with a simple boot strapping operation defined from sums of "power blocks" as follows

$$K(t;j;\Upsilon_q(Z,\xi)) \; : \; = \sum_{k=0}^{\infty} \omega_k^{j-1} t^k \, \Pi_{r=1}^m [\Xi_k(z_r, \xi_{q_r-1}^{(r)})] \, ; where,$$

$$\Xi_k(z_r, \xi_{q_r-1}^{(r)}) \; : \; = P_k(z_r) \, \Pi_{j=1}^{q_r-1} P_k(\xi_j^{(r)}) \, , \, r = 1,2,...,m, q_r = 2,3,...,$$

and $\Xi_k(z_r, \xi_{q_r-1}^{(r)}) = P_k(z_r)$ if $q_r = 1$. There is a nice feature that we shall exploit. Observe that if $q_r \geq 2$,

$$\int_{L_{\xi_j}^{(r)}} \Xi_k(z_r, \xi_{q_r-1}^{(r)}) \, P_k(\xi_j^{(r)}) \, d\xi_j^{(r)} = \omega_k^{-1} P_k(z_r) \Pi_{j=1, j\neq k}^{q_r-1} P_k(\xi_j^{(r)}) \, ;$$

whereas, if $q_r = 1$ no integration is performed. With reference to $K(t;2;z_r,\xi_u^{(r)}) = \sum_{k=0}^{\infty} \omega_k t^k P_k(z_r) P_k(\xi_u^{(r)})$, we find that for $u = 1,2,...,r,...m$,

$$\int_{L_{\xi\alpha}^{(r)}} K(t;j;\Upsilon_q(Z,\xi)) \, K(1;2;z_u,\xi\alpha^{(r)}) \, d\xi\alpha^{(r)} =$$

$$\sum_{k=0}^{\infty} \omega_k^{j-1}\, t^k P_k^2(z_u) \Pi_{r=1}^{q_m} [\Pi_{n=1,n\neq u}^{q_r-1} P_k(\xi_n^{(r)})]$$

and by induction that

$$\int_{L_{\xi_j}} \cdots \int_{L_{\xi_j}} K(t;j;\Upsilon_q(Z,\xi)) \Pi_{k=1}^{q_r-1} K(t;k;2;z_u,\xi_k^{(r)})\, d\xi_k^{(r)} =$$

$$\sum_{k=0}^{\infty} \omega_k^{j-1}\, t^k\, P_k^{q_1}(z_u)\, \Pi_{r=2}^{q_r-1} \Xi_k(z_r,\xi_{q_r-1}^{(r)}) .$$

We now complete the process

$$K(t;j;q;Z) :=$$

$$\int_{L_{\xi_1}^{(1)}} \cdots \int_{L_{\xi_1}^{(q_1-1)}} \cdots \int_{L_{\xi_m}^{(1)}} \cdots \int_{L_{\xi_m}^{(q_m-1)}} K(t;j;\Upsilon_q(Z,\xi)) \Pi_{k=1}^{m} \Pi_{r=1}^{q_k-1} K(1;2;z_r,\xi_k^{(r)})\, d\xi_k^{(r)} =$$

$$\sum_{k=0}^{\infty} \omega_k^{j-1}\, t^k\, P_k^{q_1}(z_1)\, P_k^{q_2}(z_2) \ldots P_k^{q_m}(z_m) .$$

The closed form representations of these kernels are specified in refs [8,9] as $\Gamma(1;2;z_r,\xi_u^{(r)}) = K(1;2;z_1,\xi_u^{(r)})$ and $\Gamma_q(t;j;Z;\Upsilon_q) = K(t;j;q;Z;\Upsilon_q)$. These results permit us to rewrite the kernel $K(t;j;q;Z)$ in terms of contour integrals of closed forms as

(6) $$\Gamma_q(t;j;Z) =$$

$$\int_{L_{\xi_1}^{(1)}} \cdots \int_{L_{\xi_1}^{(q_1-1)}} \cdots \int_{L_{\xi_m}^{(1)}} \cdots \int_{L_{\xi_m}^{(q_m-1)}} \Gamma_q(t;j;Z;\Upsilon_q(Z,\xi)) \Pi_{r=1}^{m} \Pi_{j=1}^{q_m-1} \Gamma(1;2;z_r,\xi_j^{(r)}) d\xi_j^{(r)}$$

which is the required form. This kernel is an analytic function on its domain of association in C^m. We remark that under the convention adopted earlier that if $q_r = 1$ for some index, the corresponding factor and integral are dropped.

The integral operators are constructed in the usual way. The ascending operator is now determined on an initial domain of definition found in ref [9] and analytically continued to its domain of association as

(7) $$F(q;Z) = T_q[f(t)](Z) := (1/2\pi i) \int_{L_t} \Gamma_q(1/t;2;Z)\; f(t)\, dt/t$$

where the associate $f(t)$ is specified in eqn (2). The descending operator requires a bit more work on account of the fact that the powers of the Legendre polynomials need not be orthogonal or linearly independent. The key is to adjust for the possible lack of orthogonality by extending the domain of $F(q;Z)$ by one dimension as $Z^+ = (z_1,\ldots,z_m,z_{m+1})$ and requiring that the corresponding factor in $F(q^+;Z^+)$ have exponent $q_{m+1} = 1$ so that $q^+ := (q_1,\ldots,q_m,1)$. Let us now define the modified kernel

$$K(t;j;q^+;Z^+) :=$$

$$\sum_{k=0}^{\infty} \omega_k^{j-1}\, t^k\, P_k^{q_1}(z_1)\, P_k^{q_2}(z_2) \ldots P_k^{q_m}(z_m)\, P_k(z_{m+1})$$

and set $\alpha_{k,q_j} := \int_{-1}^{+1} P_k^{q_j}(\varsigma)\, d\varsigma$ if $q_j \geq 2$; and, $\alpha_{k,q_j} = 1$ if $q_j = 1$. If we also temporarily boot strap the $F(q^+; Z^+)$, one computes

$$\int_{L_{z_1}} \ldots \int_{L_{z_{m+1}}} F(q^+; Z^+)\, K(t; j; q^+; Z^+)\, dZ^+ =$$

$$\sum_{k=0}^{\infty} \omega_k^{j-1}\, \alpha_{k,q_1} \ldots \alpha_{k,q_m}\, a_k\, t^k$$

with $dZ^+ := dz_1 \ldots dz_m\, dz_{m+1}$. Presuming for the moment that none of the α_{k,q_j} 's are zero, define the scaling function $\chi_q(\zeta) := \sum_{k=0}^{\infty} (\alpha_{k,q_1} \ldots \alpha_{k,q_m})^{-1} \zeta^k$ so that $f(t) = (1/2\pi i) \int_{C\epsilon} \{ \sum_{k=0}^{\infty} \omega_k^{j-1}\, \alpha_{k,q_1} \ldots \alpha_{k,q_m}\, a_k (t/\zeta)^k \}\, \chi_q(\zeta)\, d\zeta/\zeta$ where $C_\epsilon : |\zeta| = \epsilon$ is a circle of sufficiently small radius. Note, for example, that if $\alpha_{k,2} = \omega_k^{-1}$, the scaling function $\chi_q(\zeta) = [t^{1/2}\partial_t t^{1/2}(\ldots(t^{1/2}\partial_t t^{1/2}(t^{1/2}\partial_t t^{1/2})))](1-t)^{-1}$ (m-times). We are now in a position to specify the kernel for the inverse as

$$K^*(t; j; q^+; Z^+) := (1/2\pi i) \int_{C\epsilon} K(t/\zeta; j; q^+; Z^+)\chi_q(\zeta)\, d\zeta.$$

Referring to the closed form representations as before, we arrive at

$$\Gamma_q^*(t; j; q^+; Z^+) := (1/2\pi i) \int_{C\epsilon} \Gamma(t/\zeta; j; q^+; Z^+)\chi_q(\zeta)\, d\zeta.$$

The descending operator

$$(8)\quad f(t) = T_{q^+}^{-1}[F(q^+; Z^+)](t) := \int_{L_{z_1}} \ldots \int_{L_{z_{m+1}}} \Gamma_q^*(t; m+1; Z^+)\, F(q^+; Z^+)\, dZ^+$$

defines an analytic function on its domain of association. A word is in order about the null space $N(T_q^{-1})$ of $T_{q^+}^{-1}$. This set is clearly characterized by $N(T_{q^+}^{-1}) = \{F(q^+; Z^+) : \alpha_{k,q_j} = 0, k = 0, 1, 2, \ldots, \textit{and}, q_j = 2, 3, \ldots\}$. Furthermore, the set of all analytic multi-power multivariable-variable Legendre series of the form specified in eqn (1) is designated as $A_q(C^m)$. To summarize, we state the following.

Theorem 2.1. *Let $F(q^+; Z^+) \in A_{q^+}(C^{m+1})/N(T_{q^+}^{-1})$. The $T-$ transform pair linking $F(q^+; Z^+)$ and $f(t)$ on their respective domains of association are given by $F(q^+; Z^+) = T_{q^+}[f(t)](Z^+)$ and $f(t) = T_{q^+}^{-1}[F(q^+; Z^+)](t)$.*

3. The Singularities

We begin by designating the set of possible singularities of the ascending transform as $PS[F(q^+; Z^+)]$. Let the associate $f(t)$ of $F(q^+; Z^+)$ have an isolated singularity at $t = t_o \neq 0$. Following standard function theoretic format, the set $PS[F(q^+; Z^+)] = SH_{q^+}(f; t_o) \cup SE_{q^+}(f, t_o)$ where $SH_{q^+}(f; t_o)$ is the set of possible Hadamard singularities and $SE_{q^+}(f, t_o)$ is the set of possible envelope singularities. The Hadamard singularities are determined parametically from the integrand of the ascending operator as $SH_{q^+}(f; t_o) = \{(t_o, (Z^+)) : t = t_o$ *is a singularity of* $\Gamma_{q^+}(1/t;2;Z^+)$ *for some* $Z^+ = Z_o^+ \in C^{m+1}\}$. The singular manifold $\{(t_o,(Z^+)) : S_{q^+}(t;2;Z^+) = 0\}$ of $\Gamma_{q^+}(1/t;2;Z^+)$ is thereby identified parametrically from those of the constituent product kernels $\Gamma_{q^+}(1/t;2;Z^+;\Upsilon_{q^+}(Z^+,\xi))$ $\Pi_{r=1}^{m+1}\Pi_{j=1}^{q_m-1}\Gamma(1;2;z_r;\xi_j^{(r)})$ using the expressions in ref[9] (eqn (12) ff). In other words, $SH_{q^+}(f;t_o) = \{(t_o,(Z^+)) : S_{q^+}(t_o;2;Z^+) = 0,\ Z^+ = Z_o^+ \in C^{m+1}\}$. The possible envelope singularities are precisely $SE_{q^+}(f,t_o) = \{(t_o,(Z^+)) : S_{q^+}(t_o;2;Z^+) = 0,\ \partial_t S_{q^+}(t_o;2;Z^+) = 0,\ Z^+ = Z_o^+ \in C^{m+1}\}$ and are themselves deduced parametrically from the previously mentioned constituent product kernels. Noting that $F(q;Z) = F(q^+;Z^+)_{z_{m+1}=1}$, we see that the set $PS[F(q;Z)] = PS[F(q^+;Z^+)] \cap \{(t_o,Z^+) : z_{m+1} = 1\}$.

We now consider the inverse transform and proceed under the assumption that $F(q_+;Z^+)$ has a singularity at $(Z^+) = (Z_o^+)$. The set of possible singularities is $PS_{q^+}[f(t)] = SH(F_{q^+};Z_o^+) \cup SE(F_{q^+};Z_o^+) \cup SEP(F_{q^+};Z_o^+) \cup SHE(F_{q^+};Z_o^+)$. Here, additional sets of singularities must be considered. The set $SEP(F_{q^+};Z_o^+)$ of "endpinch singularities" arises from the fact that the end points of the contours of integration in T^{-1} are fixed and thus have special status in the analytic continuation of the inverse operator over its domain of association. The set $SHE(F_{q^+};Z_o^+)$ of Hadamard eliminate singularities will be described in due course.

The set of Hadamard singularities $SH(F_{q^+};Z_o^+) = \{(t,(Z_o^+)) : \Gamma_q^*(t;m+1;Z_o^+)$ *has a singularity for some* $t = t_o \in C_t\}$. The singular manifold of the kernel $\Gamma_q^*(t;m+1;Z_o^+)$ is also determined from the corresponding product kernels as in the ref [9] (eqn (12) ff) and is designated parametrically by $\{(t,(Z_o^+)) : S_{q^+}^*(t;m+1;Z_o^+) = 0, t = t_o \in C_t\}$. There is a useful property of the kernel, namely, $\Gamma_q^*(t;m+1;Z_o^+) = -1/t\ \Gamma_q^*(1/t;m+1;Z_o^+)$. Thus, one finds the possible Hadamard singularities as $SH(F_{q^+};Z_o^+) = \{(t,(Z_o^+)) : (Z_o^+)$ *is a singularity* $S_{q^+}^*(t;m+1;Z_o^+) = 0$ *for some* $t = t_o \in C_t\}$. The envelope singularities are $SE(F_{q^+};Z_o^+) = \{(t,(Z_o^+)) : S_{q^+}^*(t;m+1;Z_o^+) = 0,\ \partial_{z_k} S_{q^+}^*(t;m+1;Z_o^+) = 0, k = 1,...,m; t = t_o \in C_t\}$. And, the endpinch singularities are $SEP(F_{q^+};Z_o^+) = \{(t,Z_o^+) : S_{q^+}^*(t;m+1;Z_o^+) = 0, \partial_{z_k} S_{q^+}^*(t;m+1;Z_o^+) = 0, k = 1,2,...,m, (Z_o^+) = (\pm 1,,.....,\pm 1); t = t_o \in C_t\}$. Finally, the set of Hadamard eliminates $SHE(F_{q^+};Z_o^+) = \cup_{k=1}^{m} \cup_{j=1}^{m} \{(t,(Z_o^+)) : S_{q^+}^*(t;m+1;Z_o^+) = 0,\ E_{q^+}^{(k,j)}(t;m+1;Z_o^+) = 0, \partial_{z_j} E_{q^+}^{(k,j)}(t;m+1;Z_o^+) = 0, j \neq k; t = t_o \in C_t\}$ where $E_{q^+}^{(k)}(t;m+1;Z_o^+)$ is the eliminant set obtained for the $j-th$ variable [see ref[4] page 25ff]. We now find that the set $PS_q[f(t)] = PS_{q^+}[f(t)] \cap \{(t,Z_o^+) : z_{m+1} = 1\}$. The set of actual singularities of the transform pair $\{F,f\}$ is determined by $PS[F(q;Z)] \cap PS_q[f(t)]$. We sum-up our progress.

Theorem 3.1. *Let* $F(q^+;Z^+) \in A_{q^+}(C^{m+1})/N(T_{q^+}^{-1})$ *and let* $f(t)$ *be the* $T-$ *transform associate. Then, the singular set of the function pair* $\{F,f\}$ *is precisely* $PS[F(q;Z)] \cap PS_q[f(t)]$.

4. The Pythagorean Series

In this section, we illustrate the applicability of the T-transforms by using these operators to construct some new representation formulae. The focus is on

Pythagorean series which we refer to as the analytic sums

$$(9) \qquad P(q_\alpha, Z) = \sum_{k=0}^{\infty} \omega_k \, a_k \, \{P_k^{q_\alpha}(z_1) + ... + P_k^{q_\alpha}(z_m)\}, \; q_\alpha \geq 2.$$

These series are built up from the functions $P(q_\alpha, z_j, \xi_j) := \sum_{k=0}^{\infty} \omega_k \, a_k P_k^{q_\alpha}(z_j) P_k(\xi_j)$ as $P(q_\alpha, \Phi(z;\xi)) = \sum_{j=1}^{m} P(q_\alpha, z_j, \xi_j)$ where $\Phi(z;\xi) := \Phi[(z_1;\xi_1), ..., (z_m;\xi_m)]$. Following standard procedure, $P(q_\alpha, Z) = P(q_\alpha, \Phi(z;\xi))_Y$ where the symbol Y defines the restriction of $\Phi(z;\xi)$ as $\Phi(z;\xi)_{Y_j} : \Phi(z;\xi) \; {}_{X_j} \; where \; (z_j, \xi_j) = (z_j, 1)$, $j = 1, ..., m$. The individual terms $P(q_\alpha, z_j, \xi_j)$ are generated by placing special restrictions on the kernels of the $T-$operators. Let us set $J_{q_\alpha}(t; 2; z_j; \xi_j) := \Gamma(t; q^+; Z^+)_{X_j}$ where X_j defines the restriction: $Z^+ := (z_j, \xi_j, 1, 1, ..., 1)$ for each index j, and, q_α corresponds to the vector $q^+ := (q_\alpha, 1, 1, ..., 1)$. The scaling function is $\chi_{q_\alpha}(\zeta) := \chi_{q^+}(\zeta)_{\alpha_{k \neq j}=1}$. We find that $P(q_\alpha, z_j, \xi_j) = (1/2\pi i) \int_{L_t} J_{q_\alpha}(1/t; 2; z_j; \xi_j) \, f(t) \, dt/t \; for \; j = 1, ..., m$. Thus, the kernel we seek is $J_{q_\alpha}(t; 2; \Phi(z;\xi)) := \sum_{j=1}^{m} J_{q_\alpha}(t; 2; z_j; \xi_j)$. The ascending operator follows

$$P(q_\alpha, \Phi(z;\xi)) := Q_{q\alpha}[f(t)](\Phi(z;\xi)) =$$

$$(10) \qquad (1/2\pi i) \int_{L_t} J_{q_\alpha}(1/t; 2; \Phi(z;\xi)) f(t) \, dt/t \; .$$

We invert the operator Q_{q_α} in two steps. The kernel for the intermediate inverse operator is $J^{\#}_{q_\alpha}(t; 3; \Phi(z;\xi)) = (1/(m2^{2m-2})) \sum_{j=1}^{m} J^{\#}_{q_\alpha}(t; 3; \Phi(z_j;\xi_j))$ where $J^{\#}_{q_\alpha}(t; 3; \Phi(z_j;\xi_j))$ denotes the (reduced) closed form kernel. We now split up the product $P(q_\alpha, \Phi(z;\xi)) J^{\#}_{q_\alpha}(t; 3; \Phi(z;\xi)) = \sum_{j=1}^{m} P(q_\alpha, z_j, \xi_j) J^{\#}_{q_\alpha}(t; 3; \Phi(z_j;\xi_j)) + \sum_{j,k=1, j \neq k}^{m} P(q_\alpha, z_j, \xi_j) J^{\#}_{q_\alpha}(t; 3; \Phi(z_k;\xi_k))$ into the trace and off diagonal terms. We use the orthogonality relations

$$\int_{L_{\xi_j}} \int_{L_u} P_k(\xi_j) P_n(\xi_u) \, d\xi_j \, d\xi_u = 2\omega_k^{-1} \, \delta_{kn} \;, \; j = u; and,$$

$$\int_{L_{\xi_j}} \int_{L_u} P_k(\xi_j) P_n(\xi_u) \, d\xi_j \, d\xi_u = 4\delta_{ok}\delta_{on} \;, \; j \neq u,$$

to integrate the trace

$$\int_{L_{z_1}} \cdots \int_{L_{z_m}} \int_{L_{\xi 1}} \cdots \int_{L_{\xi_m}} \sum_{j=1}^{m} P(q_\alpha, z_j, \xi_j) J^{\#}_{q_\alpha}(t; 3; \Phi(z_j;\xi_j)) \, d\xi_1 ... d\xi_m dz_1 ... dz_m$$

$$= m2^{2m-2} \, f(t)$$

and the off diagonal terms

$$\int_{L_{z_1}} \cdots \int_{L_{z_m}} \int_{L_{\xi 1}} \cdots \int_{L_{\xi_m}} \sum_{j,k=1, j \neq k}^{m} P(q_\alpha, z_k, \xi_k) \, J^{\#}_{q_\alpha}(t; 3; \Phi(z_j;\xi_j)) \, d\xi_1 ... d\xi_m dz_1 ... dz_m$$

$$= m(m-1)2^{2m-1} \, f(0).$$

The combination of these two expressions defines a valid inverse. We shall tighten up the representation by adjusting for the off diagonal terms as follows. Set

$J^*_{q_{\alpha,k}}(t;3;\Phi(z_j;\xi_j)) := J^{\#}_{q_\alpha}(t;3;\Phi(z_j;\xi_j)) - ((m-1)/2)[1-\delta_{jk}]\delta(\xi_j)$, and, define $J^*_{q_\alpha}(t;3;\Phi(z;\xi)) := (1/m)\sum_{j,k=1}^{m} J^*_{q_{\alpha,k}}(t;3;\Phi(z_j;\xi_j))$. The inverse operator follows

$$f(t) := Q^{-1}_{q_\alpha}[P(q_\alpha, \Phi(z;\xi))](t) =$$

(11)

$$\int_{L_{z_1}} \cdots \int_{L_{z_m}} \int_{L_{\xi_1}} \cdots \int_{L_{\xi_m}} P(q_\alpha, \Phi(z;\xi))\, J^*_{q_\alpha}(t;3;\Phi(z;\xi))\; d\xi_1 ... d\xi_m dz_1 ... dz_m \;.$$

The null space $N(Q^{-1}_{q_\alpha}) = \{P(q_\alpha, \Phi(z;\xi)) : \alpha_{q_\alpha} = 0\}$. Let us summarize.

Theorem 4.1. *Let* $P(q_\alpha, \Phi(z;\xi)) \in A_{q^+}(C^{m+1})/N(Q^{-1}_{q_\alpha})$. *The* $Q-$ *transform pair linking* $P(q_\alpha, \Phi(z;\xi))$ *and* $f(t)$ *on their respective domains of association are given by* $P(q_\alpha, \Phi(z;\xi)) = Q_{q_\alpha}[f(t)](\Phi(z;\xi))$ *and* $f(t) = Q^{-1}_{q_\alpha}[P(q_\alpha, \Phi(z;\xi))](t)$.

We now find the singularities of the $Q-$associated function pair. Following notations adopted previously, set $PS[P(q_\alpha, \Phi(z;\xi))] = SH_{q^+}(f;t_o) \cup SE_{q^+}(f,t_o)$. The Hadamard singularities $SH_{q^+}(f;t_o) = \{(t_o, \Phi(z,\xi)_{X_j}) : t = t_o \text{ is a singularity of } \Gamma_{q^+}(1/t;2;\Phi(z,\xi)_{X_j}) \text{ for some } (z_j,\xi_j) \in C^2, j = 1,2,....,m\}$. In other words, the reduced singular manifold is $SH_{q^+}(f;t_o) = \cup_{j=1}^{m}\{(t_o, \Phi(z,\xi)_{X_j}) : S_{q^+}(t_o;2;\Phi(z,\xi)_{X_j}) = 0, (z_j,\xi_j) \in C^2\}$. The envelope singularities are $SE_{q^+}(f,t_o) = \cup_{j=1}^{m}\{(t_o, \Phi(z,\xi)_{X_j}) : S^*_{q^+}(t_o;2;\Phi(z,\xi)_{X_j}) = 0, \partial_t S^*_{q^+}(t_o;2;\Phi(z,\xi)_{X_j}) = 0, (z_j,\xi_j) \in C^2\}$. We thus find that the singularities of the Pythagorean sum defined by the ascending operator are located on the set $PS[P(q_\alpha, Z) := PS[P(q_\alpha, \Phi(z;\xi))] \cap \{\cup_{j=1}^{m}(t_o, \Phi(z;\xi)_{Y_j})\}$.

As we have seen previously, the analysis of the singularities of the inverse is somewhat more involved. Let us presume that $P(q_\alpha, \Phi(z;\xi))$ has a singularity at $(Z_o^+) := \{(z,\xi) : \Phi(z;\xi)_{X_j} = \Phi(z_o;\xi_o)_{X_j}, j = 1,...,m\}$ and set $PS_{q^+}[f(t)] = SH(P(q_\alpha, \Phi(z;\xi)); (Z_o^+)) \cup SE(P(q_\alpha, \Phi(z;\xi)); (Z_o^+)) \cup SEP(P(q_\alpha, \Phi(z;\xi)); (Z_o^+) \cup SHE(P(q_\alpha, \Phi(z;\xi)); (Z_o^+))$. The Hadamard singularities $SH(P(q_\alpha, \Phi(z;\xi)); (Z_o^+)) = SH(P(q_\alpha, \Phi(z;\xi)); (Z_o^+))_{trace} \cup SH(P(q_\alpha, \Phi(z;\xi)); (Z_o^+))_{offdiag}$ where the trace singularities are $SH(P(q_\alpha, \Phi(z;\xi)); (Z_o^+))_{trace} = \cup_{j=1}^{m}\{(t, \Phi(z_o;\xi_o)_{X_j}) : S^*_{q^+}(t;3;\Phi(z,\xi)_{X_j}) = 0, \Phi(z,\xi)_{X_j} = \Phi(z_o,\xi_o)_{X_j}\}$ and the off diagonal singularities are $SH(P(q_\alpha, \Phi(z;\xi)); (Z_o^+))_{offdiag} = \cup_{j,k=1, j\neq k}^{m}\{(t, \Phi(z_o;\xi_o)_{X_j}) : S^*_{q^+}(t;3;\Phi(z,\xi)_{X_j}) = 0, \Phi(z,\xi)_{X_j} = \Phi(z_o,\xi_o)_{X_k}\}$. The envelope singularities $SE(P(q_\alpha, \Phi(z;\xi)); (Z_o^+)) = SE(P(q_\alpha, \Phi(z;\xi)); (Z_o^+))_{trace} \cup SE(P(q_\alpha, \Phi(z;\xi)); (Z_o^+))_{offdiag}$ where the trace $SE(P(q_\alpha, \Phi(z;\xi)); (Z_o^+))_{trace} = \cup_{j=1}^{m}[\{(t, \Phi(z_o;\xi_o)_{X_j}) : S^*_{q^+}(t;3;\Phi(z,\xi)_{X_j}) = \partial_{z_j} S^*_{q^+}(t;3;\Phi(z,\xi)_{X_j}) = \partial_{\xi_j} S^*_{q^+}(t;3;\Phi(z,\xi)_{X_j}) = 0; \Phi(z,\xi)_{X_j} = \Phi(z_o,\xi_o)_{X_j}\}$ and the off diagonal terms $SE(P(q_\alpha, \Phi(z;\xi)); (Z_o^+))_{offdiag} = \cup_{j,k=1, j\neq k}^{m}\{\{(t, \Phi(z_o;\xi_o)_{X_j}) : S^*_{q^+}(t;3;\Phi(z,\xi)_{X_j}) = \partial_{z_j} S^*_{q^+}(t;3;\Phi(z,\xi)_{X_j}) = \partial_{\xi_j} S^*_{q^+}(t;3;\Phi(z,\xi)_{X_j}) = 0, \Phi(z,\xi)_{X_j} = \Phi(z_o,\xi_o)_{X_k}\}$. The Hadamard eliminates $SHE(P(q_\alpha, \Phi(z;\xi)); (Z_o^+)) = SHE(P(q_\alpha, \Phi(z;\xi)); (Z_o^+))_{trace} \cup SHE(P(q_\alpha, \Phi(z;\xi)); (Z_o^+))_{off}$ where $SHE(P(q_\alpha, \Phi(z;\xi)); (Z_o^+))_{trace} = \cup_{j=1}^{m}\{(t, \Phi(z_o;\xi_o)_{X_j}) : S^*_{q^+}(t;3;\Phi(z,\xi)_{X_j}) = E^{(j,j)*}_{q^+}(t;3;\Phi(z,\xi)_{X_j}) = \partial_{z_j} E^{(j,j)*}_{q^+}(t;3;\Phi(z,\xi)_{X_j}) = \partial_{\xi_j} E^{(j,j)*}_{q^+}(t;3;\Phi(z,\xi)_{X_j}) = 0, \Phi(z,\xi)_{X_j} = \Phi(z_o,\xi_o)_{X_j}\}$ where $E^{(k,j)*}_{q^+}(t_o;3;\Phi(z,\xi)_{X_j})$ is the $k-th$ eliminant set obtained for the $j-th$ variable. The set $SHE(P(q_\alpha, \Phi(z;\xi)); (Z_o^+))_{off} = \cup_{j,k=1, j\neq k}^{m}\{(t, \Phi(z_o;\xi_o)_{X_j}) : S^*_{q^+}(t;3;\Phi(z,\xi)_{X_j}) = E^{(k,j)*}_{q^+}(t;3;\Phi(z,\xi)_{X_j}) = \partial_{z_j} E^{(k,j)*}_{q^+}(t;3;\Phi(z,\xi)_{X_j}) = \partial_{\xi_k} E^{(k,j)*}_{q^+}(t;3;\Phi(z,\xi)_{X_j}) = 0, \Phi(z,\xi)_{X_j} = \Phi(z_o,\xi_o)_{X_k},$

$j \neq k\}$. The endpinch singularities $SEP(P(q_\alpha, \Phi(z;\xi)); (Z_o^+)) = SEP(P(q_\alpha, \Phi(z;\xi)); (Z_o^+))_{trace} \cup SEP(P(q_\alpha, \Phi(z;\xi)); (Z_o^+))_{offdiag}$. The trace endpinch singularities are $SEP(P(q_\alpha, \Phi(z;\xi)); (Z_o^+))_{trace} = \cup_{j=1}^m \{(t, \Phi(z_o;\xi_o)_{X_j}) : \{(t, Z_o^+) : S_{q+}^*(t;3;\Phi(z,\xi)_{X_j}) = 0, \partial_{z_j} S_{q+}^*(t;3;\Phi(z,\xi)_{X_j}) = \partial_{\xi_k} S_{q+}^*(t;3;\Phi(z,\xi)_{X_j}) = 0,\ (z,\xi)_{X_j} = (\pm 1, , \ldots\ldots, \pm 1)^2;$ $\Phi(z,\xi)_{X_j} = \Phi(z_o,\xi_o)_{X_j}, t = t_o \in C_t\}$ and the off diagonal endpinch singularities $SEP(P(q_\alpha, \Phi(z;\xi)); (Z_o^+))_{offdiag} = \cup_{j,k=1,j\neq k}^m \{(t, \Phi(z_o;\xi_o)_{X_j}) : S_{q+}^*(t;3;Z_o^+) = \partial_{z_j} S_{q+}^*(t;3;Z_o^+) = \partial_{\xi_k} S_{q+}(t;3;\Phi(z,\xi)_{X_j}) = 0, (z,\xi)_{X_j} = (\pm 1, , \ldots\ldots, \pm 1)^2, \Phi(z,\xi)_{X_j} = \Phi(z_o,\xi_o)_{X_k}\}$. One now deduces that the singularities of the inverse operator are $PS_{q_\alpha}[f(t)] = PS_{q+}[f(t)] \cap \{\cup_{j=1}^m (t, \Phi(z;\xi)_{Y_j}) : \Phi(z;\xi)_{Y_j} = \Phi(z_o;\xi_o)_{Y_j}\}$. Let us recapitulate..

Theorem 4.2. *Let $P(q_\alpha, \Phi(z;\xi)) \in A_{q+}(C^{m+1})/N(Q_{q_\alpha}^{-1})$ and let $f(t)$ be the $Q-$transform associate. Then, the singular set of the function pair $\{P(q_\alpha, \Phi(z;\xi)), f\}$ is precisely $PS[P(q_\alpha, \Phi(z;\xi))] \cap PS_{q_\alpha}[f(t)]$.*

There are a variety of interesting formulae within the domain of the $T-$transforms. One could for example consider a theory for analytic series $H(q_\beta, (z,\xi)) = \sum_{k=0}^{\infty} \omega_k a_k \{P_k^{q_{\beta_1}}(z) + \ldots + P_k^{q_{\beta_n}}(z)\}$ where the powers q_{β_j} are distinct. We remark in closing that the basic procedure and operators developed herein offer a broad potential for application to other multi-power Legendre series in C^m and provide a useful generalization of the Nehari theorem which was the basis for a number of subsequent applications in the theory of linear partial differential equations.

Acknowledgment: Funded in part by the Naval Academy Research Council and ONR grant N0001498WR20010.

References

[1] H. Begher & R.P.Gilbert, *Transmutations, Transformations and Kernel Functions,* Pitman Monographs & Surveys in Pure and Applied Math., vols 58-59, New York, 1992.

[2] R.P. Gilbert, *Singularities of three-dimensional harmonic functions,* Pacific J. Math., 10 (1960) pp. 1243-1255.

[3] R.P. Gilbert, *On harmonic functions of four variables with rational P_4 associates,* Pacific J. Math., 13 (1963) pp. 79-96.

[4] R.P. Gilbert, *Function theoretic methods in Partial Differential Equations,* Math. in Science and Engineering, 54, Academic Press, New York, 1969.

[5] R.P. Gilbert, *Constructive Methods for Elliptic Equations,* Lecture notes in Math., vol 365 (1974) Springer-Verlag, New York.

[6] E. Hille, *Analytic Function Theory,* vols 1-2, Blaisdell, New York, 1962.

[7] P.A. McCoy, *Singularities of Jacobi series on C^2,* J. Math Analysis & Appl., 128 (1987) 92-100.

[8] P.A. McCoy, *A classical theorem on the singularities of Legendre series in C^3 and associated system of hyperbolic partial differential equations,* SIAM J. Math. Anal., vol. 28, no. 3 (1997) 704-714.

[9] P.A. McCoy, *The singularities of Legendre Series in C^N,* Complex Variables, vol. 34, no.3 (1997) 231-245.

[10] Z. Nehari, *On the singularities of Legendre Series* , J. Rational Mech. Anal., 5 (1956) 987-992.

[11] G. Szego, *On the singularities of zonal harmonic expansions,* J. Rational Mech. Anal., 3 (1954) 561-564.

Mathematics Department, U.S. Naval Academy, Annapolis, MD 21402-5002, USA.

13 AN ESSAY ON THE BERGMAN METRIC AND BALANCED DOMAINS

Takeo Ohsawa

Graduate School of Mathematics,
Nagoya University, Japan
ohwawa@math.sc.nagoya-u.ac.jp

INTRODUCTION

Let Ω be a bounded pseudoconvex domain in $\mathbb{C}^n$ and let $K_\Omega(z,\overline{w})$ be the Bergman kernel function of Ω. The boundary behavior of K_Ω reflects the mass distribution of L^2 holomorphic functions on Ω through the geometry of the boundary $\partial\Omega$ in a very natural way, as one can see it from [H1] and various subsequent works (cf. [D], [P], [D'A], [O-1,2], [D-H-O], [C], [D-H], [B-S-Y], etc.). From the viewpoint of biholomorphic geometry, the Bergman metric

$$ds_\Omega^2 = \sum_{\alpha,\beta=1}^{n} \frac{\partial^2 \log K_\Omega(z,\overline{z})}{\partial z^\alpha \partial \overline{z}^\beta} dz^\alpha \otimes d\overline{z}^\beta$$

is also a natural quantity attached to Ω. Being a Hermitian metric invariant under the biholomorphic transformations, the Bergman metric is of intrinsic nature, while the values of K_Ω are not. It is well known that one can draw important information on proper holomorphic mappings from the asymptotics of K_Ω and ds_Ω^2 (see [F] and [B-N] for more precise statements). Besides such an application, the boundary behavior of the Bergman kernel and the metric is of considerable significance in the current complex analysis, because they supply questions that urge further developments of the so called L^2 method for the $\overline{\partial}$-operator. For instance, it was asked in [O-1] whether or not $K_\Omega(z,\overline{z}) \gtrsim \delta_\Omega(z)^{-2}$ if $\partial\Omega$ is C^2-smooth, $\delta_\Omega(z)$ being the distance from z to $\partial\Omega$, and this was answered affirmatively in [O-T] as a corollary of a very general extension theorem for L^2 holomorphic functions, which found even an application to algebraic geometry (cf. [A-S]).

In the present article, we shall restrict ourselves to the completeness question for the Bergman metric. Previously we obtained in [D-O] an estimate from below for the distance function $\mathrm{dist}_\Omega(\mathrm{z},\mathrm{w})$ with respect to ds_Ω^2 in terms of the euclidean distance $\delta_\Omega(z)$ to $\partial\Omega$ when $\partial\Omega$ is piecewise C^2-smooth. Namely we have proved that for each $w \in \Omega$ one can find positive constants c_1, c_2 such that

$$\mathrm{dist}_\Omega(\mathrm{z}, \mathrm{w}) > \mathrm{c}_1 \log|\log \mathrm{c}_2\, \delta_\Omega(\mathrm{z})| - 1.$$

This estimate was derived in fact as a corollary of the following more general result.

Theorem 1 (Theorem 0.1 in [D-O]). *Let $\Omega \Subset \mathbb{C}^n$ be a pseudoconvex domain on which there is a bounded plurisubharmonic C^∞ exhaustion function $\rho\colon \Omega \to [-1, 0)$ satisfying the following estimate with suitable constants $C_1, C_2 > 0$:*

$$C_1^{-1}\delta_\Omega^{C_2}(z) < -\rho(z) < C_1 \delta_\Omega^{1/C_2}(z).$$

Then there exist, for any fixed point $z_0 \in \Omega$, constants $c_3, c_4 > 0$, such that

$$\mathrm{dist}_\Omega(\mathrm{z}_0, \mathrm{z}_1) > \mathrm{c}_3 \log|\log(\mathrm{c}_4\, \delta_\Omega(\mathrm{z}_1))| - 1$$

for all $z_1 \in \Omega$.

On the other hand, there is a result by Jarnicki-Pflug [J-P-1] which asserts that ds_Ω^2 is complete if Ω is a balanced pseudoconvex domain whose Minkowski function is continuous (see also [J-P-2] for related results). Here Ω is said to be balanced if $\zeta z \in \Omega$ for all $z \in \Omega$ and $\zeta \in \mathbb{C}$ with $|\zeta| \leq 1$, and the Minkowski function of Ω is a function h on $\mathbb{C}^n$ defined by

$$h(z) := \inf\{t > 0 \mid z/t \in \Omega\}.$$

We shall establish here a quantitative version of Jarnicki-Pflug's theorem by applying Theorem 1. To realize this, the following is crucial.

Theorem 2. *Let $\Omega \subset \mathbb{C}^n$ be a bounded and balanced pseudoconvex domain with Minkowski function h. Then there exist constants $\epsilon \in (0,1)$ and $A > 0$ such that*

$$-\left(1 - h(z)^2\right)^\epsilon e^{-A|z|^2}$$

is plurisubharmonic.

Combining Theorem 1 with Theorem 2 we obtain

Theorem 3. *Let $\Omega \subset \mathbb{C}^n$ be a bounded and balanced pseudoconvex domain whose Minkowski function is Hölder continuous. Then there exist positive constants c_5, c_6 such that*

$$\mathrm{dist}_\Omega(0, \mathrm{z}) > \mathrm{c}_5 \log|\log(\mathrm{c}_6\, \delta_\Omega(\mathrm{z}))| - 1$$

for all $z \in \Omega$.

§1 A VARIATIONAL FORMULA

Let $D \subset \mathbb{C}^n$ be a Hartogs domain over a domain $D' \subset \mathbb{C}^{n-1}$, i. e.

$$D = \left\{(z, w) \in D' \times \mathbb{C} \;\middle|\; |w|^2 < e^{-\rho(z)}\right\}$$

where ρ is an upper semi-continuous function on D' with values in $[-\infty, \infty)$. As is well known, D is pseudoconvex if and only if D' is pseudoconvex and ρ is plurisubharmonic. Indeed, by a theorem of Hartogs and Oka, $-\log(e^{-\frac{1}{2}\rho(z)} - |w|)$ is plurisubharmonic if and only if D is pseudoconvex in $D' \times \mathbb{C}$. It will turn out below that $-\log(e^{-\rho(z)} - |w|^2)$ is plurisubharmonic if so is ρ. An additional property of this function which can be read off from the formula is crucial for our purpose.

Lemma .

$$\begin{aligned}&\partial\overline{\partial}\left(-\log\left(e^{-\rho}-|w|^2\right)\right) \qquad (*)\\ &= e^{-\rho}\left\{\left(e^{-\rho}-|w|^2\right)^{-2}(dw+w\partial\rho)\wedge(d\overline{w}+\overline{w}\overline{\partial}\rho)\right.\\ &\qquad\left.+\left(e^{-\rho}-|w|^2\right)^{-1}\partial\overline{\partial}\rho\right\}\end{aligned}$$

if ρ is of class C^2.

Proof.

$$\begin{aligned}&\partial\overline{\partial}\left(-\log\left(e^{-\rho}-|w|^2\right)\right)\\ &=\partial\left\{\left(e^{-\rho}-|w|^2\right)^{-1}\overline{\partial}\left(|w|^2-e^{-\rho}\right)\right\}\\ &=\left(e^{-\rho}-|w|^2\right)^{-1}\partial\overline{\partial}\left(|w|^2-e^{-\rho}\right)\\ &\quad+\left(e^{-\rho}-|w|^2\right)^{-2}\partial\left(e^{-\rho}-|w|^2\right)\wedge\overline{\partial}\left(e^{-\rho}-|w|^2\right)\\ &=\left(e^{-\rho}-|w|^2\right)^{-2}\left\{\left(e^{-\rho}-|w|^2\right)\left(dw\wedge d\overline{w}+e^{-\rho}\partial\overline{\partial}\rho-e^{-\rho}\partial\rho\wedge\overline{\partial}\rho\right)\right.\\ &\quad\left.+\left(e^{-\rho}\partial\rho+\overline{w}\,dw\right)\wedge\left(e^{-\rho}\overline{\partial}\rho+w\,d\overline{w}\right)\right\}\\ &=\left(e^{-\rho}-|w|^2\right)^{-2}\left\{e^{-\rho}\left(dw\wedge d\overline{w}+|w|^2\partial\rho\,\overline{\partial}\rho+w\,\partial\rho\wedge d\overline{w}+\overline{w}\,dw\wedge\overline{\partial}\rho\right)\right.\\ &\quad\left.+\left(e^{-\rho}-|w|^2\right)e^{-\rho}\partial\overline{\partial}\rho\right\}\\ &=e^{-\rho}\left(e^{-\rho}-|w|^2\right)^{-2}\left(dw+\overline{w}\,\partial\rho\right)\wedge\left(d\overline{w}+w\,\overline{\partial}\rho\right)+e^{-\rho}\left(e^{-\rho}-|w|^2\right)^{-1}\partial\overline{\partial}\rho.\end{aligned}$$

□

Corollary . $-\log(e^{-\rho(z)}-|w|^2)$ *is plurisubharmonic in (z,w) if $\rho(z)$ is plurisubharmonic.*

Proof. From the regularization theorem for plurisubharmonic functions (see [H-2] for instance), ρ is locally the limit of a decreasing sequence of C^∞ plurisubharmonic functions, say $\{\rho_\nu\}_{\nu=1}^{\infty}$. Since $-\log(e^{-\rho_\nu}-|w|^2)$ are plurisubharmonic functions by the above lemma and decreasing in ν, the limit $-\log(e^{-\rho}-|w|^2)$ is also plurisubharmonic. □

Remark . The equality $(*)$ has an interesting interpretation. Let $K_z(w,\zeta)$ be the Bergman kernel function of the domain $D_z=\{w\in\mathbb{C}\mid |w|^2<e^{-\rho(z)}\}$. Since

$$K_z(w,\overline{\zeta})=\frac{e^{-\rho(z)}}{\pi\left(e^{-\rho(z)}-w\overline{\zeta}\right)^2},$$

the above formula can be written as

$$\begin{aligned}&\partial\overline{\partial}\log K_z(w,\overline{w})\\ &=|w|^2\left\{\sqrt{K_z(w,\overline{w})\pi e^{\rho}}\,\partial\overline{\partial}\rho+\pi e^{\rho}K_z(w,\overline{w})\,\partial\log\left(e^{\rho}|w|^2\right)\wedge\overline{\partial}\log\left(e^{\rho}|w|^2\right)\right\}.\end{aligned}$$

Moreover, let $G_z(w,\zeta)$ be the Green function of D_z. Since

$$\log\left(e^{\rho(z)}|w|^2\right)=-2G_z(0,w),$$

we can further rewrite the formula as

$$\begin{aligned}
&\partial\overline{\partial}\log K_z(w,\overline{w})\\
&= -2e^{-2G_z(0,w)}\left(1-e^{-2G_z(0,w)}\right)^{-1}\partial\overline{\partial}G_z(0,w)\\
&\quad + 4\pi e^{-2G_z(0,w)}K_z(w,\overline{w})\,\partial G_z(0,w)\wedge\overline{\partial}G_z(0,w).
\end{aligned}$$

It is therefore natural to ask for a generalization of this relation between the Bergman kernel and the Green function to an arbitrary family of Riemann surfaces. As for the results related to this question, see [Su] and [M-Y].

Proof of Theorem 2 : By the definition of the Minkowski function, $\frac{h(z)}{|z|}$ $(z \neq 0)$ depends only on $(z_1 : z_2 : \cdots : z_n)$. Since Ω is a bounded and balanced neighbourhood of $0 \in \mathbb{C}^n$, $\left(\frac{h(z)}{|z|}\right)^{\pm 1}$ are bounded. In terms of inhomogeneous coordinates $\zeta = (\zeta_2, \dots, \zeta_n)$ with $\zeta_i = \frac{z_i}{z_1}$ $(2 \leq i \leq n)$,

$$h(z) = |z_1|\cdot\left(1+|\zeta|^2\right)^{\frac{1}{2}}\lambda(\zeta) \qquad (z_1 \neq 0)$$

for some upper semicontinuous function λ. Since Ω is pseudoconvex, $\rho(\zeta) := \log\{(1+|\zeta|^2)^{\frac{1}{2}}\lambda(\zeta)\}$ is plurisubharmonic. Suppose for a moment that λ is of class C^2. Then we have

$$-\partial\overline{\partial}\log\left(1-h^2\right) = -\partial\overline{\partial}\log\left(e^{-\rho}-|z_1|^2\right) - \partial\overline{\partial}\rho. \tag{1}$$

By the lemma, we have

$$-\partial\overline{\partial}\log\left(e^{-\rho}-|z_1|^2\right) \geq e^{-\rho}\left(e^{-\rho}-|z_1|^2\right)^{-1}\partial\overline{\partial}\rho.$$

Here and in what follows, we identify the complex Hessian of a function φ with $\partial\overline{\partial}\varphi$.

Therefore, the negative term $-\partial\overline{\partial}\rho$ in the right hand side of (1) is absorbed into the preceding term. Hence $\varphi(z) := -\log(1-h(z)^2) + A_0|z|^2$ is strictly plurisubharmonic on Ω for any positive constant A_0. Furthermore, again by the lemma,

$$\partial\overline{\partial}\varphi \geq e^{-\rho}\left(e^{-\rho}-|z_1|^2\right)^{-2}(dz_1+z_1\partial\rho)\otimes(d\overline{z}_1+\overline{z}_1\overline{\partial}\rho) + A_0\partial\overline{\partial}|z|^2. \tag{2}$$

Since

$$\partial\varphi = \frac{\partial h^2}{1-h^2} + A_0\,\partial|z|^2 = \frac{\overline{z}_1(dz_1+z_1\partial\rho)}{e^{-\rho}-|z_1|^2} + A_0\,\partial|z|^2,$$

we obtain from (2) that

$$\partial\overline{\partial}\varphi \geq \epsilon\,\partial\varphi\otimes\overline{\partial}\varphi$$

for sufficiently small positive constant ϵ. Note that ϵ depends also on the diameter of Ω.

Hence we obtain

$$-\partial\overline{\partial}e^{-\epsilon\varphi} = \epsilon e^{-\epsilon\varphi}(\partial\overline{\partial}\varphi - \epsilon\,\partial\varphi\otimes\overline{\partial}\varphi) \geq 0.$$

By the regularization theorem, this conclusion is also valid for the general case. □

Notes . A complex manifold M is said to be hyperconvex if it admits a strictly plurisubharmonic bounded exhaustion function. In several complex variables, this notation was first brought under consideration by J.-L. Stehlé [St] to study a problem raised by J.-P. Serre whether or not analytic fiber bundles over Stein manifolds with Stein fibers are Stein. Stehlé proved that Serre's conjecture is valid if the fibers are hyperconvex. It was then found by J.-L. Ermine [E] that tube domains with bounded convex bases and bounded pseudoconvex Reinhardt domains are both hyperconvex. In the same article he almost proves Theorem 2, because he notices that bounded pseudoconvex Hartogs domains over hyperconvex domains are hyperconvex. However, more extensive meaning of hyperconvexity was realized only after the work of Diederich-Fornaess [D-F]. They discovered that, for any bounded pseudoconvex domain $\Omega \subset \mathbb{C}^2$ with C^2-smooth boundary, the defining function ρ of Ω has a property that $-(-\rho)^\epsilon e^{-A|z|^2}$ is strictly plurisubharmonic for $0 < \epsilon \ll 1$ and $A \gg 1$. This observation has had a great influence on the developments of the $\overline{\partial}$-Neumann regularity theories. (See [O-S-1, 2] for very recent instances.)

Let us also note that a link between hyperconvexity and the Bergman kernel was found in [O-3, 4]. In this direction, whether or not any bounded hyperconvex domain in $\mathbb{C}^n$ is complete with respect to the Bergman metric remains still an open question.

§2 KOBAYASHI'S CHARACTERIZATION OF THE BERGMAN METRIC

For the convenience of the reader, we recall here a characterization of the Bergman metric which is due to S. Kobayashi [K]. Estimates as in Theorem 1 for the Bergman metric by the L^2 method have been built on this foundation.

Let M be a connected complex manifold of dimension n and let K_M be the canonical line bundle over M. In terms of the holomorphic local coordinates $(z_1, \dots, z_n)$, local expressions of the sections of K_M are

$$f\, dz_1 \wedge \cdots \wedge dz_n,$$

where f is a (locally defined) function. For any measurable section ω of K_M over M, with local expressions $f dz_1 \wedge \cdots \wedge dz_n$,

$$\omega \wedge \overline{\omega} = |f|^2\, dz_1 \wedge \cdots \wedge dz_n \wedge d\overline{z}_1 \wedge \cdots \wedge d\overline{z}_n$$

is a globally well defined (n,n)-form.

Hence we put

$$||\omega||^2 = (-1)^{\frac{n(n-1)}{2}} \left(\frac{\sqrt{-1}}{2}\right)^n \int_M \omega \wedge \overline{\omega} \qquad (\in [0, \infty])$$

and call $||\omega||$ the L^2 norm of ω. By Cauchy's estimate for holomorphic functions, it is easy to see that $A^2(K_M)$, the set of holomorphic sections of K_M with finite L^2 norm, is a Hilbert space. Let $\{\omega_\nu\}_{\nu=1}^{\infty}$ be a complete orthonormal system of $A^2(K_M)$. Then the series

$$\sum_{\nu=1}^{\infty} \omega_\nu(x) \wedge \overline{\omega_\nu(y)}$$

converges uniformly on compact subsets of $M \times M$ by Cauchy's estimate. We put

$$\kappa_M(x, \overline{y}) = \sum_{\nu=1}^{\infty} \omega_\nu(x) \wedge \overline{\omega_\nu(y)}.$$

Clearly κ_M does not depend on the choice of the complete orthonormal system. In terms of the local coordinates, $\kappa_M(x, y)$ is expressed as

$$\kappa_M(x, \overline{y}) = K_M(z, \overline{w})\, dz_1 \wedge \cdots \wedge dz_n \wedge d\overline{w}_1 \wedge \cdots \wedge d\overline{w}_n.$$

There will be no occasion of confusing the above K_M with the canonical bundle. $K_M(z, \overline{w})$ are holomorphic in $z = (z_1, \ldots, z_n)$ and antiholomorphic in $w = (w_1, \ldots, w_n)$.

We assume that $A^2(K_M)$ contains for each point $x \in M$ an element ω such that $\omega(x) \neq 0$. Bounded domains in $\mathbb{C}^n$ satisfy this assumption for instance. We are then allowed to introduce a $(1,1)$-form

$$\partial\overline{\partial} \log K_M(z, \overline{z})$$

on M. Note that $\partial\overline{\partial} \log K_M(z, \overline{z})$ is well defined because

$$K_M(z, \overline{z}) = K_M(w, \overline{w}) \left| \frac{\partial(w_1, \ldots, w_n)}{\partial(z_1, \ldots, z_n)} \right|^2.$$

We put

$$\partial\overline{\partial} \log K_M(z, \overline{z}) = \sum_{\alpha,\beta=1}^{n} g_{\alpha\overline{\beta}}(z)\, dz_\alpha \wedge d\overline{z}_\beta.$$

If the Hermitian metrics $(g_{\alpha\overline{\beta}})$ are positive definite, we shall call the Hermitian metric $ds_M^2 := \sum_{\alpha,\beta=1}^{n} g_{\alpha\overline{\beta}}(z)\, dz_\alpha \wedge d\overline{z}_\beta$ associated to $\partial\overline{\partial} \log K_M(z, \overline{z})$ the Bergman metric of M. We shall call M a Bergman manifold if M admits the Bergman metric.

For any Bergman manifold M, a complete orthonormal system $\{\omega_\nu\}_{\nu=1}^{\infty}$ yields a holomorphic map

$$\begin{array}{cccc} \iota: & M & \longrightarrow & (A^2(K_M) \setminus \{0\}) \,/\, \mathbb{C}^* \\ & \Psi & & \Psi \\ & x & \mapsto & (\omega_1(x) : \omega_2(x) : \cdots : \omega_\nu(x) : \cdots). \end{array}$$

Here $\mathbb{C}^* = \mathbb{C} \setminus \{0\}$. Let ds_{FS}^2 be the Hermitian metric on $(A^2(K_M) \setminus \{0\}) \,/\, \mathbb{C}^*$ induced from $\partial\overline{\partial} \log \|\omega\|^2$ $(\omega \in A^2(K_M))$. Then we have

$$ds_M^2 = \iota^* ds_{\mathrm{FS}}^2.$$

Therefore the length of a path in M is equal to the length of its image by the map ι. Since the length of a path with respect to ds_{FS}^2 is at most the distance between the end points, this characterization of the Bergman metric provides a natural way to estimate it from below.

Given $x, y \in M$, let $|x, y|$ be the distance between $\iota(x)$ and $\iota(y)$ with respect to ds_{FS}^2. We shall express $|x, y|$ in terms of the values of $K_M(z, \overline{w})$, by fixing coordinates z, w around x, y. We shall write $K(x, \overline{y})$ instead of $K_M(z(x), \overline{z(y)})$, for brevity.

Proposition .

$$|x, y| = \mathrm{Arctan} \frac{\sqrt{\mathrm{K}(\mathrm{x}, \overline{\mathrm{x}})\mathrm{K}(\mathrm{y}, \overline{\mathrm{y}}) - |\mathrm{K}(\mathrm{x}, \overline{\mathrm{y}})|^2}}{|\mathrm{K}(\mathrm{x}, \overline{\mathrm{y}})|}. \qquad (\star)$$

Proof. We define $B_i \in A^2(K_M)$ $(i = 1, 2)$ by

$$B_1 = K(x, \overline{x})^{-\frac{1}{2}} K(u, \overline{x})\, du_1 \wedge \cdots \wedge du_n$$

and

$$B_2 = K(y, \overline{y})^{-\frac{1}{2}} K(u, \overline{y})\, du_1 \wedge \cdots \wedge du_n.$$

If $B_1 = B_2$, then the both sides of $(\star)$ are 0. So let us suppose otherwise. Since we are allowed to use any complete orthonormal system, we may choose $\{\omega_\nu\}_{\nu=1}^\infty$ as

$$\omega_1 = B_1$$

and

$$\omega_2 = \frac{B_2 - (B_2, B_1)B_1}{||B_2 - (B_2, B_1)B_1||},$$

where (,) denotes the inner product. Then $|x, y|$ is equal to the distance between 0 and $\frac{\omega_2(x)}{\omega_1(x)}$ in $\mathbb{C} \cup \{\infty\}$ with respect to the metric (associated to) $\partial\bar{\partial}\log(1 + |z|^2)$. Namely,

$$|x, y| = \int_0^{\left|\frac{\omega_2(x)}{\omega_1(x)}\right|} \frac{dr}{1 + r^2} = \text{Arctan}\left|\frac{\omega_2(\mathrm{x})}{\omega_1(\mathrm{x})}\right|. \tag{3}$$

On the other hand, we have

$$(B_2, B_1) = \frac{K(x, \overline{y})}{\sqrt{K(x, \overline{x})K(y, \overline{y})}},$$

so that

$$||B_2 - (B_2, B_1)B_1||^2 = 1 - |(B_2, B_1)|^2 = 1 - \frac{|K(x, \overline{y})|^2}{K(x, \overline{x})K(y, \overline{y})}.$$

Therefore

$$\begin{aligned}
\left|\frac{\omega_2(x)}{\omega_1(x)}\right| &= \frac{|B_2(y) - (B_2, B_1)B_1(y)|}{|B_1(y)| \cdot ||B_2 - (B_2, B_1)B_1||} \qquad (4)\\
&= \frac{K(x, \overline{x})K(y, \overline{y}) - |K(x, \overline{y})|^2}{\sqrt{K(x, \overline{x})K(y, \overline{y})}\,|K(x, \overline{y})| \left(1 - \frac{|K(x,\overline{y})|}{K(x,\overline{x})K(y,\overline{y})}\right)^{\frac{1}{2}}}\\
&= \frac{\sqrt{K(x, \overline{x})K(y, \overline{y}) - |K(x, \overline{y})|^2}}{|K(x, \overline{y})|}.
\end{aligned}$$

Combining (3) and (4) we obtain $(\star)$.

Note . In [D-O], for the proof of Theorem 1 we used an inequality

$$|x, y| \geq \min\left(\frac{1}{2}, \sup\left\{\frac{|\omega(x)|^2}{K(x, \overline{x})} \;\middle|\; \omega \in A^2(K_M),\ ||\omega|| = 1,\ \omega(y) = 0\right\}\right)$$

instead of the above precise formula, since it was already sufficient for our purpose. Nevertheless we preferred to present the formula here. Perhaps it is not only from an aesthetic reason.

References

[A-S] Angehrn, U. and Siu, Y. T. *Effective freeness and separation of points for disjoint bundles*, Invent. Math. **122** (1995), 291–308.

[B-N] Bell, S. R. and Narasimhan, R. *Proper holomorphic mappings of complex spaces*, Complex manifolds, S. R. Bell et al. Springer, 1997.

[B-S-Y] Boas, H. P, Straube, E. J. and Yu, J. *Boundary limits of the Bergman kernal and metric*, Michigan Math. J. **42** (1995), 449–461.

[C] Catlin, D. *Estimates of invariant metrics on pseudoconvex domains of dimension two*, Math. Z. **200** (1989), 429–466.

[D'A] D'Angelo, J. *A note on the Bergman kernel*, Duke. Math. J. **45** (1978), 259–265.

[D] Diederich, K. *Das Randverhalten der Bergmanschen Kernfunktion und Metrik in streng pseudokonvexen Gebieten*, Math. Ann. **187** (1970), 9–36.

[D-F] Diederich, K. and Fornaess, J. E. *Pseudoconvex domains : Bounded strictly plurisubharmonic exhaustion functions*, Inv. Math. **39** (1977), 129–141.

[D-H] Diederich, K. and Herbort, G. *Extension of holomorphic L^2-functions with weighted growth conditions*, Nagoya Math. J. **126** (1992), 141–157.

[D-H-O] Diederich, K, Herbort, G. and Ohsawa, T. *The Bergman kernel on uniformly extendable pseudoconvex domains*, Math. Ann. **273** (1986), 471–478.

[D-O] Diederich, K. and Ohsawa, T. *An estimate for the Bergman distance on pseudoconvex domains*, Ann. Math. **141** (1995), 181–190.

[E] Ermine, J.-L. *Conjecture de Sérre et espaces hyperconvexes*, Lecture Notes in Mathematics 670, Springer, 1978, 124–139.

[F] Fefferman, C. *The Bergman kernel and biholomorphic mappings of pseudoconvex domains*, Invent. Math. **26** (1974), 1–65.

[H-1] Hörmander, L. *L^2 estimates and existence theorems for the $\bar{\partial}$-operator*, Acta Math. **113** (1965), 89–152.

[H-2] ——— *An introduction to complex analysis in several variables*, North Holland, 1990.

[J-P-1] Jarnicki, M. and Pflug, P. *Bergman completeness of complete circular domains*, Ann. Pol. Math. **50** (1989), 219–222.

[J-P-2] ——— *Invariant distances and metrics in complex analysis*, de Gruyter expositions in math. 9, 1993.

[K] Kobayashi, S. *Geometry of bounded domains*, Trans. Amer. Math. Soc. **92** (1959), 267–290.

[M-Y] Maitani, F. and Yamaguchi, H. *Variation of three metrics on the moving Riemann surfaces*, preprint.

[O-1] Ohsawa, T. *A remark on the completeness of the Bergman metric*, Proc. Jap. Acad. **57** (1981), 238–240.

[O-2] ——— *Boundary behavior of the Bergman kernel function on pseudoconvex domains*, Publ. RIMS, Kyoto Univ. **20** (1984), 897–902.

[O-3] ——— *On the Bergman kernel of hyperconvex domains*, Nagoya. Math. J. **129** (1993), 43–52.

[O-4] ——— *Addendum to "On the Bergman kernel of hyperconvex domains"*, Nagoya. Math. J. **137** (1995), 145–148.

[O-S-1] Ohsawa, T. and Sibony, N. *Bounded P. S. H. functions and pseudoconvexity in Kähler manifolds*, to appear in Nagoya Math. J.

[O-S-2] ——— *Nonexistence of Levi-flat hypersurfaces in $\mathbb{P}^k$*, preprint.

[O-T] Ohsawa, T. and Takegoshi, K. *On the extension of L^2 holomorphic functions*, Math. Z. **195** (1987), 197–204.

[P] Pflug, P. *Quadratintegrable holomorphe Funktionen und die Serre Vermutung*, Math. Ann. **216** (1975), 285–288.

[St] Stehlé, J.-L. *Fonctions plurisousharmoniques et convexité holomorphe de certain fibrés analytiques*, Lecture Notes in Mathematics 474, 155–180.

[Su] Suita, N. *Capacities and kernels on Riemann surfaces*, Arch. Rational Mech. Anal. **46** (1972), 212–217.

14 INTEGRAL TRANSFORMS INVOLVING SMOOTH FUNCTIONS

Saburou Saitoh[1]
Masahiro Yamamoto[2]

[1]Department of Mathematics, Faculty of Engineering
Gunma University, Kiryu 376-8515, Japan
ssaitoh@eg.gunma-u.ac.jp

[2]Department of Mathematical Sciences
The University of Tokyo
Komaba, Tokyo 153-8914, Japan
myama@ms.u-tokyo.ac.jp

Abstract: We present a general method for obtaining isometrical identities and inversion formulas for integral transforms involving smooth functions. We illustrate our method using Fourier transforms with weights as well as for Weierstrass, Laplace, and Mellin transforms.

INTRODUCTION

In [7, 8], the first author obtained inversion formulas for integral transforms using the theory of reproducing kernels involving Hilbert spaces and developed various applications. See [12] for a survey article and [9, 13] for the research books for its various applications.

Let $\mathcal{H}$ be a Hilbert (possibly finite-dimensional) space. Let E be an abstract set. Let $\mathbf{h}(p)$ be a fixed $\mathcal{H}$-valued function on E such that

$$\mathbf{h} : E \longrightarrow \mathcal{H}.$$

Then, we consider a linear mapping L defined by

$$(L\mathbf{f})(p) = (\mathbf{f}, \mathbf{h}(p))_{\mathcal{H}}, \quad \mathbf{f} \in \mathcal{H}, \quad p \in E$$

from $\mathcal{H}$ into a linear space comprising complex-valued functions on E. We shall denote the function $(L\mathbf{f})(p)$ on E by $f(p)$ as follows:

$$f(p) = (\mathbf{f}, \mathbf{h}(p))_{\mathcal{H}}, \quad \mathbf{f} \in \mathcal{H}. \tag{1}$$

Then, the linear mapping (1) is a general and abstract form of integral transforms in Hilbert space.

To investigate the linear transform (1), we form the Hermitian form $K(p,q)$ on $E \times E$ defined by

$$K(p,q) = (\mathbf{h}(q), \mathbf{h}(p))_{\mathcal{H}}. \tag{2}$$

Note that $K(p,q)$ is a positive matrix in the sense of Moore-Aronszajn (cf. [2] and [13]), that is, any finite points $\{p_j\}_j$ of E and for any complex numbers $\{c_j\}_j$,

$$\sum_{j,j'} c_j \overline{c_{j'}} K(p_{j'}, p_j) \geqq 0.$$

Then, by the fundamental theorem of Moore-Aronszajn, there exists a uniquely determined Hilbert space $H_K(E)$ comprising complex-valued functions f on E satisfying

$$K(\cdot, q) \in H_K(E) \quad \text{for any} \quad q \in E \tag{3}$$

and, for any $f \in H_K(E)$ and for any $q \in E$,

$$(f(\cdot), K(\cdot, q))_{H_K(E)} = f(q). \tag{4}$$

$K(p,q)$ and $H_K(E)$ are called a reproducing kernel and a reproducing kernel Hilbert space admitting the reproducing kernel $K(p,q)$, respectively. The reproducing kernel $K(p,q)$ is uniquely determined by the reproducing properties (3) and (4) in the Hilbert space $H_K(E)$. Note that a Hilbert space $H(E)$ comprising complex-valued functions on E admits a reproducing kernel if and only if for any fixed $p \in E$ the point evaluation $f(p), f \in H(E)$, is continuous. For some general properties of reproducing kernels, see Aronszajn [2], Schwartz [14] and Saitoh [13].

Consider now the following three fundamental theorems (see [7]-[9], [12] and [13] for the details):

(I) The images f of $\mathbf{f} \in \mathcal{H}$ by the linear transform (1) form precisely the reproducing kernel Hilbert space $H_K(E)$ and we have the inequality

$$\|f\|_{H_K(E)} \leqq \|\mathbf{f}\|_{\mathcal{H}}. \tag{5}$$

Furthermore, for any $f \in H_K(E)$ there exists a uniquely determined member $\mathbf{f}^*$ of $\mathcal{H}$ satisfying

$$f(p) = (\mathbf{f}^*, \mathbf{h}(p))_{\mathcal{H}}, \quad p \in E \tag{6}$$

and

$$\|f\|_{H_K(E)} = \|\mathbf{f}^*\|_{\mathcal{H}}. \tag{7}$$

Here, $\{\mathbf{h}(p); p \in E\}$ is complete in $\mathcal{H}$, (that is for any $\mathbf{f} \in \mathcal{H}$, if $(\mathbf{f}, \mathbf{h}(p))_{\mathcal{H}} = 0$ on E, then $\mathbf{f} = 0$), if and only if the mapping (1) is isometrical from $\mathcal{H}$ onto $H_K(E)$.

(II) By using the image space $H_K(E)$, we can obtain an inversion formula of the form

$$f \longrightarrow \mathbf{f}^* : H_K(E) \longrightarrow \mathcal{H}$$

in $\mathcal{H}$ for the member $\mathbf{f}^*$ defined by (6) and (7). See Section 4 in [12] and [13] for some methods.

(III) Conversely, suppose that a reproducing kernel Hilbert space $H_K(E)$ on E admitting the reproducing kernel K on E and a Hilbert space $\mathcal{H}$ are given. Furthermore, suppose that there exists an isometrical mapping $\widetilde{L}$ from $H_K(E)$ onto $\mathcal{H}$. Then by using the isometrical mapping $\widetilde{L}$ from $H_K(E)$ onto $\mathcal{H}$ and by using the reproducing kernel $K(p,q)$, we can determine the linear transform function $\mathbf{h}(p)$ which is a Hilbert $\mathcal{H}$-valued function on E and which transforms $\mathcal{H}$ onto $H_K(E)$ in the form

$$f(p) = (\mathbf{f}, \mathbf{h}(p))_{\mathcal{H}}, \quad \mathbf{f} \in \mathcal{H}$$

with

$$\|f\|_{H_K(E)} = \|\mathbf{f}\|_{\mathcal{H}}$$

as follows:

$$\mathbf{h}(q) = \widetilde{L}K(\cdot, q).$$

For a general linear transform (1) involving Hilbert spaces we shall consider in this paper (1) with smooth functions $\mathbf{f}$. Here, "smooth" functions are as members of some reproducing kernel Hilbert space. For example, although the usual Hilbert space $L_2(a,b)$ does not admit a reproducing kernel, the Sobolev Hilbert space $H^1(a,b)$ does admit a reproducing kernel and comprises absolutely continuous functions on (a,b) with the inner product

$$(f,g)_{H^1(a,b)} = \int_a^b f(x)\overline{g(x)}dx + \int_a^b f'(x)\overline{g'(x)}dx.$$

For various reproducing kernel Hilbert spaces, see [13].

In the inversion formula (II) involving Hilbert spaces, the inverse $\mathbf{f}^*$ will be generally determined by the strong convergence in $\mathcal{H}$ (Theorem 2.1.5 and Theorem 2.1.6 in [13]). In reproducing kernel Hilbert spaces strong convergences induce pointwise convergences because the point evaluations are continuous on reproducing kernel Hilbert spaces. Hence, when we consider a reproducing kernel Hilbert space as an input function space $\mathcal{H}$ in the inversion formula (II), we can obtain pointwise convergence. Of course, in applied mathematics, pointwise convergence is more favourable than strong convergences. Furthermore, an inversion formula with uniform convergence is desirable if possible. We can obtain such formulas if we take a reproducing kernel Hilbert space as an input function space. In this paper we shall examine the linear transform (1) for generating reproducing kernel Hilbert space (as $\mathcal{H}$).

The purposes of this paper are two fold:

(i) We derive an isometrical identity and an inversion formula for a general linear transform (1) where $\mathbf{f}$ varies in a reproducing kernel Hilbert space (§2).

(ii) On the basis of the general method given in §2, we apply our results to Fourier transforms with weights, the Weierstrass transform, the Laplace transform and the Mellin transform of functions in Sobolev spaces, not in L^2-space (§3 ∼ §6).

The important aspect of §2 is the ability to derive new isometrical identities and inversion formulas for various integral transforms.

GENERAL FORMULATION FOR LINEAR TRANSFORM

As stated in §1, we shall examine linear transforms for reproducing kernel Hilbert spaces of type (1) involving the reproducing kernel Hilbert space $H_K(E)$ as given in §1. Furthermore we shall use (2) involving $\mathcal{H}$-valued functions $\mathbf{h}$ on E. Hence, properties (I) and (II) are valid.

Let $H(E)$ be a Hilbert space comprising complex-valued functions on E and containing the members of $H_K(E)$ as $H_K(E) \subset H(E)$. Let $\hat{E}$ be an abstract set. Let $h(p, \hat{p})$ be a complex-valued function on $E \times \hat{E}$ satisfying

$$h(\cdot, \hat{p}) \in H(E) \quad \text{for any fixed} \quad \hat{p} \in \hat{E}.$$

Then, we shall consider linear transforms for $H_K(E)$ defined by

$$\hat{f}(\hat{p}) = (f(\cdot), h(\cdot, \hat{p}))_{H(E)}, \quad f \in H_K(E) \tag{8}$$

which gives complex-valued functions $\hat{f}$ on $\hat{E}$. The linear transform defined by (8) is generally different from the linear transform (1), since, in general, $H(E) \neq H_K(E)$.

Our purposes are to establish an isometrical identity in the linear transform (8) and an inversion formula. For these purposes, we define another reproducing kernel and its reproducing kernel Hilbert space.

Let us assume that

$$\overline{\mathbf{h}(\cdot)} \in H(E) \tag{9}$$

and form the positive matrix

$$\hat{K}(\hat{p}, \hat{q}) = \left((h(\cdot, \hat{q}), \overline{\mathbf{h}(\cdot)})_{H(E)}, (h(\cdot, \hat{p}), \overline{\mathbf{h}(\cdot)})_{H(E)}\right)_{\mathcal{H}} \quad on \quad \hat{E} \times \hat{E}. \tag{10}$$

We denote the reproducing kernel Hilbert space admitting the reproducing kernel (10) by $H_{\hat{K}}(\hat{E})$. Furthermore, if we define a linear map $\hat{L}: \mathbf{f} \longrightarrow \hat{f}$ from $\mathcal{H}$ to $H_{\hat{K}}(\hat{E})$ by

$$\hat{f}(\hat{p}) = (\mathbf{f}, (h(\cdot,\hat{p}), \overline{\mathbf{h}(\cdot)})_{H(E)})_{\mathcal{H}}, \quad \mathbf{f} \in \mathcal{H}, \tag{11}$$

then by the fundamental theorem (II) stated in §1 (where we take $(h(\cdot,\hat{p}), \overline{\mathbf{h}(\cdot)})_{H(E)}$ as $\mathbf{h} = \mathbf{h}(p)$) we obtain the inversion formula

$$\mathbf{f} = \underset{N\to\infty}{\mathrm{l.i.m.}}\, \hat{L}_N^{-1} \hat{f} \tag{12}$$

with strong convergence in $\mathcal{H}$ for $\hat{f} \in H_{\hat{K}}(\hat{E})$. Having given the preliminary definitions, we now state the following theorem involving isometrical identities and inversion formulas employing the linear transform (8):

Theorem 1 *Let*

$$((\mathbf{f}, \mathbf{h}(p))_{\mathcal{H}}, h(p,\hat{p}))_{H(E)} = \left(\mathbf{f}, (h(p,\hat{p}), \overline{\mathbf{h}(p)})_{H(E)}\right)_{\mathcal{H}} \quad \text{on} \quad E \times \hat{E} \quad \text{for} \quad \mathbf{f} \in \mathcal{H}, \tag{13}$$

and

$$\{\mathbf{h}(p); p \in E\} \quad \text{and} \quad \{h(p,\hat{p}); \hat{p} \in \hat{E}\} \tag{14}$$

be complete in $\mathcal{H}$ and $H(E)$, respectively. Then, the transform (8) from $H_K(E)$ onto $H_{\hat{K}}(\hat{E})$ gives the isometrical identity

$$\|\hat{f}\|_{H_{\hat{K}}(\hat{E})} = \|f\|_{H_K(E)} \tag{15}$$

and the inversion formula

$$f(p) = \left(\hat{f}(\hat{p}), (K(\cdot,p), h(\cdot,\hat{p}))_{H(E)}\right)_{H_{\hat{K}}(\hat{E})}. \tag{16}$$

Furthermore, the limit

$$f(p) = \lim_{N\to\infty} (\hat{L}_N^{-1}\hat{f}, \mathbf{h}(p))_{\mathcal{H}} \tag{17}$$

converges uniformly on a subset of E such that $K(p,p)$ is bounded.

Proof. By (8), (1) and (13), we have

$$\begin{aligned}\hat{f}(\hat{p}) &= (f(p), h(p,\hat{p}))_{H(E)} \\ &= ((\mathbf{f}, \mathbf{h}(p))_{\mathcal{H}}, h(p,\hat{p}))_{H(E)} \\ &= \left(\mathbf{f}, (h(p,\hat{p}), \overline{\mathbf{h}(p)})_{H(E)}\right)_{\mathcal{H}}.\end{aligned}$$

From assumptions (14), the mapping from $\mathbf{f}$ to f by (1) and the mapping from f to $\hat{f}$ by (8) are one to one, respectively. Therefore, the mapping from $\mathbf{f}$ to $\hat{f}$ by (11) is also one to one. Hence, we have the isometrical identity between $\mathcal{H}$ and $H_{\hat{K}}(\hat{E})$

$$\|\hat{f}\|_{H_{\hat{K}}(\hat{E})} = \|\mathbf{f}\|_{\mathcal{H}}.$$

Hence, the desired isometrical identity (15)

$$\|\hat{f}\|_{H_{\hat{K}}(\hat{E})} = \|\mathbf{f}\|_{\mathcal{H}} = \|f\|_{H_K(E)},$$

where we have used isometry by (1).

Using the reproducing property of $K(\cdot, p)$ in $H_K(E)$ and isometry (15) between $H_K(E)$ and $H_{\hat{K}}(\hat{E})$, we have the following inversion formula (16)

$$f(p) = (f(\cdot), K(\cdot, p))_{H_K(E)} = \left(\hat{f}(\hat{p}), (K(\cdot, p), h(\cdot, \hat{p}))_{H(E)}\right)_{H_{\hat{K}}(\hat{E})}.$$

Equation (17) then follows from the inversion formula (12). An uniform convergence condition may be computed taking the absolute of (17) or

$$|f(p)| \le \lim_{N\to\infty} \|\hat{L}_N^{-1}\hat{f}\|_{\mathcal{H}}\|\mathbf{h}(p)\|_{\mathcal{H}} = \|f\|_{H_K(E)}K(p,p)^{\frac{1}{2}}.$$

Note that assumption (13) corresponds to Fubini's theorem on exchanging the orders of integrals. If the norms in the Hilbert spaces $H(E)$ and $\mathcal{H}$ are given in terms of some L_2 spaces on E and $\hat{E}$, respectively, then the assumption will be a natural one.

Theorem 1 is very simple. It applies to any integral transform with smooth functions which have members of suitable reproducing kernel Hilbert spaces. Thus Theorem 1 gives a unified way for finding new isometrical identities and inversion formulas involving special functions, integral transforms and reproducing kernels.

Although we must still find reproducing kernels for a particular special function, the point here is that we now have a method for finding such identities and inversion formulas. Once we obtain these formulas, we should be able to prove the results by standard methods of integral transforms.

In Sections 3 ~ 6, we illustrate Theorem 2.1 on various integral transforms.

FOURIER TRANSFORMS WITH WEIGHTS

In this section we illustrate our results using the Fourier transform with a weight function. To this end, for any fixed $a > 0$ and $b > 0$, we set $E = \hat{E} = (-\infty, \infty), \mathcal{H} = L^2(E; \frac{dt}{a^2t^2+b^2})$ with the inner product

$$(f,g)_{\mathcal{H}} = \int_{-\infty}^{\infty} f(t)\overline{g(t)}\frac{dt}{a^2t^2+b^2},$$

and $\mathbf{h} = \frac{1}{\sqrt{2\pi}} e^{ixt}, x \in E, t \in \hat{E}$. In this case the integral transform (1) is

$$F(x) = \frac{1}{\sqrt{2\pi}} \int_{-\infty}^{\infty} e^{-ixt} \frac{f(t)}{a^2t^2 + b^2} dt \tag{18}$$

for real-valued functions f satisfying

$$\int_{-\infty}^{\infty} f(t)^2 \frac{dt}{a^2t^2 + b^2} < \infty. \tag{19}$$

From now on, we shall write f in place of $\mathbf{f}$. See [9, 11, 12, 13].

Let us examine the integral transform (18) for functions f in the Sobolev Hilbert space $H_{A,B}$ comprising real-valued and absolutely continuous functions $f(\xi)$ with the inner product

$$(f_1, f_2)_{H_{A,B}} = \int_{-\infty}^{\infty} (A^2 f_1'(\xi) f_2'(\xi) + B^2 f_1(\xi) f_2(\xi)) d\xi, \tag{20}$$

for any fixed $A, B > 0$. Here we note that

$$H_{A,B} \subset L^2\left((-\infty, \infty), \frac{dt}{a^2t^2 + b^2}\right).$$

and that the reproducing kernel

$$G_{A,B}(\xi, \eta) = \frac{1}{2AB} e^{-\frac{B}{A}|\xi - \eta|} = \frac{1}{2\pi} \int_{-\infty}^{\infty} \frac{e^{i\tau(\xi - \eta)}}{A^2\tau^2 + B^2} d\tau \tag{21}$$

for the space $H_{A,B}$ is the Green function of the differential equation

$$-A^2 \partial_\xi^2 G_{A,B}(\xi, \eta) + B^2 G_{A,B}(\xi, \eta) = \delta(\xi - \eta) \tag{22}$$

which satisfies the following conditions

$$\lim_{\xi \to \infty} G_{A,B}(\xi, \eta) = \lim_{\xi \to -\infty} G_{A,B}(\xi, \eta) = 0$$

([11]). Hence, from (21), the integral transform

$$f(t) = \frac{1}{2\pi} \int_{-\infty}^{\infty} \frac{e^{i\tau t}}{A^2\tau^2 + B^2} \hat{F}(\tau) d\tau \tag{23}$$

has the isometrical identity

$$\|f\|_{H_{A,B}}^2 = \frac{1}{2\pi} \int_{-\infty}^{\infty} \frac{\hat{F}(\tau)^2}{A^2\tau^2 + B^2} d\tau \tag{24}$$

by the fundamental theorem (I) stated in §1. The same results also follow directly by using Fourier transforms.

From (18) and (23), we have that

$$
\begin{aligned}
F(x) &= \frac{1}{2\pi}\frac{\sqrt{2\pi}}{2ab}\int_{-\infty}^{\infty}\hat{F}(\tau)e^{-\frac{b}{a}|\tau-x|}\frac{d\tau}{A^2\tau^2+B^2} \\
&= \frac{1}{\sqrt{2\pi}}\int_{-\infty}^{\infty}\hat{F}(\tau)G_{a,b}(\tau,x)\frac{d\tau}{A^2\tau^2+B^2}. \qquad (25)
\end{aligned}
$$

To examine the integral transform (25) for functions $\hat{F}$ satisfying (24), we form the positive matrix

$$
\begin{aligned}
K(x,y) &= \frac{1}{2\pi}\frac{2\pi}{4a^2b^2}\int_{-\infty}^{\infty}e^{-\frac{b}{a}|\tau-x|}e^{-\frac{b}{a}|\tau-y|}\frac{d\tau}{A^2\tau^2+B^2} \\
&= \int_{-\infty}^{\infty}G_{a,b}(\tau,x)G_{a,b}(\tau,y)\frac{d\tau}{A^2\tau^2+B^2}. \qquad (26)
\end{aligned}
$$

From the property of the Green function $G_{a,b}(\tau,x)$, we have from (22) that

$$
-a^2\partial_x^2K(x,y)+b^2K(x,y)=\frac{G_{a,b}(x,y)}{A^2x^2+B^2}
$$

or

$$
(A^2x^2+B^2)(-a^2\partial_x^2+b^2)K(x,y)=G_{a,b}(x,y). \qquad (27)
$$

Hence, applying (22)

$$
(-a^2\partial_x^2+b^2)(A^2x^2+B^2)(-a^2\partial_x^2+b^2)K(x,y)=\delta(x-y). \qquad (28)
$$

This identity implies that for a smooth function F on $(-\infty,\infty)$

$$
\begin{aligned}
F(y) &= \int_{-\infty}^{\infty}F(x)\delta(x-y)dx \\
&= \int_{-\infty}^{\infty}F(x)\left\{(-a^2\partial_x^2+b^2)(A^2x^2+B^2)(-a^2\partial_x^2+b^2)K(x,y)\right\}dx. \qquad (29)
\end{aligned}
$$

For any C_0^2 function $F(x)$,

$$
\begin{aligned}
F(y) = \int_{-\infty}^{\infty} &\{a^4F''(x)\partial_x^2K(x,y)+2a^2b^2F'(x)\partial_xK(x,y) \\
&+b^4F(x)K(x,y)\}(A^2x^2+B^2)dx \qquad (30)
\end{aligned}
$$

by repeated integration by parts.

By the usual completion procedure, we see that the reproducing kernel Hilbert space H_K admitting the reproducing kernel $K(x,y)$ in (26) is composed of all absolutely continuous functions with finite norms

$$\|F\|^2_{H_K} = \int_{-\infty}^{\infty} \{a^4 F''(x)^2 + 2a^2b^2F'(x)^2 + b^4F(x)^2\}(A^2x^2+B^2)dx. \quad (31)$$

In order to use the inversion formula (16), we calculate

$$\begin{aligned}
&\left(G_{A,B}(\hat{t},t), h(\hat{t},x)\right)_{H(E)} \\
&= \frac{1}{\sqrt{2\pi}} \int_{-\infty}^{\infty} \left(\frac{1}{2\pi} \int_{-\infty}^{\infty} \frac{e^{i\tau(\hat{t}-t)}}{A^2\tau^2+B^2} d\tau \right) \frac{e^{-ix\hat{t}}}{a^2\hat{t}^2+b^2} d\hat{t} \\
&= \frac{1}{\sqrt{2\pi}} \int_{-\infty}^{\infty} \frac{e^{-i\tau t} G_{a,b}(\tau,x)}{A^2\tau^2+B^2} d\tau.
\end{aligned}$$

Hence, we obtain the inversion formula

$$\begin{aligned}
f(t) = &\frac{1}{\sqrt{2\pi}} \int_{-\infty}^{\infty} \int_{-\infty}^{\infty} \{a^4 F''(x) \partial_x^2 G_{a,b}(\tau,x) \\
&+ 2a^2b^2 F'(x) \partial_x G_{a,b}(\tau,x) + b^4 F(x) G_{a,b}(\tau,x)\} \\
&\cdot \frac{e^{i\tau t} d\tau}{A^2\tau^2+B^2} (A^2x^2+B^2) dx. \qquad (32)
\end{aligned}$$

Furthermore, by (II), we have the inversion formula of (25)

$$\begin{aligned}
\hat{F}(\tau) = &\operatorname*{l.i.m.}_{N\to\infty} \sqrt{2\pi} \int_{-N}^{N} \{a^4 F''(x) \partial_x^2 G_{a,b}(\tau,x) \\
&+ 2a^2b^2 F'(x) \partial_x G_{a,b}(\tau,x) + b^4 F(x) G_{a,b}(\tau,x)\}(A^2x^2+B^2)dx, \qquad (33)
\end{aligned}$$

with strong convergence in $L^2(\mathbb{R}, 1/(A^2\tau^2+B^2)d\tau)$. See [9], [12] or [13] for the details.

For this result (33), the following formal identities will be instructive. From the reproducing property in H_K and from (26),

$$\begin{aligned}
F(y) &= (F(x), K(x,y))_{H_K} \\
&= \left(F(x), \int_{-\infty}^{\infty} G_{a,b}(\tau,x) G_{a,b}(\tau,y) \frac{d\tau}{A^2\tau^2+B^2}\right)_{H_K} \\
&= \int_{-\infty}^{\infty} (F(x), G_{a,b}(\tau,x))_{H_K} G_{a,b}(\tau,y) \frac{d\tau}{A^2\tau^2+B^2} \\
&= \frac{1}{\sqrt{2\pi}} \int_{-\infty}^{\infty} \hat{F}(\tau) G_{a,b}(\tau,y) \frac{d\tau}{A^2\tau^2+B^2}. \qquad (34)
\end{aligned}$$

Hence, using (23), we would have the desired inversion formula

$$f(t) = \frac{1}{\sqrt{2\pi}} \int_{-\infty}^{\infty} e^{i\tau t}(F(x), G_{a,b}(\tau, x))_{H_K} \frac{d\tau}{A^2\tau^2 + B^2}.$$

Note however that

$$G_{a,b}(\tau, x) \notin H_K$$

so that

$$(F(x), G_{a,b}(\tau, x))_{H_K}$$

is considered in the same manner as the right side of (33). More precisely, we obtain the inversion formula

$$\begin{aligned} f(t) = \lim_{N\to\infty} \frac{1}{\sqrt{2\pi}} \int_{-\infty}^{\infty} \int_{-N}^{N} &\{a^4 F''(x)\partial_x^2 G_{a,b}(\tau, x) \\ &+ 2a^2b^2 F'(x)\partial_x G_{a,b}(\tau, x) + b^4 F(x) G_{a,b}(\tau, x)\} \\ &\cdot (A^2x^2 + B^2)dx \frac{e^{i\tau t} d\tau}{A^2\tau^2 + B^2}. \end{aligned} \tag{35}$$

Because the reproducing kernel $G_{A,B}(x, x)$ is bounded on $\mathbb{R}$, the limit in (35) converges uniformly on $\mathbb{R}$ by Theorem 1.

WEIERSTRASS TRANSFORM

The Weierstrass transform, for any fixed $t > 0$, is

$$u(x, t) = \frac{1}{\sqrt{4\pi t}} \int_{-\infty}^{\infty} F(\xi) e^{-\frac{(x-\xi)^2}{4t}} d\xi. \tag{36}$$

Thus, by integration by parts

$$\partial_x u(x, t) = \frac{1}{\sqrt{4\pi t}} \int_{-\infty}^{\infty} F'(\xi) e^{-\frac{(x-\xi)^2}{4t}} d\xi. \tag{37}$$

Hence, for the integral transform (36) in the Sobolev space $H_{A,B}$, we immediately have the following isometrical identity for the $L_2(-\infty, \infty)$ space:

$$\begin{aligned} \int_{-\infty}^{\infty} (A^2 F'(\xi)^2 + B^2 F(\xi)^2) d\xi = A^2 \sum_{n=0}^{\infty} \frac{(2t)^n}{n!} \int_{-\infty}^{\infty} (\partial_x^{n+1} u(x, t))^2 dx \\ + B^2 \sum_{n=0}^{\infty} \frac{(2t)^n}{n!} \int_{-\infty}^{\infty} (\partial_x^n u(x, t))^2 dx \end{aligned} \tag{38}$$

(see, [5], [13]). Hence, by using the inversion formula (16) and

$$\hat{G}_{A,B}(x, \xi) := \frac{1}{\sqrt{4\pi t}} \int_{-\infty}^{\infty} G_{A,B}(\hat{\xi}, \xi) e^{-\frac{(\hat{\xi}-x)^2}{4t}} d\hat{\xi} = \frac{1}{2\pi} \int_{-\infty}^{\infty} \frac{e^{-\tau^2 t} e^{i\tau(x-\xi)}}{A^2\tau^2 + B^2} d\tau,$$

we have the inversion formula (36)

$$\begin{aligned}F(\xi) &= \sum_{n=0}^{\infty} \frac{(2t)^n}{n!} \int_{-\infty}^{\infty} (A^2 \partial_x^{n+1} u(x,t) \partial_x^{n+1} \overline{\hat{G}_{A,B}(x,\xi)} \\ &\quad + B^2 \partial_x^n u(x,t) \partial_x^n \overline{\hat{G}_{A,B}(x,\xi)}) dx \\ &= \sum_{n=0}^{\infty} \frac{(2t)^n}{n!} \int_{-\infty}^{\infty} u(x,t) \{(-1)^{n+1} A^2 \partial_x^{2n+2} \overline{\hat{G}_{A,B}(x,\xi)} \\ &\quad + (-1)^n B^2 \partial_x^{2n} \overline{\hat{G}_{A,B}(x,\xi)}\} dx. \end{aligned} \tag{39}$$

The convergence criteria are easily obtained. We can also find the corresponding complex inversion formulas. See [13], pp. 128-132.

LAPLACE TRANSFORM

From the definition of the Laplace transform

$$f(z) = \int_0^{\infty} e^{-zt} F(t) dt, \tag{40}$$

we have the isometrical identity

$$\int_0^{\infty} |F(t)|^2 dt = \lim_{x \to +0} \frac{1}{2\pi} \int_{-\infty}^{\infty} |f(x+iy)|^2 dy = \frac{1}{2\pi} \int_{-\infty}^{\infty} |f(iy)|^2 dy. \tag{41}$$

Here, $f(z)$ belongs to the Szegö space in the right half plane $\{Re(z) > 0\}$ and $f(z)$ has Fatou's nontangential boundary values belonging to the L_2 space in (41) along the imaginary axis. See, for example, [13], pp. 97-98.

For the reproducing kernel Hilbert space $H_{\min(t,t')}$ admitting the reproducing kernel $\min(t,t')$ on $t, t' \geq 0$, the members F of $H_{\min(t,t')}$ are composed of all absolutely continuous functions on $[0,\infty)$ satisfying $F(0) = 0$ and having the inner product

$$(F_1, F_2)_{H_{\min(t,t')}} = \int_0^{\infty} F_1'(t) \overline{F_2'(t)} dt$$

([7], [13]). Then, by integration by parts, we obtain the isometrical identity

$$\begin{aligned}\int_0^{\infty} |F'(t)|^2 dt &= \lim_{x \to +0} \frac{1}{2\pi} \int_{-\infty}^{\infty} |(x+iy) f(x+iy)|^2 dy \\ &= \frac{1}{2\pi} \int_{-\infty}^{\infty} |f(iy)|^2 y^2 dy, \end{aligned} \tag{42}$$

directly from (41). Hence, by using the inversion formula (16) and

$$\int_0^{\infty} e^{-zt} \min(t,t') dt = \frac{1}{z^2}(1 - e^{-zt'}),$$

we obtain the inversion formula

$$F(t) = \lim_{x \to +0} \frac{1}{2\pi} \int_{-\infty}^{\infty} (x+iy) f(x+iy) \overline{\frac{1}{x+iy}(1-e^{-t(x+iy)})} dy$$
$$= \frac{-1}{2\pi} \int_{-\infty}^{\infty} f(iy)(1-e^{iyt}) dy. \tag{43}$$

For some real inversion formulas involving the Laplace transform, see [13], Chapter 5, §2.

MELLIN TRANSFORM

We shall examine the Mellin transform

$$f(z) = \int_0^\infty F(t) t^{z-1} dt \tag{44}$$

which has the reproducing kernel Hilbert space $H_{\min(t,t')}$ admitting the reproducing kernel $\min(t,t')$ on $t, t' \geqq 0$. Because

$$\min(t,t') = \frac{2}{\pi} \int_0^\infty \frac{\sin \xi t \sin \xi t'}{\xi^2} d\xi, \tag{45}$$

$F \in H_{\min(t,t')}$ is expressible as

$$F(t) = \frac{2}{\pi} \int_0^\infty \frac{\sin \xi t \hat{F}(\xi)}{\xi^2} d\xi \tag{46}$$

for any uniquely determined function $\hat{F}$ satisfying

$$\int_0^\infty \frac{|\hat{F}(\xi)|^2}{\xi^2} d\xi < \infty. \tag{47}$$

The isometrical identity is

$$\int_0^\infty |F'(t)|^2 dt = \frac{2}{\pi} \int_0^\infty \frac{|\hat{F}(\xi)|^2}{\xi^2} d\xi. \tag{48}$$

From (44) and (46) we have that

$$f(z) = \Gamma(z) \sin \frac{\pi z}{2} \frac{2}{\pi} \int_0^\infty \xi^{-z} \frac{\hat{F}(\xi)}{\xi^2} d\xi \tag{49}$$

for $-1 < Re(z) < 1$ (for the Mellin transform of $\sin \xi t$, see [4] and [6]).

To examine the integral transform

$$f_1(z) = \frac{2}{\pi}\int_0^\infty \xi^{-z}\frac{\hat{F}(\xi)}{\xi^2}d\xi \tag{50}$$

for functions $\hat{F}$ satisfying (47), we calculate the positive matrix

$$\begin{aligned} K(z,\overline{u}) &= \frac{2}{\pi}\int_0^\infty \xi^{-z}\xi^{-\overline{u}}\frac{d\xi}{\xi^2} \\ &= \frac{2}{\pi}\frac{1}{z+\overline{u}+1} \quad \text{on} \quad Re(z), Re(u) > -\frac{1}{2}. \end{aligned} \tag{51}$$

Because

$$\frac{1}{2\pi(z+\overline{u}+1)} \tag{52}$$

is the Szegö kernel for the Hilbert space $H_S(Re(z) > -\frac{1}{2})$ comprising all analytic functions $f(z)$ on $Re(z) > -\frac{1}{2}$ and

$$\sup_{x>-\frac{1}{2}}\int_{-\infty}^\infty |f(x+iy)|^2 dy < \infty, \tag{53}$$

the members $f(z)$ of $H_S(Re(z) > -\frac{1}{2})$ have Fatou's nontangential boundary values on $x = -\frac{1}{2}$ a.e. and the norm is given by

$$\|f\|^2_{H_S(Re(z)>-\frac{1}{2})} = \sup_{x>-\frac{1}{2}}\int_{-\infty}^\infty |f(x+iy)|^2 dy = \int_{-\infty}^\infty \left|f(-\frac{1}{2}+iy)\right|^2 dy. \tag{54}$$

Hence, for the integral transform (50) we have the isometrical identity

$$\frac{1}{4}\int_{-\infty}^\infty \left|f_1(-\frac{1}{2}+iy)\right|^2 dy = \frac{2}{\pi}\int_0^\infty \frac{|\hat{F}(\xi)|^2}{\xi^2}d\xi. \tag{55}$$

From the identities

$$\left|\Gamma(-\frac{1}{2}+iy)\right|^2 = \frac{\pi}{(\frac{1}{4}+y^2)\cosh \pi y} \quad ([1], 256\text{p.})$$

and

$$\left|\sin\frac{\pi}{2}(-\frac{1}{2}+iy)\right|^2 = \frac{1}{2}+\sinh^2\frac{\pi y}{2},$$

we have the desired isometrical identity

$$\int_0^\infty |F'(t)|^2 dt = \frac{1}{4\pi}\int_{-\infty}^\infty \left|f(-\frac{1}{2}+iy)\right|^2 \frac{(\frac{1}{4}+y^2)\cosh \pi y}{\frac{1}{2}+\sinh^2\frac{\pi y}{2}}dy. \tag{56}$$

Note that (49) is valid on $\{Re(z) > -\frac{1}{2}\}$ following the principle of invariance of analytic relationship. Hence, by using the inversion formula (16) and

$$\int_0^\infty \min(\hat{t}, t)\hat{t}^{z-1}d\hat{t} = \frac{1}{z+1}t^{z+1} \quad \text{for} \quad \text{Im}\,(z) > -1,$$

we have the inversion formula

$$F(t) = \frac{1}{4\pi}\int_{-\infty}^{\infty} f(-\frac{1}{2}+iy)\overline{\frac{1}{\frac{1}{2}+iy}t^{\frac{1}{2}+iy}}\frac{(\frac{1}{4}+y^2)\cosh\pi y}{\frac{1}{2}+\sinh^2\frac{\pi y}{2}}dy.$$

The authors wish to express their sincere thanks Professor Dean G. Duffy for her careful readings and improvement of the expression.

References

[1]. M. Abramowitz and L. A. Stegun, *Handbook of Mathematical Functions, with Formulas, Graphs, and Mathematical Tables*, U.S. Department of Commerce, AMS 55, National Bureau of Standards Applied Mathematics Series 55, 1972.

[2]. N. Aronszajn, *Theory of reproducing kernels*, Trans. Amer. Math. Soc. **68**, pp.337-404, 1950.

[3]. D. -W. Byun and S. Saitoh, *A real inversion formula for the Laplace transform*, Zeitschrift für Analysis und ihre Anwendungen **12**, pp.597-603, 1993.

[4]. A. Erdélyi, W. Magnus, F. Oberhettinger, F. G. Tricomi, *Tables of Integral Transforms, Volume I*, McGraw-Hill Book Company, Inc. 1954.

[5]. N. Hayashi and S. Saitoh, *Analyticity and smoothing effect for the Schrödinger equation*, Ann. Inst. Henri Poincaré **52**, pp.163-173, 1990.

[6]. F. Oberhettinger, *Tables of Mellin Transforms*, Springer-Verlag, 1974.

[7]. S. Saitoh, *Integral transforms in Hilbert spaces*, Proc. Japan Acad. **58**, pp.361-364, 1982.

[8]. ———, *Hilbert spaces induced by Hilbert space valued functions*, Proc. Amer. Math. Soc. **89**, pp.74-78, 1983.

[9]. ———, *Theory of reproducing kernels and its applications*, Pitman Res. Notes in Math. Series **189**, Longman Scientific & Technical, England, 1988.

[10]. ———, *Representations of the norms in Bergman-Selberg spaces on strips and half planes*, Complex Variables **19**, pp.231-241, 1992.

[11]. ———, *Inequalities in the most simple Sobolev space and convolutions of L_2 functions with weights*, Proc. Amer. Math. Soc. **118**, pp.515-520, 1993.

[12]. ———, *One approach to some general integral transforms and its applications*, Integral Transforms and Special Functions **3**, pp.49-84, 1995.

[13]. ———, *Integral Transforms, Reproducing Kernels and Their Applications*, Pitman Res. Notes in Math. Series **369**, Addison Wesley Longman, 1997.

[14]. L. Schwartz, *Sous-espaces hilbertiens d'espaces vectoriels topologiques et noyaux associès (noyaux reproduisants)*, J. Analyse Math. **13**, pp.115-256, 1964.

References

15 APPLICATIONS OF THE GENERAL THEORY OF REPRODUCING KERNELS

Saburou Saitoh

Department of Mathematics, Faculty of Engineering
Gunma University, Kiryu 376-8515, Japan
ssaitoh@eg.gunma-u.ac.jp

Abstract: In this paper, we present a survey of the contents of the research note [[Sa2]] which was published recently. For the following general applications of the general theory of reproducing kernels :

a new characterization of the adjoint L-kernel of Szegö type,

nonharmonic integral transforms,

and

interpolation problems of Pick-Nevanlinna type,

see the previous research note [[Sa1]], which was also dealt with the history of reproducing kernels and the classical reproducing kernels in one complex analysis. {The publication of this survey article was permitted by Addison Wesley Longman Ltd in connection with the original book [[Sa 2]].}

A GENERALIZED ISOPERIMETRIC INEQUALITY

In 1976, the author obtained the generalized isoperimetric inequality in his thesis [Sa1]:

For a bounded regular region G in the complex $z = x + iy$ plane surrounded by a finite number of analytic Jordan curves and for any analytic functions $\varphi(z)$ and $\psi(z)$ on $\overline{G} = G \cup \partial G$,

$$\frac{1}{\pi} \iint_G |\varphi(z)\psi(z)|^2 dxdy \leqq \frac{1}{2\pi} \int_{\partial G} |\varphi(z)|^2 |dz| \frac{1}{2\pi} \int_{\partial G} |\psi(z)|^2 |dz|.$$

The crucial point in this thesis was to determine completely the case when equality holds in the inequality.

In order to prove this simple inequality, surprisingly enough, we must apply the historical results of

G.F.B.Riemann (1826–1866); F.Klein (1849–1925); S.Bergman; G.Szegö; Z.Nehari; M.M.Schiffer; P.R.Garabedian; D.A.Hejhal (1972, thesis).

In particular, a profound result of D. A. Hejhal, which establishes the fundamental interrelationship between the Bergman and the Szegö reproducing kernels of G ([[He]]), must be applied. Furthermore, we must use the general theory of reproducing kernels by N. Aronszajn ([Ar]) described in 1950. These circumstances have not changed, since the paper [Sa1] was published about 20 years ago.

This thesis is a milestone in the development of the theory of reproducing kernels. In the thesis, the author realized that miscellaneous applications of the general theory of reproducing kernels are possible for many concrete problems. See also [[Sa1]] for the details. It seems that the general theory of reproducing kernels was, in a strict sense, not active in the theory of concrete reproducing kernels until the publication of the thesis. Indeed, after the publication of the thesis, we derived miscellaneous fundamental norm inequalities containing quadratic norm inequalities in matrices which have been described in over 20 papers. Furthermore, we obtained a general idea for linear transforms essentially by using the general theory of reproducing kernels, which is the main tool in [[Sa2]].

LINEAR TRANSFORMS OF HILBERT SPACES

In 1982 and 1983, we published the very simple theorems in [Sa2] and [Sa3], respectively. Certainly the results are very simple mathematically, but they appear to be extremely fundamental and widely applicable for general linear transforms. Moreover, the results gave rise to several new ideas for linear transforms themselves.

We shall formulate a *'linear transform'* as follows:

$$f(p) = \int_T F(t)\overline{h(t,p)}dm(t), \quad p \in E. \tag{1}$$

Here, the input $F(t)$ (source) is a function on a set T, E is an arbitrary set, $dm(t)$ is a σ-finite positive measure on the dm measurable set T, and $h(t,p)$ is a complex-valued function on $T \times E$ which determines the transform of the system.

This formulation will give a generalized form of a linear transform L:

$$L(aF_1 + bF_2) = aL(F_1) + bL(F_2).$$

Indeed, following the Schwartz kernel theorem (cf. [[Tr]]), we see that very general linear transforms are realized as integral transforms as in (1) above by using generalized functions as the integral kernels $h(t,p)$.

We shall assume that $F(t)$ is a member of the Hilbert space $L_2(T,dm)$ satisfying

$$\int_T |F(t)|^2 dm(t) < \infty. \tag{2}$$

The space $L_2(T,dm)$ whose norm gives an energy integral will be the most fundamental space as the input function space. In other spaces we shall modify them in order to comply with our situation. As a prototype case, we shall consider primarily or, as the first stage, the linear transform (1) in our situation.

As a natural result of our basic assumption (2), we assume that

$$\text{for any fixed} \quad p \in E, \quad h(\cdot,p) \in L_2(T,dm) \tag{3}$$

for the existence of the integral in (1).

We shall consider the following two typical linear transforms:

We take $E = \{1,2\}$ and let $\{\mathbf{e}_1, \mathbf{e}_2\}$ be some orthonormal vectors of $\mathbb{R}^2$. Then, we shall consider the linear transform from $\mathbb{R}^2$ to $\{x_1, x_2\}$ as follows:

$$\mathbf{x} \longrightarrow \begin{cases} x_1 = (\mathbf{x}, \mathbf{e}_1) \\ x_2 = (\mathbf{x}, \mathbf{e}_2). \end{cases} \tag{4}$$

For $F \in L_2(\mathbb{R}, dx)$ we shall consider the integral transform

$$u(x,t) = \frac{1}{\sqrt{4\pi t}} \int_{-\infty}^{\infty} F(\xi) e^{-\frac{(x-\xi)^2}{4t}} d\xi, \tag{5}$$

which gives the solution $u(x,t)$ of the heat equation

$$u_{xx}(x,t) = u_t(x,t) \quad \text{on} \quad \mathbb{R} \times \{t > 0\} \tag{6}$$

subject to the initial condition

$$u(x,0) = F(x) \quad \text{on} \quad \mathbb{R}. \tag{7}$$

IDENTIFICATION OF THE IMAGES OF LINEAR TRANSFORMS

We formulated linear transforms as the integral transforms (1) satisfying (2) and (3) in the framework of Hilbert spaces. In this general situation, we can identify the space of output functions $f(p)$ and we can completely characterize the output functions $f(p)$. Regarding this fundamental idea, it seems that the mathematical community still does not realize its importance, since the papers [Sa2] and [Sa3] were published about fourteen years ago.

One reason why we have no idea of the identification of the images of linear transforms is based on the definition of linear transforms themselves. A linear transform is, in general, a linear mapping from a linear space into a linear space, and so the image space of the linear mapping will be considered as an, a priori, given one. For this, our idea will show that the image spaces of linear transforms, in our situation, form the uniquely determined and intuitive ones which are, in general, different from the image spaces stated in the definition of linear transforms.

Another reason is that the very fundamental theory of reproducing kernels by Aronszajn is still not widely known. The general theory seems to be a very fundamental one in mathematics, as in the theory of Hilbert spaces.

Recall the paper of [Schw] for this fact, which extended globally the theory of Aronszajn. Our basic idea for linear transforms is very simple mathematically, but it was initially deduced from the theory of Schwartz using the direct integrals of reproducing kernel Hilbert spaces. See [[Sa1]] for the details.

In order to identify the image space of the integral transform (1), we consider the Hermitian form

$$K(p,q) = \int_T h(t,q)\overline{h(t,p)}dm(t) \quad \text{on} \quad E \times E. \tag{8}$$

The kernel $K(p,q)$ is apparently a positive matrix on E in the sense of

$$\sum_{j=1}^{n}\sum_{j'=1}^{n} C_j \overline{C_{j'}} K(p_{j'}, p_j) \geqq 0$$

for any finite points $\{p_j\}$ of E and for any complex numbers $\{C_j\}$. Then, following the fundamental theorem of Aronszajn–Moore ([Ar]), there exists a uniquely determined Hilbert space H_K comprising functions $f(p)$ on E satisfying

$$\text{for any fixed} \quad q \in E, K(p,q) \quad \text{belongs to} \quad H_K \quad \text{as a function in} \quad p, \tag{9}$$

and

$$\text{for any} \quad q \in E \quad \text{and for any} \quad f \in H_K \quad (f(\cdot), K(\cdot,q))_{H_K} = f(q). \tag{10}$$

Then, the point evaluation $f(p) (p \in E)$ is continuous on H_K and, conversely, a functional Hilbert space such that the point evaluation is continuous admits the reproducing kernel $K(p,q)$ satisfying (9) and (10). Then, we obtain

Theorem 1 *The images $f(p)$ of the integral transform (1) for $F \in L_2(T, dm)$ form precisely the Hilbert space H_K admitting the reproducing kernel $K(p,q)$ in (8).*

In Example (4), using Theorem 1 we can naturally deduce that

$$\|\{x_1, x_2\}\|_{H_K} = \sqrt{x_1^2 + x_2^2}$$

for the image.

In Example (5), we naturally deduce the very surprising result that the image $u(x, t)$ is extended analytically onto the entire complex $z = x + iy$ plane and when we denote its analytic extension by $u(z, t)$, we have

$$\|u(z,t)\|_{H_K}^2 = \frac{1}{\sqrt{2\pi t}} \iint_{\mathbb{C}} |u(z,t)|^2 e^{-\frac{y^2}{2t}} dx dy \tag{11}$$

([Sa4]).

The images $u(x, t)$ of (5) for $F \in L_2(\mathbb{R}, dx)$ are characterized by (11); that is, $u(x, t)$ are entire functions in the form $u(z, t)$ with finite integrals (11).

In 1989 ([HS1]), we deduced that (11) equals

$$\sum_{j=0}^{\infty} \frac{(2t)^j}{j!} \int_{-\infty}^{\infty} (\partial_x^j u(x,t))^2 dx \tag{12}$$

using the property that $u(x, t)$ is the solution of the heat equation (6) with (7) (see [[Sa2]], Theorem 3.2.2 for a direct proof). Hence, we see that the images $u(x, t)$ of (5) are also characterized by the property that $u(x, t) \in C^\infty$ with finite integrals (12).

RELATIONSHIP BETWEEN THE MAGNITUDES OF INPUT AND OUTPUT FUNCTIONS – A GENERALIZED PYTHAGORAS THEOREM

Our second theorem is

Theorem 2 *In the integral transform (1), we have the inequality*

$$\|f\|_{H_K}^2 \leqq \int_T |F(t)|^2 dm(t).$$

Furthermore, there exist functions F^* with the minimum norms satisfying (1), and we have the isometrical identity

$$\|f\|_{H_K}^2 = \int_T |F^*(t)|^2 dm(t).$$

In Example (4), we have, surprisingly enough, the Pythagoras theorem

$$\|\mathbf{x}\|^2 = x_1^2 + x_2^2.$$

In Example (5), we have the isometrical identity

$$\int_{-\infty}^{\infty} |F(\xi)|^2 d\xi = \frac{1}{\sqrt{2\pi t}} \iint_{\mathbb{C}} |u(z,t)|^2 e^{-\frac{y^2}{2t}} dx dy, \tag{13}$$

whose integrals are independent of $t > 0$. At this moment, we will be able to say that by the general principle (Theorems 1 and 2) for linear transforms we can prove the Pythagoras (572–492 B.C.) theorem apart from the idea of *'orthogonality'*, and we can understand Theorem 2 as a generalized theorem of Pythagoras in our general situation of linear transforms.

By using the general principle, we derived miscellaneous Pythagoras type theorems in over 30 papers, whose typical results were given in Chapter 3 of [[Sa2]]. We shall refer to one typical example.

For the solution $u(x,t)$ of the simplest wave equation

$$u_{tt}(x,t) = c^2 u_{xx}(x,t) \quad (c > 0 : \text{constant})$$

subject to the initial conditions

$$u_t(x,t)|_{t=0} = F(x), \quad u(x,0) = 0 \quad \text{on} \quad \mathbb{R}$$

for $F \in L_2(\mathbb{R}, dx)$, we obtain the isometrical identity

$$\frac{1}{2}\int_{-\infty}^{\infty} |F(x)|^2 dx = \frac{2\pi c}{t} \int_{-\infty}^{\infty} \left| l.i.m._{N\to\infty} \int_{-N}^{N} u(x,t) \exp\left(\frac{ix\xi}{2\pi ct}\right) dx \right|^2 \frac{d\xi}{\left(\frac{\sin\frac{\xi}{2}}{\frac{\xi}{2}}\right)^2}, \tag{14}$$

whose integrals are independent of $t > 0$.

Recall here the *conservative law of energy*

$$\frac{1}{2}\int_{-\infty}^{\infty} |F(x)|^2 dx = \frac{1}{2}\int_{-\infty}^{\infty} (u_t(x,t)^2 + c^2 u_x(x,t)^2) dx. \tag{15}$$

To compare the two integrals (14) and (15) will be very interesting, because (15) contains the derived functions $u_t(x,t)$ and $u_x(x,t)$; meanwhile (14) contains the values $u(x,t)$ only.

In the viewpoint of the conservative law of energy in (14) and (15), could we give some physical interpretation to the isometrical identities in (13) and (12) whose integrals are independent of $t > 0$ in the heat equation?

INVERSION FORMULAS FOR LINEAR TRANSFORMS

In our **Theorem 3**, we establish the inversion formula

$$f \longrightarrow F^* \tag{16}$$

of the integral transform (1) in the sense of Theorem 2.

The basic idea to derive the inversion formula (16) is, first, to represent $f \in H_K$ in the space H_K in the form

$$f(q) = (f(\cdot), K(\cdot, q))_{H_K},$$

secondly, to consider as follows:

$$\begin{aligned} f(q) &= (f(\cdot), \int_T h(t,q)\overline{h(t,\cdot)}dm(t))_{H_K} \\ &= \int_T (f(\cdot), \overline{h(t,\cdot)})_{H_K} \overline{h(t,q)}dm(t) \\ &= \int_T F^*(t)\overline{h(t,q)}dm(t) \end{aligned}$$

and, finally, to deduce that

$$F^*(t) = (f(\cdot), \overline{h(t,\cdot)})_{H_K}. \tag{17}$$

However, in these arguments the integral kernel $h(t,p)$ does not generally belong to H_K as a function of p and therefore (17) is generally not valid.

For this reason, we shall realize the norm in H_K in terms of a σ-finite positive measure $d\mu$ in the form

$$\|f\|_{H_K}^2 = \int_E |f(p)|^2 d\mu(p).$$

Then, for some suitable exhaustion $\{E_N\}$ of E, we obtain, in general, the inversion formula

$$F^*(t) = s - \lim_{N\to\infty} \int_{E_N} f(p)h(t,p)d\mu(p) \tag{18}$$

in the sense of strong convergence in $L_2(T, dm)$.

Note that F^* is a member of the visible component of $L_2(T, dm)$ which is the orthocomplement of the null space (the invisible component)

$$\left\{F_0 \in L_2(T, dm); \int_T F_0(t)\overline{h(t,p)}dm(t) = 0 \quad \text{on} \quad E\right\}$$

of $L_2(T, dm)$. Therefore, our inversion formula (16) will be considered as a very natural one.

By our Theorem 3, for example, in Example (5) we can establish the inversion formulas

$$u(x,t) \longrightarrow F(x)$$

and

$$u(x,t) \longrightarrow F(x)$$

for any fixed $t > 0$.

Our inversion formula will present a new viewpoint and a new method for Fredholm integral equations of the first kind which are fundamental in the theory of integral equations. The characteristics of our inversion formula are as follows:

(i) Our inversion formula is given in terms of the reproducing kernel Hilbert space H_K which is intuitively determined as the image space of the integral transform (1).
(ii) Our inversion formula gives the visible component F^* of F with the minimum $L_2(T, dm)$ norm.
(iii) The inverse F^* is, in general, given in the sense of strong convergence in $L_2(T, dm)$.
(iv) Our integral equation (1) is, in general, an ill-posed problem, but our solution F^* is given as the solution of a well-posed problem in the sense of Hadamard (1902, 1923) ([H1-2]).

At this point, we can see why we meet ill-posed problems; that is, because we do not consider the problems in the natural image spaces H_K, but in some artificial spaces.

Nowday it is considered that the general theory of integral equations of the first kind has not been formed yet ([[Bi]], Preface and see also [[Gr]]). Our method will give a general theory for the integral equations in the framework of Hilbert spaces.

For a general reference and for a historical background to integral equations, see [[Co]] with its references. See also [[Kon]], [[Tri]] and [[Yo]].

DETERMINATION OF THE SYSTEM BY INPUT AND OUTPUT FUNCTIONS

In our **Theorem 4,** we can construct the integral kernel $h(t,p)$ conversely, in terms of the isometrical mapping $\widetilde{L}$ from a reproducing kernel Hilbert space H_K onto $L_2(T, dm)$ and the reproducing kernel $K(p,q)$, in the form

$$h(t,p) = \widetilde{L}K(\cdot, p), \tag{19}$$

which were discussed in Chapter 2, Section 5 of [[Sa]].

GENERAL APPLICATIONS

Our basic assumption for the integral transform (1) is (3). When this assumption is not valid, we will be able to apply the following techniques to comply with our assumption (3).

(a) We restrict the sets E or T, or we exchange the set E.
(b) We multiply a positive continuous function ρ in the form $L_2(T, \rho dm)$.

For example, in the Fourier transform

$$\int_{-\infty}^{\infty} F(t) e^{-itx} dt, \tag{20}$$

we consider the integral transform with the weighted function such that

$$\int_{-\infty}^{\infty} F(t) e^{-itx} \frac{dt}{1+t^2}.$$

(c) We integrate the integral kernel $h(t,p)$.

For example, in the Fourier transform (20), we consider the integral transform

$$\int_{-\infty}^{\infty} F(t) \left(\int_0^{\hat{x}} e^{-itx} dx \right) dt = \int_{-\infty}^{\infty} F(t) \left(\frac{e^{-it\hat{x}} - 1}{-it} \right) dt.$$

By these techniques we can apply our general method even for integral transforms with integral kernels of generalized functions. Furthermore, for the integral transforms with the integral kernels of

miscellaneous Green's functions,
Cauchy's kernel,

and

Poisson's kernel

and even for cases of the Fourier transform and the Laplace transform, we were able to derive novel results. We gave the miscellaneous isometrical identities and inversion formulas in Chapter 3 of [[Sa2]].

Recall the Whittaker–Kotel'nikov–Shannon sampling theorem:
In the integral transform

$$f(t) = \frac{1}{\sqrt{2\pi}} \int_{-\pi}^{\pi} F(\omega) e^{i\omega t} d\omega$$

for functions $F(\omega)$ satisfying

$$\int_{-\pi}^{\pi} |F(\omega)|^2 d\omega < \infty,$$

we have the expression

$$f(t) = \sum_{n=-\infty}^{\infty} f(n) \frac{\sin \pi(n-t)}{\pi(n-t)} \quad \text{on} \quad (-\infty, \infty).$$

All the signals $f(t)$ are expressible in terms of the discrete data $f(n)$ (n: integers) only, therefore many scientists are interested in this theorem. Thus, this theorem is applied in miscellaneous fields. Furthermore, very interesting relationships between fundamental theorems and formulas of signal analysis, of analytic number theory and of applied analysis have been described recently in [Klu].

In our general situation (1), the essence of the sampling theorem is given clearly and simply as follows:

For a sequence of points $\{p_n\}$ of E, if $\{h(t, p_n)\}_n$ is a complete orthonormal system in $L_2(T, dm)$, then for any $f \in H_K$, we have the sampling theorem

$$f(p) = \sum_n f(p_n) K(p, p_n) \quad \text{on} \quad E.$$

These results and related topics were discussed in Chapter 4, Section 2.1 of [[Sa2]], in detail. See also [[Hi]], [Hig] and [J] for sampling theory.

Meanwhile, the theory of wavelets which was created by [MAFG] and [Mo] about fourteen years ago is developing rapidly in both the mathematical sciences and pure mathematics. See [[D]] and [[Chu]] for the details. The theory is applicable to signal analysis, numerical analysis and many other fields, as in Fourier transforms. Since the theory is that of integral transforms in the framework of Hilbert spaces, our general theory for integral transforms will be applicable to wavelet theory, globally, and in particular, our method will give a good unified understanding of the wavelet transform, frames, multiresolution analysis and sampling theory in the theory of wavelets. For the typical Meyer wavelets, we examined the isometrical identities and inversion formulas in Chapter 3, Section 5 of [[Sa2]].

ANALYTIC EXTENSION FORMULAS

The equality of the two integrals (11) and (12) means that a C^∞ function $g(x)$ with a finite integral

$$\sum_{j=0}^{\infty} \frac{(2t)^j}{j!} \int_{-\infty}^{\infty} |\partial_x^j g(x)|^2 dx < \infty,$$

is extended analytically onto $\mathbb{C}$ and when we denote its analytic extension by $g(z)$, we have the identity

$$\sum_{j=0}^{\infty} \frac{(2t)^j}{j!} \int_{-\infty}^{\infty} |\partial_x^j g(x)|^2 dx = \frac{1}{\sqrt{2\pi t}} \iint_{\mathbb{C}} |g(z)|^2 \exp\{-\frac{y^2}{2t}\} dx dy. \tag{21}$$

In this way, we have derived miscellaneous analytic extension formulas in over 15 papers with H. Aikawa and N. Hayashi, and the analytic extension formulas are applied to the investigation of analyticity of solutions of nonlinear partial differential equations. See, for example [HS1], [HS2], [Hay1], [Hay2], [HK] and [BHK].

One typical result of another type is obtained from the integral transform

$$v(x,t) = \frac{1}{t}\int_0^t F(\xi)\frac{x\exp[\frac{-x^2}{4(t-\xi)}]}{2\sqrt{\pi}(t-\xi)^{\frac{3}{2}}}\xi d\xi$$

in connection with the heat equation

$$u_t(x,t) = u_{xx}(x,t) \quad \text{for} \quad u(x,t) = tv(x,t)$$

satisfying the conditions

$$u(0,t) = tF(t) \quad \text{on} \quad t \geqq 0$$

and

$$u(x,0) = 0 \quad \text{on} \quad x \geqq 0.$$

Then, we obtain:

Let $\Delta(\frac{\pi}{4})$ denote the sector $\{|\arg z| < \frac{\pi}{4}\}$. For any analytic function $f(z)$ on $\Delta(\frac{\pi}{4})$ with a finite integral

$$\iint_{\Delta(\frac{\pi}{4})} |f(z)|^2 dxdy < \infty,$$

we have the identity

$$\iint_{\Delta(\frac{\pi}{4})} |f(z)|^2 dxdy = \sum_{j=0}^{\infty} \frac{2^j}{(2j+1)!}\int_0^{\infty} x^{2j+1}|\partial_x^j f(x)|^2 dx. \tag{22}$$

Conversely, for any smooth function $f(x)$ with a finite integral in (22) on $(0,\infty)$, there exits an analytic extension $f(z)$ onto $\Delta(\frac{\pi}{4})$ satisfying (22) ([AHS]).

We discussed miscellaneous analytic extension formulas in Chapter 3, Sections 1, 2, 4, 5 and Chapter 5, Section 1 of [[Sa2]].

BEST APPROXIMATION FORMULAS

As shown, when we consider linear transforms in the framework of Hilbert spaces, we naturally have the idea of reproducing kernel Hilbert spaces. As a natural extension of our theorems, we have the fundamental theorems for approximations of functions in the framework of Hilbert spaces.

For a function F on a set X, we shall look for a function which is nearest to F among some family of functions $\{f\}$. In order to formulate the *'nearest'* precisely, we shall consider F as a member of some Hilbert space $H(X)$ comprising functions on X. Meanwhile, as the family $\{f\}$ of approximation functions, we shall consider some reproducing kernel Hilbert space H_K comprising functions f on, in general, a set E containing the set X. Here the reproducing kernel Hilbert space H_K as a family of approximation functions will be considered as a natural one, since the point evaluation $f(p)$ is continuous on H_K.

We shall assume that for the relation between the two Hilbert spaces $H(X)$ and H_K:

$$\text{for the restriction } f|_X \text{ of the members } f \text{ of } H_K \text{ to the set } X, \quad f|_X \text{ belongs to the Hilbert space } H(X), \tag{23}$$

and

$$\text{the linear operator } Lf = f|_X \text{ is continuous from } H_K \text{ into } H(X). \tag{24}$$

In this natural situation, we can discuss the best approximation problem

$$\inf_{f \in H_K} \|Lf - F\|_{H(X)}$$

for a member F of $H(X)$.

For the sake of the nice properties of the restriction operator L and its adjoint L^*, we can obtain *'algorithms'* to decide whether best approximations f^* of F in the sense of

$$\inf_{f \in H_K} \|Lf - F\|_{H(X)} = \|Lf^* - F\|_{H(X)}$$

exist. Further, when the best approximations f^* exist, we can obtain constructive *'algorithms'* for them. Moreover, we can give the representations of f^* in terms of the given function F and the reproducing kernel $K(p,q)$. Meanwhile, when the best approximations f^* do not exist, we can construct the minimizing or approximating sequence $\{f_n\}$ of H_K satisfying

$$\inf_{f \in H_K} \|Lf - F\|_{H(X)} = \lim_{n \to \infty} \|Lf_n - F\|_{H(X)}.$$

As an example, for an $L_2(\mathbb{R}, dx)$ function $h(x)$, we shall approximate it by the family of functions $u_F(x,t)$ for any fixed $t > 0$ which are the solutions of the heat equation (6) with (7) for $F \in L_2(\mathbb{R}, dx)$.

Then, we can see that a member F of $L_2(\mathbb{R}, dx)$ exists such that

$$u_F(x,t) = h(x) \quad \text{on} \quad \mathbb{R}$$

if and only if

$$\iint_{\mathbb{C}} \left| \int_{-\infty}^{\infty} h(\xi) \exp\left\{-\frac{\xi^2}{8t} + \frac{\xi z}{4t}\right\} d\xi \right|^2 \exp\left\{\frac{-3x^2 + y^2}{12t}\right\} dxdy < \infty.$$

If this condition is not satisfied for h, then we can construct the sequence $\{F_n\}$ satisfying

$$\lim_{n \to \infty} \int_{-\infty}^{\infty} |u_{F_n}(x,t) - h(x)|^2 dx = 0;$$

that is, for any $L_2(\mathbb{R}, dx)$ function $h(x)$, we can construct the initial functions $\{F_n\}$ whose heat distributions $u_{F_n}(x,t)$ of time t later converge to $h(x)$.

We discussed these problems in Chapter 4 of [[Sa2]] based on [BS2, 3, 4].

APPLICATIONS TO RANDOM FIELDS ESTIMATIONS

We assume that the random field is of the form

$$u(x) = s(x) + n(x),$$

where $s(x)$ is the useful signal and $n(x)$ is noise. Without loss of generality, we can assume that the mean values of $u(x)$ and $n(x)$ are zero. We assume that the covariance functions

$$R(x,y) = \overline{u(x)u(y)}$$

and

$$f(x,y) = \overline{u(x)s(y)}$$

are known. We shall consider the general form of a linear estimation $\hat{u}$ of u in the form

$$\hat{u}(x) = \int_T u(t)h(x,t)dm(t)$$

for an $L_2(T, dm)$ space and for a function $h(x,t)$ belonging to $L_2(T, dm)$ for any fixed $x \in E$. For the desired information As for a linear operator A of s, we wish to determine the function $h(x,t)$ satisfying

$$\inf \overline{(\hat{u} - As)^2}$$

which gives the minimum of the variance by the least squares method. Many topics in filtering and estimation theory in signal and image processing, underwater acoustics, geophysics, optical filtering, etc., which were initiated by N. Wiener (1894–1964), will be presented in this framework. Then, we see that the linear transform $h(x,t)$ is given by the integral equation

$$\int_T R(x',t)h(x,t)dm(t) = f(x',x)$$

[[Ra1]]. Therefore, our random fields estimation problems will be reduced to finding the inversion formula

$$f(x', x) \longrightarrow h(x, t)$$

in our framework. So, our general method for integral transforms will be applied to these problems. For this situation and other topics and methods for the inversion formulas, see [[Ra1]] for details.

APPLICATIONS TO INVERSE PROBLEMS

Inverse problems will be considered as the problems of determining unobservable quantities from observable quantities. These problems are miscellaneous and are, in general, difficult. In many cases, the problems are reduced to certain Fredholm integral equations of the first kind and then our method will be applicable to these equations. Meanwhile, in many cases, the problems will be reduced to determining the inverse F^* from the data $f(p)$ on some subset of E in our integral transform (1). See, for example, [[Ra2]] and [[Gr]].

In each case, we shall state a typical example.

We shall consider the Poisson equation

$$\Delta u = -\rho(\mathbf{r}) \quad \text{on} \quad \mathbb{R}^3 \tag{25}$$

for a real-valued $L_2(\mathbb{R}^3, d\mathbf{r})$ source function ρ whose support is contained in a sphere $r < a(|\mathbf{r}| = r)$. By using our method for the integral transform

$$u(\mathbf{r}) = \frac{1}{4\pi} \int_{r'<a} \frac{1}{|\mathbf{r} - \mathbf{r}'|} \rho(\mathbf{r}') d\mathbf{r}',$$

we can obtain the characteristic property and natural representation of the potential u on the outside of the sphere $\{r < a\}$. Furthermore, we can obtain the surprisingly simple representation of ρ^* in terms of u on any sphere (a', θ', φ') $(a < a')$, which has the minimum $L_2(\mathbb{R}^3, dr)$ norm among ρ satisfying (25) on $r > a$, in the form:

$$\begin{aligned}
\rho^*(r, \theta, \varphi) = & \frac{1}{4\pi} \sum_{n=0}^{\infty} \frac{(2n+1)^2(2n+3)}{a^{2n+3}} r^n a'^{n+1} \\
& \times \sum_{m=0}^{n} \frac{\varepsilon_m (n-m)!}{(n+m)!} P_n^m(\cos\theta) \int_0^{\pi} \int_0^{2\pi} u(a', \theta', \varphi') \\
& \times P_n^m(\cos\theta') \cos m(\varphi' - \varphi) \sin\theta' d\theta' d\varphi'.
\end{aligned}$$

Here, ε_m is the Neumann factor $\varepsilon_m = 2 - \delta_{m0}$.

In Chapter 6 of [[Sa2]], we examined inverse source problems in the Poisson and the Helmholtz equations. For the details and the related topics, see also [Sa6], [[Che]], [[I]] and [[Bert]].

Next, we shall consider an analytical real inversion formula for the Laplace transform

$$f(p) = \int_0^\infty e^{-pt} F(t) dt, \quad p > 0;$$
$$\int_0^\infty |F(t)|^2 dt < \infty.$$

For the polynomial of degree $2N+2$

$$P_N(\xi) = \sum_{0 \leqq \nu \leqq n \leqq N} \frac{(-1)^{\nu+1}(2n)!}{(n+1)!\nu!(n-\nu)!(n+\nu)!} \xi^{n+\nu} \cdot \left\{ \frac{2n+1}{n+\nu+1} \xi^2 - \left(\frac{2n+1}{n+\nu+1} + 3n + 1 \right) \xi + n(n+\nu+1) \right\},$$

we set

$$F_N(t) = \int_0^\infty f(p) e^{-pt} P_N(pt) dp.$$

Then, we have

$$\lim_{N \to \infty} \int_0^\infty |F(t) - F_N(t)|^2 dt = 0.$$

Furthermore, an estimation of the error of $F_N(t)$ is also given, based on [BS1].

In Chapter 5 of [[Sa2]], we derived this formula in a more general form.

The application to the stability of Lipschitz type in the determination of initial heat distribution of the real inversion formula of the Laplace transform was discussed in Appendix 3 of [[Sa2]] based on [SY1].

Compare our formula with ([BW], 1940 and [[Wi]], Chapter VII), and with ([[Ra2]], p.221, 1992):

For the Laplace transform

$$\int_0^b e^{-pt} F(t) dt = f(p),$$

we have

$$F(t) = \frac{2tb^{-1}}{\pi} \frac{d}{du} \int_0^u \frac{G(v)}{(u-v)^{\frac{1}{2}}} dv \bigg|_{u=t^2b^{-2}};$$
$$G(v) = v^{-\frac{1}{2}} \frac{2}{\pi} \int_0^\infty dy \cos(y \cosh^{-1} v^{-1}) \cosh \pi y \times \int_0^\infty dz \cos(zy)(\cosh z)^{-\frac{1}{2}} \int_0^\infty dp f(p) J_0 \left(p \frac{b}{(\cosh z)^{\frac{1}{2}}} \right).$$

Unfortunately, in this very complicated formula, the characteristic properties of both the functions F and f making the inversion formula hold are not given.

Some characteristics of the strong singularity of the polynomial $P_N(\xi)$ as $N \to \infty$ and some effective algorithms for the real inversion formula are examined by [KT1, 2] and [T] for computers.

NONHARMONIC TRANSFORMS

In our general transform (1), suppose that $\varphi(t,p)$ is near to the integral kernel $h(t,p)$ in the following sense:

For any $F \in L_2(T, dm)$,

$$\left\| \int_T F(t)\overline{(h(t,p) - \varphi(t,p))}dm(t) \right\|_{H_K}^2 \leqq \omega^2 \int_T |F(t)|^2 dm(t)$$

where $0 < \omega < 1$ and ω is independent of $F \in L_2(T, dm)$.

Then, we can see that for any $f \in H_K$, there exists a function F_φ^* belonging to the visible component of $L_2(T, dm)$ in (1) such that

$$f(p) = \int_T F_\varphi^*(t)\overline{\varphi(t,p)}dm(t) \quad \text{on} \quad E \tag{26}$$

and

$$(1-\omega)^2 \int_T |F_\varphi^*(t)|^2 dm(t) \leqq \|f\|_{H_K}^2$$
$$\leqq (1+\omega)^2 \int_T |F_\varphi^*(t)|^2 dm(t).$$

The integral kernel $\varphi(t,p)$ will be considered as a perturbation of the integral kernel $h(t,p)$. When we look for the inversion formula of (26) following our general method, we must calculate the kernel form

$$K_\varphi(p,q) = \int_T \varphi(t,q)\overline{\varphi(t,p)}dm(t) \quad \text{on} \quad E \times E.$$

We will, however, in general, not be able to calculate this kernel.

Suppose that the image $f(p)$ of (26) belongs to the known space H_K. Then, we can construct the inverse F_φ^* by using our inversion formula in H_K repeatedly and by constructing some approximation of F_φ^* by our inverses.

In particular, for the reproducing kernel $K(p,q) \in H_K (q \in E)$ we construct (or we obtain, by some other method or directly) the function $\hat{\varphi}(t,p)$ satisfying

$$K(p,q) = \int_T \hat{\varphi}(t,q)\overline{\varphi(t,p)}dm(t) \quad \text{on} \quad E \times E,$$

where $\hat{\varphi}(t,q)$ belongs to the visible component of $L_2(T, dm)$ in (26) for any fixed $q \in E$. Then, we have the idea of a *'nonharmonic integral transform'* and we can formulate the inversion formula of (26) in terms of the kernel $\hat{\varphi}(t, q)$ and the space H_K, globally ([[Sa1]], Chapter 7).

NONLINEAR TRANSFORMS

Our generalized isoperimetric inequality will mean that for an analytic function $\varphi(z)$ on $\overline{G}$ satisfying

$$\int_{\partial G} |\varphi(z)|^2 |dz| < \infty,$$

the image of the simplest nonlinear transform

$$\varphi(z) \longrightarrow \varphi(z)^2$$

belongs to the space of analytic functions satisfying

$$\iint_G |\varphi(z)|^4 dxdy < \infty$$

and we have the norm inequality

$$\frac{1}{\pi} \iint_G |\varphi(z)^2|^2 dxdy \leqq \left\{ \frac{1}{2\pi} \int_{\partial G} |\varphi(z)|^2 |dz| \right\}^2 .$$

We established in Theorem 1 the method of identification of the images of linear transforms, and we will also be able to look for some Hilbert spaces containing the image spaces of some general nonlinear transforms. In these cases, however, the spaces will be too large for the image spaces, as in the generalized isoperimetric inequality. So, the inversion formulas for nonlinear transforms will be, in general, extremely involved. However, in many nonlinear transforms of reproducing kernel Hilbert spaces, some norm inequalities exist, as in the generalized isoperimetric inequality. We shall state one example in the strongly nonlinear transform

$$f(x) \quad \longrightarrow \quad e^{f(x)}$$

which was recently obtained in [Sa5]:

For an absolutely continuous real-valued function f on $[a, b)$ $(a > 0)$ satisfying

$$f(a) = 0,$$
$$\int_a^b f'(x)^2 x dx < \infty,$$

we obtain the inequality

$$1 + a \int_a^b |(e^{f(x)})'|^2 dx \leqq e^{\int_a^b f'(x)^2 x dx}.$$

Here, we should note that the equality holds for many functions $f(x)$.

For nonlinear transforms, we shall need other treatments different from linear transforms and so we referred to nonlinear transforms in Appendix 2 of [[Sa2]] based on [Sa7].

REPRESENTATIONS OF INVERSE FUNCTIONS

By considering transforms of reproducing kernels we can obtain some general principles for representations of the inverse φ^{-1} for an arbitrary mapping $p = \varphi(\hat{p})$

$$\varphi : \widehat{E} \longrightarrow E$$

from an abstract set $\widehat{E}$ into an abstract set E. Of course, the inverse φ^{-1} is, in general, multivalued. We can, however, obtain representations of φ^{-1} in terms of φ for many concrete mappings φ by a unified principle. For example, we have the expression

$$\sqrt[n]{x} = \frac{2}{\pi}\int_0^\infty \int_0^\infty \frac{\cos(\xi^n t)\sin xt}{t} dt d\xi$$

which was given in Appendix 1, (31) of [[Sa2]] based on [Sa8].

LINEAR TRANSFORMS BY SMOOTH FUNCTIONS

For a general linear transform (1) in the framework of Hilbert spaces we shall consider (1) for some smooth functions F. Here, for 'some smooth' functions we shall consider them as members of some reproducing kernel Hilbert space.

For a typical example, compare the usual Hilbert space $L_2(a,b)$ which does not admit a reproducing kernel with the Sobolev Hilbert space $H^1(a,b)$ comprising absolutely continuous functions on (a,b) and equipped with the inner product

$$(f,g)_{H^1(a,b)} = \int_a^b f(x)\overline{g(x)}dx + \int_a^b f'(x)\overline{g'(x)}dx,$$

admitting a reproducing kernel.

For various reproducing kernel Hilbert spaces, see Chapter 1, §3 of [[Sa2]].

In Chapter 1, §1, 5 of [[Sa2]], we examined the linear transform (1) for a reproducing kernel Hilbert space and we established the corresponding isometrical identity and inversion formula based on [SY2].

LINEAR INTEGRO-DIFFERENTIAL EQUATIONS AND REPRODUCING KERNELS

We shall formulate Volterra-type integral equations of the first kind in the following way:

$$\int_a^t F(\xi)h(\xi,t)d\xi = f(t). \tag{27}$$

We shall assume that

$$F \in L_2(a,\infty) \tag{28}$$

and

$$h(\cdot,t) \in L_2(a,\infty) \quad for \quad t > a, \tag{29}$$

in the framework of Hilbert spaces. Then, note that by using the characteristic function

$$\chi(\xi;(a,t)) = \begin{cases} 1 & \text{on} \quad (a,t) \\ 0 & \text{on} \quad t < \xi, \end{cases}$$

the integral equation (27) of Volterra type can be reduced to the integral equation of Fredholm type as follows:

$$\int_a^\infty F(\xi)h(\xi,t)\chi(\xi;(a,t))d\xi = f(t). \tag{30}$$

As examples, see Theorem 3.3.4 and Theorem 3.3.10 in [[Sa2]].

In general, integral equations of Volterra type can be transformed to those of Fredholm type by using the characteristic function as in (30) from the viewpoint of our fundamental theorems of linear transforms.

Next, we shall formulate Fredholm-type integral equations of the second kind in the form

$$F(t) + \int_T F(\xi)\overline{h(\xi,t)}dm(\xi) = f(t), \tag{31}$$

where we shall assume that F belongs to some reproducing kernel Hilbert space $H_{\mathbb{K}}$ on E and for the reproducing kernel $\mathbb{K}(t,t')$

$$\mathbb{K}(t,t') = \int_{\hat{E}} \hat{h}(\hat{\xi},t')\overline{\hat{h}(\hat{\xi},t)}d\hat{m}(\hat{\xi}), \quad \text{on} \quad E \times E \tag{32}$$

and

$$\left\{\hat{h}(\hat{\xi},t); t \in E\right\} \quad \text{is complete in} \quad L_2(\hat{E},d\hat{m}). \tag{33}$$

Further we assume that

$$h(\cdot,t) \in L_2(T,dm) \quad \text{on} \quad E$$

and

$$H_{\mathbb{K}} \subset L_2(T,dm).$$

Then, any member $F \in H_{\mathbb{K}}$ is expressible in the form

$$F(t) = \int_{\hat{E}} \hat{F}(\hat{\xi})\overline{\hat{h}(\hat{\xi},t)}d\hat{m}(\hat{\xi}) \tag{34}$$

and we have the isometrical identity

$$\|F\|^2_{H_{\mathbb{K}}} = \int_{\hat{E}} |\hat{F}(\hat{\xi})|^2 d\hat{m}(\hat{\xi}). \tag{35}$$

Now, from (31) and (34) we have

$$\int_{\hat{E}} \hat{F}(\hat{\xi}) \left\{ \overline{\hat{h}(\hat{\xi},t)} + \int_T \overline{\hat{h}(\hat{\xi},\xi)}\, \overline{h(\xi,t)} dm(\xi) \right\} d\hat{m}(\hat{\xi}) = f(t), \tag{36}$$

where we assume that the order of integrals is exchangeable. If

$$\int_T \mathbb{K}(t,t) dm(t) < \infty,$$

then this assumption will be satisfied by Fubini's theorem.

We thus see that following this procedure integral equations of the second kind can be transformed to those of the first kind. The circumstances are similar for integral equations of Fredholm type of the third kind. We shall state the procedure in the following more general integro-differential equations.

We shall formulate integro-differential equations as follows. For real intervals T and E,

$$\begin{aligned} a_0(t)F(t) + a_1(t)F'(t) + \cdots + a_n(t)F^{(n)}(t) \\ + \int_T F(\xi)\overline{h(\xi,t)} dm(\xi) = f(t), \quad \text{on} \quad E \end{aligned} \tag{37}$$

where

$$h(\cdot,t) \in L_2(T,dm) \quad \text{for} \quad t \in E,$$

and $\{a_j\}_{j=0}^n$ are *arbitrary* complex-valued functions on E.

From the form (37), we shall assume that F belongs to some reproducing kernel Hilbert space $H_{\mathbb{K}}$ on E satisfying (32), (33) (and so, satisfying (34) and (35)). From (37), we shall assume furthermore that

$$\frac{\partial^{j+j'} \mathbb{K}(t,t')}{\partial t^j \partial t'^{j'}} \quad (j,j' = 0,1,2,\cdots,n) \quad \text{on} \quad E \times E \tag{38}$$

are continuously differentiable. Then, we see that any member F of $H_{\mathbb{K}}$ belongs to the C^n-class and we have the expression

$$F^{(j)}(t) = \int_{\hat{E}} \hat{F}(\hat{\xi}) \overline{\frac{\partial^j}{\partial t^j} \hat{h}(\hat{\xi},t)} d\hat{m}(\hat{\xi}), \quad \text{on} \quad E. \tag{39}$$

From (37) and (39) we have

$$\begin{aligned} \int_{\hat{E}} \hat{F}(\hat{\xi}) \Big\{ a_0(t)\overline{\hat{h}(\hat{\xi},t)} + \cdots + a_n(t)\overline{\partial_t^n \hat{h}(\hat{\xi},t)} \\ + \int_T \overline{\hat{h}(\hat{\xi},\xi)}\, \overline{h(\xi,t)} dm(\xi) \Big\} d\hat{m}(\hat{\xi}) = f(t), \quad \text{on} \quad E. \end{aligned} \tag{40}$$

Here we assume that the order of the integrals is exchangeable. Again, if

$$\int_E \mathbb{K}(t,t)dm(t) < \infty,$$

it is assured by Fubini's theorem. Our procedure implies that the integro-differential equation (37) can be transformed to the Fredholm-type integral equation of the first kind (Chapter 2, §1, 6) based on ([Sa9]).

The difficulty of solving the integro-differential equation (37) with variable coefficients will be transformed to that of the complicated form in the integral kernel in (40). In the integral equation (40) of the first kind we can apply also various numerical constructions of the solutions. See [[Gr]], [[TA]], [NW1-3], [Stg1-2] and [W1-2].

References

Books

[[Bert]]. Bertero, M., *Linear inverse and ill-posed problems*, INFNITC-8812, 19, Gennaio (1988).

[[Bi]]. Bitsadze, A. V., *Integral equations of first kind*, World Scientific, Singapore (1995).

[[Che]]. Cherednichenko, V. G., *Inverse Logarithmic Potential Problem*, VSP, Utrecht, The Netherlands (1996).

[[Chu]]. Chui, C. K., *An introduction to wavelets*, Academic Press, New York (1992).

[[Co]]. Corduneanu, C., *Integral equations and applications*, Cambridge University Press. (1991).

[[D]]. Daubechies, I., *Ten lectures on wavelets*, Society for Industrial and Applied Mathematics, Philadelphia (1992).

[[Gr]]. Groetsch, C. W., *Inverse problems in the mathematical sciences*, Vieweg Mathematics for Scientists and Engineers, Vieweg (1993).

[[He]]. Hejhal, D. A., *Theta functions, kernel functions and Abel integrals*, Memoirs of Amer. Math. Soc., Providence, R. I. **129** (1972).

[[Hi]]. Higgins, J. R., *Sampling theory in Fourier and signal analysis foundations*, Oxford Science Publications (1996).

[[I]]. Isakov, V., *Inverse source problems*, Mathematical Surveys and Monographs, Number 34, American Mathematical Society, Providence, R. I. (1990).

[[Kon]]. Kondo, J., *Integral equations*, Kodanshia, Tokyo (1991) and Clarendon Press, Oxford (1991).

[[Ra]]. Ramm, A.G.

1. *Random fields estimation theory*, Pitman Monographs and Surveys in Pure and Applied Mathematics, **48**, Longman Scientific & Technical, UK (1990).

2. *Multidimensional inverse scattering problems*, Pitman Monographs and Surveys in Pure and Applied Mathematics, **51**, Longman Scientific & Technical, UK (1992).

[[Sa]]. Saitoh, S.

1. *Theory of reproducing kernels and its applications*, Pitman Research Notes in Mathematics Series, **189**, Longman Scientific & Technical, UK (1988).
2. *Integral transforms, reproducing kernels and their applications*, Pitman Research Notes in Mathematics Series, **369**, Addison Wesley Longman, UK (1997).

[[TA]]. Tikhonov, A. N., Arsenin, V. Y., *Solution of ill-posed problems*, John Wiley & Sons, Inc., New York (1977).

[[Tr]]. Treves, F., *Topological vector spaces, distributions and kernels*, Academic Press, New York (1967).

[[Tri]]. Tricomi, F. G., *Integral equations*, Pure and Applied Mathematics Vol. 5, New York (1957).

[[Wi]]. Widder, D. V., *The Laplace transform*, Princeton University Press, Princeton, N. J. (1972).

[[Yo]]. Yoshida, K., *Lectures on differential and integral equations*, Pure and Applied Mathematics, Interscience Publishers, New York (1960).

Articles

[AHS]. Aikawa, H., Hayashi, N., Saitoh, S., *The Bergman space on a sector and the heat equation*, Complex Variables **15** (1990), 27-36.

[Ar]. Aronszajn, N., *Theory of reproducing kernels*, Trans. Amer. Math. Soc. **68** (1950), 337-404.

[BHK]. Bouard, A. de, Hayashi, N., Kato, K., *Gevrey regularizing effect for the (generalized) Korteweg-de Vries equation and nonlinear Schrödinger equations*, Ann. Henri Inst. Poincare Analyse nonlinear **12** (1995), 673-725.

[BS]. Byun, D.-W., Saitoh, S.

1. *A real inversion formula for the Laplace transform*, Zeitschrift für Analysis und ihre Anwendungen **12** (1993), 597-603.
2. *Approximation by the solutions of the heat equation*, J. Approximation Theory **78** (1994), 226-238.
3. *Best approximation in reproducing kernel Hilbert spaces*, Proc. of the 2th International Colloquium on Numerical Analysis, VSP-Holland (1994), 55-61.
4. *Analytic extensions of functions on the real line to entire functions*, Complex Variables **26** (1995), 277-281.

[BW]. Boas, R. P., Widder, D. V., *An inverse formula for the Laplace integral*, Duke Math. J. **6** (1940), 1-26.

[H]. Hadamard, J.

1. *Sur les problème aux dérivées partielles et leur signification physique*, Bull. Univ. Princeton **13** (1902).

2. *Lectures on Cauchy's Problems in Linear Partial Differential Equations*, New Haven, Yale University Press. **348** (1923).

[Hay]. Hayashi, N.

1. *Global existence of small analytic solutions to nonlinear Schrödinger equations*, Duke Math. J. **60** (1990), 717-727.
2. *Solutions of the (generalized) Korteweg-de Vries equation in the Bergman and the Szegö spaces on a sector*, Duke Math. J. **62** (1991), 575-591.

[Hig]. Higgins, J. R., *Five short stories about the cardinal series*, Bull. of Amer. Math. Soc. **12** (1985), 45-89.

[HK]. Hayashi, N., Kato, K., *Regularity of solutions in time to nonlinear Schrödinger equations*, J. Funct. Anal. **128** (1995), 255-277.

[HS]. Hayashi, N., Saitoh, S.

1. *Analyticity and smoothing effect for the Schrödinger equation*, Ann. Inst. Henri Poincaré **52** (1990), 163-173.
2. *Analyticity and global existence of small solutions to some nonlinear Schrödinger equation*, Commun. Math. Phys. **139** (1990), 27-41.

[J]. Jerri, A. J., *The Shannon sampling theorem – its various extensions and applications*, a tutorial review, Proceedings of the IEEE **65** (1977), 1565-1596.

[Klu]. Klusch, D., *The sampling theorem, Dirichlet series and Hankel transforms*, J. of Computational and Applied Math. **44** (1992), 261-273.

[KT]. Kajiwara, J., Tsuji, M.

1. *Program for the numerical analysis of inverse formula for the Laplace transform*, Proceedings of the Second Korean-Japanese Colloquium on Finite or Infinite Dimensional Complex Analysis, (1994), 93-107.
2. *Inverse formula for Laplace transform*, Proceedings of the 5th International Colloquium on Differential Equations, VSP-Holland (1995), 163-172.

[MAFG]. Morlet, J., Arens, G., Fourgeau, I., Giard, D., *Wave propagation and sampling theory*, Geophysics **47** (1982), 203-236.

[Mo]. Morlet, J., *Sampling theory and wave propagation*, C. H. Chen, ed., Issues in Acoustic Signal Image Processing and Recognition, NATO ASI Series, Vol. 1, Springer-Verlag, Berlin (1983), 233-261.

[NW]. Nashed, M. Z., Wahba, G.

1. *Convergence rates of approximate least squares solutions of linear integral and operator equations of the first kind*, Math. Computation **28** (1974), 69-80.
2. *Regularization and approximation of linear operator equations and reproducing kernel spaces*, Bull. of Amer. Math. Soc. **80** (1974), 1213-1218.
3. *Generalized inverses in reproducing kernel spaces: an approach to regularization of linear operator equations*, SIAM J. Math. Anal. **5** (1974), 974-987.

[Sa]. Saitoh, S.

1. *The Bergman norm and the Szegö norm*, Trans. Amer. Math. Soc. **249** (1979), 261-279.
2. *Integral transforms in Hilbert spaces*, Proc. Japan Acad. **58** (1982), 361-364.

3. *Hilbert spaces induced by Hilbert space valued functions*, Proc. Amer. Math. Soc. **89** (1983), 74-78.
4. *The Weierstrass transform and an isometry in the heat equation*, Applicable Analysis **16** (1983), 1-6.
5. *An integral inequality of exponential type for real-valued functions*, World Scientific Series in Applicable Analysis 3, Inequalities and Applications, World Scientific (1994), 537-541.
6. *Inverse source problems in Poisson's equation*, Suri Kaiseki Kenkyu Jo, Koukyu Roku **890** (1994), 19-30 , J. of Inverse and Ill-posed Problems **5** (1997), 477-486.
7. *Natural norm inequalities in nonlinear transforms*, General Inequalities 7, Birkhäuser Verlag, Basel, Boston (1997).
8. *Representations of inverse functions*, Proc. Amer. Math. Soc. **125** (1997), 3633-3639.
9. *Linear integro-differential equations and the theory of reproducing kernels*, Proceedings of Volterra Centennial Symposium, Marcel Dekker, New York (to appear).

[Schw]. Schwartz, L., *Sous-espaces hilbertiens d'espaces vectoriels topologiques et noyaux associés (noyaux reproduisants)*, J. Analyse Math. **13** (1964), 115-256.

[Stg]. Stenger, F.
1. *Numerical methods based on Whittaker cardinal, or sinc functions*, SIAM Review **23** (1981), 165-225.
2. *Numerical Methods Based on Sinc and Analytic Functions*, Springer-Verlag, New York (1993).

[SY]. Saitoh, S., Yamamoto, M.
1. *Stability of Lipschitz type in determination of initial heat distribution*, J. of Inequalities and Applications **1** (1997), 73-83.
2. *Integral transforms involving smooth functions*, This Proceedings.

[T]. Tsuji, K., *An algorithm for sum of floating-point numbers without rounding error*, In Abstracts of the Third International Colloquium on Numerical Analysis, Bulgaria (1994).

[W]. Wahba, G.
1. *On the approximate solution of Fredholm integral equations of the first kind*, Tech. Summ. Rep. 990, Mathematics Research Center, Univ. of Wisconsin-Madison (1969).
2. *Convergence rates of certain approximate solutions to Fredholm integral equations of the first kind*, J. of Approx. Theory **7** (1973), 167-185.

16 A SURVEY OF THE EXTENDED INTERPOLATION

Sechiko Takahashi

Department of Mathematics, Faculty of Science,
Nara Women's University, Japan
sechiko@cc.nara-wu.ac.jp

Abstract: In this paper, we consider the extended interpolation problem, which combines Schur's coefficient problem and Pick's interpolation problem, and make a survey of our results on this problem, which were already shown.

INTRODUCTION

In 1911, Carathéodory [6] developed the theory of power series with positive real part converging in the open unit disc $D = \{z \in \mathbb{C} : |z| < 1\}$ and Toeplitz [42] supplemented in an important respect. We may say that this is the origin of the theory of bounded analytic functions.

In 1917, replacing the right half plane by $\overline{D}$ by means of a conformal mapping, Schur formulated the theorem in the following form and established, in order to prove it, the so-called Schur's algorithm in each step of which one uses carefully Schwarz's lemma [32].

Let $\mathcal{B}$ denote the set of holomorphic functions f in D such that $|f| \leq 1$.

Schur's coefficient theorem. A function $f(z) = \sum_{n=0}^{\infty} c_n z^n$ belongs to $\mathcal{B}$ if and only if the Hermitian matrix

$$\begin{bmatrix} 1 & & & \\ & 1 & & \\ & & \ddots & \\ & & & 1 \end{bmatrix} - \begin{bmatrix} c_0 & & & \\ c_1 & c_0 & & \\ \vdots & \ddots & \ddots & \\ c_{n-1} & \cdots & c_1 & c_0 \end{bmatrix} \begin{bmatrix} \overline{c}_0 & \overline{c}_1 & \cdots & \overline{c}_{n-1} \\ & \ddots & \ddots & \vdots \\ & & \overline{c}_0 & \overline{c}_1 \\ & & & \overline{c}_0 \end{bmatrix}$$

is positive semidefinite for all $n \in \mathbb{N}$.

On the other hand, in 1915, after restating Schwarz's lemma in terms of the non-euclidien distance in an invariant form under Möbius transformations, Pick [28] showed a crucial interpolation theorem in many versions. We state here his theorem in the disc version.

Pick's interpolation theorem. Let $z_1, \ldots, z_n$ be distinct points in D and let $w_1, \ldots, w_n$ be complex numbers. Then there exists a function $f \in \mathcal{B}$ such that $f(z_i) = w_i$ $(i = 1, \cdots, n)$ if and only if the Hermitian matrix

$$\begin{bmatrix} \dfrac{1 - w_1\overline{w}_1}{1 - z_1\overline{z}_1} & \cdots & \dfrac{1 - w_1\overline{w}_n}{1 - z_1\overline{z}_n} \\ \cdots & \cdots & \cdots \\ \dfrac{1 - w_n\overline{w}_1}{1 - z_n\overline{z}_1} & \cdots & \dfrac{1 - w_n\overline{w}_n}{1 - z_n\overline{z}_n} \end{bmatrix}$$

is posotive semidefinite.

Differently from the viewpoint of Pick's matrix representation, Nevanlinna investigated in 1919 Pick's interpolation problem by means of Schur's algorithm, giving another formulation of the existence theorem [23]. In 1929, he expressed the set of solutions in a explicit manner [24], which is called Nevanlinna parametrization.

In this paper, we consider the extended interpolation problem, which combines Schur's coefficient problem and Pick's interpolation problem, and make a survey of our results on this problem, which were already shown.

First of all, in §1, we shall formulate the extended interpolation problem on Riemann surfaces. This coordinate-free definition is found in [41]. After having prepared in §2 some tools of great use in our studies, we shall deal with the problem in the open unit disc and state our main results obtained in [38] and [39]; notably, the existence theorem in §3, the invariance under Möbius transformations in §4 and the Nevanlinna parametrization in §5. In §6, we shall give a sufficient condition for Nevanlinna parametrizations, which is established in [41] and will be used in §8.

We shall proceed to finitely connected domains and give two generalizations to our extended intepolation problem; in §7, the Abrahamse theorem in finitely connected domains with analytic boundaries, [1] and [40], and in §8, the Heins theorem in doubly connected Riemann surfaces, [19] and [41].

§1. A FORMULATION OF EXTENDED INTERPOLATION

Let X be a Riemann surface, i.e. a connected 1-dimensional complex manifold. For each $x \in X$, $\mathcal{O}_x$ denotes the ring of germs of holomorphic functions at x. Consider for each $x \in X$ a nonzero ideal $\mathcal{I}_x$ of $\mathcal{O}_x$ and an element $\mathfrak{c}_x$ of the quotient ring $\mathcal{O}_x/\mathcal{I}_x$. The collection $(\mathcal{I}, \mathfrak{c})$, where $\mathcal{I} = (\mathcal{I}_x)_{x \in X}$ and $\mathfrak{c} = (\mathfrak{c}_x)_{x \in X}$, will be called *extended interpolation problem* on X. Let $\mathcal{B}$ denote the set of all holomorphic functions f on X such that $|f| \leq 1$ on X. Our problem is to find a function $f \in \mathcal{B}$ which satisfies the condition

$$(*) \qquad f_x + \mathcal{I}_x = \mathfrak{c}_x \qquad (\forall x \in X),$$

where f_x is the germ at x represented by f and $f_x + \mathcal{I}_x$ is the coset of f_x modulo $\mathcal{I}_x$. Such a function will be called *solution* in $\mathcal{B}$ of the problem $(\mathcal{I}, \mathfrak{c})$. Let

$$\mathcal{E} = \{ f \in \mathcal{B} : f \text{ satisfies } (*) \}$$

denote the set of solutions in $\mathcal{B}$ of the problem $(\mathcal{I}, \mathfrak{c})$.

Consider the set $\sigma = \{x \in X : \mathcal{I}_x \neq \mathcal{O}_x\}$. Suppose $x \in \sigma$. Then, as $\{0\} \neq \mathcal{I}_x \neq \mathcal{O}_x$, there is a unique positive integer n_x such that $\mathcal{I}_x = \mathfrak{m}_x^{n_x}$, where $\mathfrak{m}_x$ is the maximal ideal of the local ring $\mathcal{O}_x$. Associating to x a local coordinate z such that $z(x) = 0$, we obtain

$$\mathfrak{c}_x = (c_0 + c_1 z + \cdots + c_{n_x - 1} z^{n_x - 1})_x + \mathcal{I}_x ,$$

where the n_x and the constants $c_0, c_1, \cdots, c_{n_x-1}$ are uniquely determined by $\mathfrak{c}_x$ and z. To give an extended interpolation problem $(\mathcal{I}, \mathfrak{c})$ is thus to give for each $x \in \sigma$ a positive integer n_x, a local coordinate z at x, and first n_x Taylor coefficients $c_0, c_1, \cdots, c_{n_x-1}$ with respect to z.

The case where σ is nonempty and has no limit points in X, i.e. σ is a discrete closed set of X, is of prime importance. For example, when $\mathcal{E}$ has at least two elements, σ has no limit points in X. In such a case, we can express the situation of our problem in terms of sheaves as follows:

$$\mathcal{O} = \bigcup_{x \in X} \mathcal{O}_x : \text{ structure sheaf of } X;$$

$$\mathcal{I} = \bigcup_{x \in X} \mathcal{I}_x : \text{ coherent analytic sheaf of } \mathcal{O};$$

$$\theta : H^0(X, \mathcal{O}) \longrightarrow H^0(X, \mathcal{O}/\mathcal{I}) : \text{ canonical homomorphism};$$

$$\mathfrak{c} \in H^0(X, \mathcal{O}/\mathcal{I});$$

$$f \in \mathcal{E} \Longleftrightarrow f \in H^0(X, \mathcal{O}) \cap \mathcal{B} \text{ and } \theta(f) = \mathfrak{c}.$$

Let X, Y be Riemann surfaces and let $\varphi : X \longrightarrow Y$ be a holomorphic mapping. For each $x \in X$, setting $y = \varphi(x)$, we have the canonical ring homomorphism $\varphi_x^* : \mathcal{O}_y \longrightarrow \mathcal{O}_x$ defined by

$$\varphi_x^*(g_y) = (g \circ \varphi)_x ,$$

where g is a holomorphic function at y on Y. Moreover, when an ideal $\mathcal{I}_x$ of $\mathcal{O}_x$ and an ideal $\mathcal{I}_y$ of $\mathcal{O}_y$ are given in such a way that $\varphi_x^*(\mathcal{I}_y) \subset \mathcal{I}_x$, we have a canonical ring homomorphism $\mathcal{O}_y/\mathcal{I}_y \longrightarrow \mathcal{O}_x/\mathcal{I}_x$, which will be denoted by the same symbol φ_x^*. Let $\theta_x : \mathcal{O}_x \longrightarrow \mathcal{O}_x/\mathcal{I}_x$ be the canonical homomorphism. Then

$$\theta_x \circ \varphi_x^* = \varphi_x^* \circ \theta_y \,.$$

Let Z be a Riemann surface and let $\psi : Y \longrightarrow Z$ be a holomorphic mapping. Put $z = \psi(y)$ and let $\mathcal{I}_z$ be an ideal of $\mathcal{O}_z$ such that $\psi_y^*(\mathcal{I}_z) \subset \mathcal{I}_y$. Then

$$(\psi \circ \varphi)_x^* = \varphi_x^* \circ \psi_y^* \,.$$

If φ is invertible at x, that is, φ is locally homeomorphic at x, then its inverse φ^{-1} is holomorphic at y and we have

$$(\varphi_x^*)^{-1} = (\varphi_y^{-1})^* : \mathcal{O}_x \longrightarrow \mathcal{O}_y$$

and, under the assumption $\varphi_x^*(\mathcal{I}_y) = \mathcal{I}_x$,

$$(\varphi_x^*)^{-1} = (\varphi_y^{-1})^* : \mathcal{O}_x/\mathcal{I}_x \xrightarrow{\sim} \mathcal{O}_y/\mathcal{I}_y \,.$$

Let X and Y be Riemann surfaces and φ be a holomorphic mapping of a neighborhood of x to Y. Put $y = \varphi(x)$ and take local coordinates z and w of X and Y at x and y, with $z(x) = 0$ and $w(y) = 0$, respectively. With respect to these coordinates, φ may be expressed in a convergent Taylor series

$$w = a_\mu z^\mu + a_{\mu+1} z^{\mu+1} + \cdots \qquad (a_\mu \neq 0)\,,$$

where $\mu \geq 1$ is the order of φ at x. Take two positive integer m and n such that $\mu n \geq m$. Then we have $\varphi_x^*(\mathfrak{m}_y^n) \subset \mathfrak{m}_x^m$ and hence the ring homomorphism $\varphi_x^* : \mathcal{O}_y/\mathfrak{m}_y^n \longrightarrow \mathcal{O}_x/\mathfrak{m}_x^m$.

Let $\mathfrak{c} \in \mathcal{O}_y/\mathfrak{m}_y^n$. $\mathfrak{c}$ is represented by a function g holomorphic at y :

$$\mathfrak{c} = g_y + \mathfrak{m}_y^n.$$

With respect to w, g is expressed in its Taylor expansion

$$c_0 + c_1 w + c_2 w^2 + \cdots \,.$$

The first n coefficients $c_0, c_1, \cdots, c_{n-1}$ depend only on $\mathfrak{c}$, but not on the choice of g. The mapping

$$\mathfrak{c} \mapsto {}^t(c_0 \; c_1 \; \cdots \; c_{n-1})$$

gives a linear isomorphism $\rho_y : \mathcal{O}_y/\mathfrak{m}_y^n \longrightarrow \mathbb{C}^n$. Similarly we have a linear isomorphism $\rho_x : \mathcal{O}_x/\mathfrak{m}_x^m \longrightarrow \mathbb{C}^m$. Finally, φ_x^* induces a unique linear mapping $\bar{\varphi}_x : \mathbb{C}^n \longrightarrow \mathbb{C}^m$ such that

$$\bar{\varphi}_x \circ \rho_y = \rho_x \circ \varphi_x^* \,.$$

Calculating the Taylor expansion

$$d_0 + d_1 z + d_2 z^2 + \cdots$$

of $g \circ \varphi$ at x with respect to z, one sees that the linear mapping $\widetilde{\varphi}_x$ is represented by the matrix Ω_{mn} defined as follows:

When $\mu \geq 2$, we put $a_1 = \cdots = a_{\mu-1} = 0$. Take an integer k such that $k \geq m$ and $k \geq n$. Put

$$\Phi = \begin{bmatrix} a_1 & 0 & \cdots & 0 \\ a_2 & a_1 & & \\ \vdots & \ddots & \ddots & 0 \\ a_k & \cdots & a_2 & a_1 \end{bmatrix}.$$

and

$$\Omega = \sum_{\alpha=0}^{k-1} \Phi^\alpha \cdot E_\alpha ,$$

where $\Phi^0 = I_k$(the unit $k \times k$ matrix), $\Phi^\alpha = \Phi^{\alpha-1} \cdot \Phi$ $(\alpha = 1, 2, \cdots)$, and E_α is the $k \times k$ matrix whose $(\alpha+1, \alpha+1)$-entry only is 1 and other entries are all 0. The matrix Ω is of the form

$$\Omega = \begin{bmatrix} 1 & & & & \\ 0 & a_1 & & & \\ \cdot & * & a_1^2 & & \\ \vdots & \vdots & \ddots & \ddots & \\ 0 & * & \cdots & * & a_1^{k-1} \end{bmatrix}.$$

(See the next §2. Φ, E_α and Ω defined here correspond to Φ_k, $E_k^{(\alpha)}$ and $\Omega(\varphi; x; k)$ in §2, respectively.) The required matrix Ω_{mn} is the $m \times n$ matrix composed of the first m rows and the first n columns of Ω. Note that Ω_{mn} does not depend on the choice of k. We have thus

$${}^t(d_0 \; d_1 \; \cdots \; d_{m-1}) = \Omega_{mn} \cdot {}^t(c_0 \; c_1 \; \cdots \; c_{n-1}) .$$

§2. COEFFICIENT MATRICES AND TRANSFORMATION FORMULAS

In the coefficient problem, Schur's triangular matrix was a very powerful tool [32]. In our extended case, we introduce a rectangular matrix M for a function of two variables. This matrix made it possible to unify Schur's coefficient theorem and Pick's interpolation theorem [38].

1. To a function

$$f(z) = \sum_{\alpha=0}^{\infty} c_\alpha (z - z_0)^\alpha$$

holomorphic at z_0 and to a positive integer $n \in \mathbb{N}$, we assign a triangular $n \times n$ matrix

$$\Delta(f; z_0; n) = \begin{bmatrix} c_0 & & & \\ c_1 & c_0 & & \\ \vdots & \ddots & \ddots & \\ c_{n-1} & \cdots & c_1 & c_0 \end{bmatrix}.$$

Let $g(z)$ be another function holomorphic at z_0, we see immediately

$$\begin{aligned} \Delta(f+g; z_0; n) &= \Delta(f; z_0; n) + \Delta(g; z_0; n), \\ \Delta(fg; z_0; n) &= \Delta(f; z_0; n) \cdot \Delta(g; z_0; n) \\ &= \Delta(g; z_0; n) \cdot \Delta(f; z_0; n), \\ \Delta(1; z_0; n) &= I_n \quad \text{(the unit matrix of order } n). \end{aligned}$$

To a function

$$F(z, \zeta) = \sum_{\alpha,\beta=0}^{\infty} a_{\alpha\beta}(z - z_0)^{\alpha}\overline{(\zeta - \zeta_0)}^{\beta}$$

holomorphic w.r.t. $(z, \overline{\zeta})$ at (z_0, ζ_0) and to $(m, n) \in \mathbb{N} \times \mathbb{N}$, we associate an $m \times n$ matrix

$$\mathrm{M}(F; z_0, \zeta_0; m, n) = \begin{bmatrix} a_{00} & \cdots & a_{0n-1} \\ \cdots & \cdots & \cdots \\ a_{m-10} & \cdots & a_{m-1n-1} \end{bmatrix}$$

For another function $G(z, \zeta)$ holomorphic w.r.t. $(z, \overline{\zeta})$ at (z_0, ζ_0), we have

$$\mathrm{M}(F+G; z_0, \zeta_0; m, n) = \mathrm{M}(F; z_0, \zeta_0; m, n) + \mathrm{M}(G; z_0, \zeta_0; m, n).$$

Moreover, for functions $f(z)$ and $g(\zeta)$, holomorphic at z_0 and ζ_0 respectively, we have the *product formula*

$$\text{(PF)} \quad \mathrm{M}(fF\bar{g}; z_0, \zeta_0; m, n) = \Delta(f; z_0; m) \cdot \mathrm{M}(F; z_0, \zeta_0; m, n) \cdot \Delta(g; \zeta_0; n)^*,$$

where $\Delta^* = {}^t\overline{\Delta}$.

2. For a transformation $z = \varphi(x)$ holomorphic at x_0 with $z_0 = \varphi(x_0)$ and for $m \in \mathbb{N}$, we define the *transformation matrix* $\Omega(\varphi; x_0; m)$ as follows: Write

$$\begin{aligned} \varphi(x) &= z_0 + (x - x_0)\varphi_1(x), \\ \Phi_m &= \Delta(\varphi_1; x_0; m), \\ E_m^{(\alpha)} &= \mathrm{M}((z - z_0)^{\alpha}\overline{(\zeta - \zeta_0)}^{\alpha}; z_0, \zeta_0; m, m) \quad (\alpha = 0, \cdots, m-1) \end{aligned}$$

and put

$$\Omega(\varphi; x_0; m) = \sum_{\alpha=0}^{m-1} \Phi_m^{\alpha} E_m^{(\alpha)}.$$

If $\varphi_1(x_0) = \varphi'(x_0) \neq 0$, then $\Omega(\varphi; x_0; m)$ is an invertible matrix.

Using this transformation matrix Ω, we showed in [38] and [40] the following

Transformation formulas:

(a) *Let $F(z, \zeta)$ be a function holomorphic w.r.t. $(z, \overline{\zeta})$ at (z_0, ζ_0). Let $z = \varphi(x)$ and $\zeta = \psi(\xi)$ be functions holomorphic at x_0 and ξ_0, with $z_0 = \varphi(x_0)$ and $\zeta_0 = \psi(\xi_0)$, respectively. Put*

$$G(x, \xi) = F(\varphi(x), \psi(\xi)) .$$

Then, for $(m, n) \in \mathbb{N} \times \mathbb{N}$, we have

$$\text{(1)} \qquad \mathrm{M}(G; x_0, \xi_0; m, n) = \Omega(\varphi; x_0; m) \cdot \mathrm{M}(F; z_0, \zeta_0; m, n) \cdot \Omega(\psi; \xi_0; n)^* .$$

(b) *Let*

$$f(z) = \sum_{\alpha=0}^{\infty} c_\alpha (z - z_0)^\alpha$$

be a function holomorphic at z_0 and let φ be a function holomorphic at x_0 with $z_0 = \varphi(x_0)$. Set

$$g(x) = f(\varphi(x)) = \sum_{\alpha=0}^{\infty} d_\alpha (x - x_0)^\alpha .$$

Then we have for $n \in \mathbb{N}$

$$\text{(2)} \qquad \begin{bmatrix} d_0 & & & \\ d_1 & d_0 & & \\ \vdots & \ddots & \ddots & \\ d_{n-1} & \cdots & d_1 & d_0 \end{bmatrix} \Omega(\varphi; x_0; n) = \Omega(\varphi; x_0; n) \begin{bmatrix} c_0 & & & \\ c_1 & c_0 & & \\ \vdots & \ddots & \ddots & \\ c_{n-1} & \cdots & c_1 & c_0 \end{bmatrix},$$

and

$$\text{(3)} \qquad {}^t(d_0\ d_1\ \cdots\ d_{n-1}) = \Omega(\varphi; x_0; n) \cdot {}^t(c_0\ c_1\ \cdots\ c_{n-1}) .$$

We see easily that, for two transformations φ and ψ holomorphic at x_0 and at $\varphi(x_0)$ respectively, we have for any $m \in \mathbb{N}$

$$\Omega(\psi \circ \varphi; x_0; m) = \Omega(\varphi; x_0; m) \cdot \Omega(\psi; \varphi(x_0); m)$$

and that, if φ is invertible locally at x_0, then $\Omega(\varphi; x_0; m)$ is a regular matrix and

$$\Omega^{-1}(\varphi; x_0; m) = \Omega(\varphi^{-1}; \varphi(x_0); m) .$$

§3. EXISTENCE THEOREM IN THE UNIT DISC

In this section, we deal with the extended interpolation in the open unit disc $D = \{z : |z| < 1\}$. In this case, the problem has been considered in much more general situations with a great deal of fruitful results ([4],[8],[9],[13]). But, we shall here restrict ourselves to the classical situation of ordinary holomorphic functions, because our results concern only this simple situation. In the unit disc D, we can express our problem in a more explicit form.

1. Let $\sigma = \{z_1, z_2, \cdots, z_k\}$ be a nonempty set of k distinct points in D and, for each point z_i, let be given n_i complex numbers $c_{i0}, \cdots, c_{in_i-1}$. Our problem is to find a holomorphic function f in D, satisfying $|f| \leq 1$ and the conditions

$$\text{(EI)} \qquad f(z) = \sum_{\alpha=0}^{n_i-1} c_{i\alpha}(z-z_i)^\alpha + O((z-z_i)^{n_i}) \qquad (i = 1, \cdots, k).$$

From the given data $\{z_i\}$ and $\{c_{i\alpha}\}$ $(1 \leq i \leq k; 0 \leq \alpha \leq n_i - 1)$, we construct an Hermitian matrix A of order $n = \sum_{i=1}^k n_i$, which will be called *criterion matrix* of the problem (EI).

Write

$$C_i = \begin{bmatrix} c_{i0} & & & \\ c_{i1} & c_{i0} & & \\ \vdots & \ddots & \ddots & \\ c_{in_i-1} & \cdots & c_{i1} & c_{i0} \end{bmatrix}, \qquad C = \begin{bmatrix} C_1 & & \\ & \ddots & \\ & & C_k \end{bmatrix},$$

$$\Gamma_{ij} = \mathrm{M}\left(\frac{1}{1-z\bar{\zeta}}; z_i, z_j; n_i, n_j\right), \qquad \Gamma = \begin{bmatrix} \Gamma_{11} & \cdots & \Gamma_{1k} \\ \cdots & \cdots & \cdots \\ \Gamma_{k1} & \cdots & \Gamma_{kk} \end{bmatrix},$$

$$A_{ij} = \Gamma_{ij} - C_i \cdot \Gamma_{ij} \cdot C_j^*, \qquad A = \begin{bmatrix} A_{11} & \cdots & A_{1k} \\ \cdots & \cdots & \cdots \\ A_{k1} & \cdots & A_{kk} \end{bmatrix}.$$

Then we have

$$A = \Gamma - C \cdot \Gamma \cdot C^*.$$

Let $\mathcal{B}$ be the set of holomorphic functions f with $|f| \leq 1$ in D and let $\mathcal{E}$ denote the set of all solutions of (EI) in $\mathcal{B}$.

Theorem 1. *There exists an $f \in \mathcal{E}$ if and only if $A \geq 0$ (positive semidefinite). If so, there is a Blaschke product of degree at most n in $\mathcal{E}$.*

As special cases,

(i) in the case where $k = 1$ and $z_1 = 0$, since

$$\Gamma_{11} = \mathrm{M}\left(\frac{1}{1-z\bar{\zeta}}; 0, 0; n_1, n_1\right) = I_{n_1}, \qquad A = I_{n_1} - C_1 \cdot C_1^*,$$

Theorem 1 reduces to that of Schur; and

(ii) in the case where $n_1 = \cdots = n_k = 1$, that is, where only the values of solutions on σ are prescribed, this is Pick's theorem.

Theorem 2. *For the problem* (EI), *the following conditions are equivalent:*

(a) *The set $\mathcal{E}$ has just one element.*

(b) *Some finite Blaschke product of degree $r < n$ is in $\mathcal{E}$.*

(c) $A \geq 0$ *and* $\det A = 0$.

If one of, therefore all of, these conditions are satisfied, then $r = \operatorname{rank} A$.

The proof of these theorems given in [38] was based on Marshall's method by induction in [21], which makes use of Schur's algorithm.

In the case where the solution is not unique, that is, where $A > 0$ (positive definite), how many solutions are there ? With respect to this question, as an easy consequence of the above theorem, we have the following

Theorem 3. *Suppose* $A > 0$.

(a) *Let* $z_0 \in D$, $z_0 \neq z_i$ $(i = 1, \cdots, k)$. *The set*

$$W(z_0) = \{f(z_0) : f \in \mathcal{E}\}$$

is a nondegenerate closed disc in D. *For any* $w \in \partial W(z_0)$, *there is a unique* $f \in \mathcal{E}$ *such that* $f(z_0) = w$.

(b) *For each* z_i $(i = 1, \cdots, k)$, *the set*

$$W'(z_i) = \{\, f^{(n_i)}(z_i) : f \in \mathcal{E}\}$$

is a nondegenerate compact disc in $\mathbb{C}$. *For any* $c \in \partial W'(z_i)$, *there is a unique* $f \in \mathcal{E}$ *such that* $f^{(n_i)}(z_i) = c$.

As an application of the existence theorem and the transformation formulas, we can give a criterion matrix of the extended interpolation problem in the case where the domain D is a simply connected domain in the Riemann sphere having at least two boundary points. In this case, the range W is a closed disc in the Riemann sphere. When W contains the point at infinity, we have some meromorphic interpolation, modifying the conditions (EI) appropriately [40].

2. Let us proceed to the case where σ is countably infinite.

Let $\sigma = \{z_i\}_{i\in\mathbb{N}}$ be an infinite sequence of distinct points in D. For each $z_i \in \sigma$, let $c_{i\alpha}$ $(0 \le \alpha \le n_{i-1})$ be n_i complex numbers. Our problem is to find an $f \in \mathcal{B}$ such that

$$\text{(EI)} \qquad f(z) = \sum_{\alpha=0}^{n_i-1} c_{i\alpha}(z-z_i)^\alpha + O((z-z_i)^{n_i}) \qquad (\forall z_i \in \sigma).$$

For each finite section $\sigma_k = \{z_1, z_2, \cdots, z_k\}$ of σ $(k = 1, 2, \cdots)$, we have an associated extended interpolation problem

$$\text{(EI)}_k \qquad f(z) = \sum_{\alpha=0}^{n_i-1} c_{i\alpha}(z-z_i)^\alpha + O((z-z_i)^{n_i}) \qquad (i = 1, \cdots, k).$$

Set

$$\mathcal{E} = \{f \in \mathcal{B} : f \text{ satisfies (EI)}\} \quad \text{and} \quad \mathcal{E}_k = \{f \in \mathcal{B} : f \text{ satisfies (EI)}_k\} \quad (k \in \mathbb{N}).$$

Clearly, $\mathcal{E}_k \supset \mathcal{E}_{k+1}$ and $\bigcup_{k\in\mathbb{N}} \mathcal{E}_k = \mathcal{E}$.

Let A_k be the criterion matrix of the problem (EI)_k. A normal families argument shows that

$$\mathcal{E} \ne \emptyset \Longleftrightarrow A_k \ge 0 \quad \text{for all } k \in \mathbb{N}\,.$$

§4. INVARIANCE UNDER THE MÖBIUS TRANSFORMATIONS

The positivity and the rank of the criterion matrix A of the problem (EI) are invariant under the Möbius tansformations of the variable as well as those of the function [38]. This invariance was shown by virtue of the notion of local solution, which is convenient to avoid complicated calculations with the coefficients, and by means of the transformtion formula given in §2. We shall outline the proof here.

A function f holomorphic in some neighborhood of the set $\{z_1, z_2, \cdots, z_k\}$ is said to be *local solution* of (EI) if the condition (EI) are satisfied. Evidently a global solution is a local solution and there exist always many local solutions.

If f is a local solution of (EI), then putting,

$$F(z,\zeta) = \frac{1 - f(z)\overline{f(\zeta)}}{1 - z\bar{\zeta}} = \frac{1}{1 - z\bar{\zeta}} - f(z) \cdot \frac{1}{1 - z\bar{\zeta}} \cdot \overline{f(\zeta)},$$

we have the fundamental relation for the matrix A, defined in §3,

$$A_{ij} = \Gamma_{ij} - C_i \cdot \Gamma_{ij} \cdot C_j^* = \mathrm{M}(F; z_i, z_j; n_i, n_j)$$

1. Consider a Möbius transformation

$$\varphi(x) = \lambda \frac{x-c}{1-\bar{c}x} \qquad (|\lambda| = 1, \quad |c| < 1)$$

and set $x_i = \varphi^{-1}(z_i)$. Take a local solution f of (EI). Then the function $g(x) = f(\varphi(x))$ is holomorhic in a neighborhood of the set $\{x_1, x_2, \cdots, x_k\}$ and is a local solution of another extended interpolation problem

$$\text{(EI)}^* \qquad g(x) = \sum_{\alpha=0}^{n_i-1} d_{i\alpha}(x-x_i)^\alpha + O((x-x_i)^{n_i}) \qquad (i = 1, \cdots, k),$$

$$(d_{i0} \ \cdots \ d_{in_i-1}) = (c_{i0} \ \cdots \ c_{in_i-1}) \cdot {}^t\Omega_i,$$

where $\Omega_i = \Omega(\varphi; x_i; n_i)$ is the transformation matrix defined in §2. Set

$$G(x,\xi) = \frac{1-g(x)\overline{g(\xi)}}{1-x\bar{\xi}} = (1-|c|^2) \cdot \frac{1}{1-\bar{c}x} \cdot \frac{1-f(\varphi(x))\overline{f(\varphi(\xi))}}{1-\varphi(x)\overline{\varphi(\xi)}} \cdot \frac{1}{1-c\bar{\xi}} \,.$$

We have by the formulas (PF) and (1) in §2

$$\mathrm{M}(G; x_i, x_j; n_i, n_j) = (1-|c|^2) M_i \cdot \Omega_i \cdot A_{ij} \cdot \Omega_j^* \cdot M_j,$$

where $M_i = \Delta(1/(1-\bar{c}x); x_i, n_i)$ and Ω_i are invertible matrices. Hence the criterion matrix of the propblem (EI)* for g is

$$(1-|c|^2) S \cdot A \cdot S^*, \quad \text{with} \quad S = \begin{bmatrix} M_1\Omega_1 & & \\ & \ddots & \\ & & M_k\Omega_k \end{bmatrix}, \quad |c| < 1,$$

which asserts the invariance under the Möbius transformation φ.

2. Take a local solution f of (EI) and put

$$g(z) = \lambda \frac{f(z)-c}{1-\bar{c}f(z)} \qquad (|\lambda| = 1, \quad |c| < 1)\,.$$

Then g is a local solution of another extended interpolatioin problem determined by

$$\Delta(g; z_i; n_i) = \lambda(C_i - cI_{n_i})/(I_{n_i} - \bar{c}C_i)^{-1}\,.$$

Write

$$G(z,\zeta) = \frac{1-g(z)\overline{g(\zeta)}}{1-z\bar{\zeta}} = (1-|c|^2) \cdot \frac{1}{1-\bar{c}f(z)} \cdot \frac{1-f(z)\overline{f(\zeta)}}{1-z\bar{\zeta}} \cdot \frac{1}{1-c\overline{f(\zeta)}}\,.$$

Then the formula (PF) gives the relation

$$\mathrm{M}(G; z_i, z_j; n_i, n_j) = (1 - |c|^2) N_i \cdot A_{ij} \cdot N_j^* ,$$

where $N_i = \Delta(1/(1 - \bar{c}f); z_i, n_i)$ is an invertible matrix. Hence the criterion matrix corresponding to this problem is

$$(1 - |c|^2) N \cdot A \cdot N^*, \quad \text{with} \quad N = \begin{bmatrix} N_1 & & \\ & \ddots & \\ & & N_k \end{bmatrix}, \quad |c| < 1,$$

which asserts the invariance under the Möbius transformation of the function.

The proof of main theorems in §3, as mentioned there, is based on Marshall's method of induction. Using the fact stated above, we reduce the problem (EI) to the case, where (i) $c_{10} = 0$ and (ii) $z_1 = 0$, by a Möbius transformation of the function f and one of the variable z respectively. After these reductions, the correspondence (iii) $f \longleftrightarrow \dfrac{f}{z}$ makes it possible to perform our induction on the total order $n = \sum_{i=1}^{k} n_i$.

§5. NEVANLINNA PARAMETRIZATIONS

As in §3, we treat in this section the extended interpolation problem in the open unit disc D. In this case, analysing Nevanlinna's method [24], we define axiomatically the notion of Nevanlinna parametrization of the solutions. The existence, some properties and some applications of such paramertizations were shown in [39].

Let $\mathcal{B} = \{f : \text{ holomorphic and } |f| \leq 1 \text{ in } D\}$ as in §3.

Let $\sigma = \{z_i\}$ be a nonempty finite or infinite sequence of distinct points in D and, for each point $z_i \in \sigma$, let be given n_i complex numbers $c_{i0}, \cdots, c_{in_i-1}$.

Let $\mathcal{E}$ denote the set of all functions $f \in \mathcal{B}$, satisfying the extended interpolation condition

$$\text{(EI)} \qquad f(z) = \sum_{\alpha=0}^{n_i-1} c_{i\alpha}(z - z_i)^\alpha + O((z - z_i)^{n_i}) \qquad (\forall z_i \in \sigma).$$

1. A bijection $\pi : \mathcal{B} \longrightarrow \mathcal{E}$ is called a *Nevanlinna parametrization* of $\mathcal{E}$ if there exist four holomorphic functions P, Q, R and S in D such that

$$\pi(g) = \frac{Pg + Q}{Rg + S}, \qquad Rg + S \not\equiv 0 \qquad (\forall g \in \mathcal{B}).$$

We shall say that the *quadruple* (P, Q, R, S) *represents* π.

Assume that

(H) $\mathcal{E}$ has at least two elements.

By virtue of Schur-Nevanlinna's algorithm, there exists a Nevanlinna parametrization of $\mathcal{E}$ if and only if (H) is fulfilled.

To the problem (EI), we associate the Blaschke product

$$B(z) = \prod_{z_i \in \sigma} \left(\lambda_i \frac{z - z_i}{1 - \overline{z_i}\, z} \right)^{n_i}, \qquad \lambda_i = \begin{cases} 1 & (\text{if } z_i = 0) \\ -|z_i|/z_i & (\text{if } z_i \neq 0) \end{cases}.$$

Note that, when σ is infinite, the above hypothesis (H) implies the Blaschke condition $\sum_{z_i \in \sigma} n_i(1 - |z_i|) < \infty$ and hence the convergence of this infinite product.

We now show some basic properties of Nevanlinna parametrizations.

Proposition 1. *Let* (P, Q, R, S) *represents a Nevanlinna parametrization* π *of* $\mathcal{E}$. *Then these functions have the following properties:*

(a) $S \not\equiv 0$.

(b) $|P/S| < 1$, $|Q/S| < 1$, $|R/S| < 1$ *in* D.

(c) $Q/S \in \mathcal{E}$.

(d) *We may write*

$$\frac{PS - QR}{S^2} = U \cdot B,$$

where $U \in \mathcal{B}$, $U \neq 0$ *in* D *and* B *is a Blaschke product defined above.*

This proposition allows us to assume $S \equiv 1$. One observes that if two quadruples (P, Q, R, S) and $(\widehat{P}, \widehat{Q}, \widehat{R}, \widehat{S})$ represent the same π, then there exists a meromorphic function M in D such that $(\widehat{P}, \widehat{Q}, \widehat{R}, \widehat{S}) = M(P, Q, R, S)$. Therefore,

(i) each Nevanlinna parametrization of $\mathcal{E}$ is represented by one and only one quadruple of the form $(P, Q, R, 1)$ and

(ii) the function U in (d) is independent of the representative of π.

2. How many parameters has the totality of Nevanlinna parametrization ? The following is our answer of this question.

We define $\mathcal{P}$ to be the set of all Nevanlinna parametrizations of $\mathcal{E}$. Let G denote the group of all Möbius transformations

$$\tau(z) = \lambda \frac{z + a}{1 + \overline{a} z} \qquad (|\lambda| = 1,\ |a| < 1),$$

regarded as analytic automorphisms of the closed unit disc $\overline{D} = D \bigcup \partial D$. The group G operates on $\mathcal{P}$ in the following way:

$$(\tau^*(\pi))(g) = \pi(\tau \circ g) \qquad (\tau \in G,\ \pi \in \mathcal{P},\ g \in \mathcal{B}).$$

We have the following

Theorem 4. *Let π, $\pi_0 \in \mathcal{P}$. Then there exists one and only one $\tau \in G$ such that $\pi = \tau^*(\pi_0)$.*

This theorem shows that, if we fix one $\pi_0 \in \mathcal{P}$ as a base point, $\tau \mapsto \tau^*(\pi_0)$ gives a bijective mapping of G onto $\mathcal{P}$. Hence the totality $\mathcal{P}$ of Nevanlinna parametrizations has three real parameters.

3. For each $z \in D$, let

$$W(z) = \{f(z) : f \in \mathcal{E}\}$$

denote the set of values taken at z by all solutions of (EI) in $\mathcal{B}$. Clearly $W(z) \subset D$. To $\pi \in \mathcal{P}$ and to $z \in D$, we shall associate the mapping $\pi_z : \overline{D} \longrightarrow W(z)$ defined by

$$\pi_z(\zeta) = \pi(\zeta)(z) \qquad (\zeta \in \overline{D}),$$

where ζ is regarded as a constant function.

If π is represented by (P, Q, R, S), then

$$W(z) = \left\{ \frac{P(z)\zeta + Q(z)}{R(z)\zeta + S(z)} : \zeta \in \overline{D} \right\}.$$

If $z \in D \setminus \sigma$, then π_z is bijective and $W(z)$ is a nondegenerate closed disc in D whose center and radius are

$$\frac{Q(z)\overline{S(z)} - P(z)\overline{R(z)}}{|S(z)|^2 - |R(z)|^2} \qquad \text{and} \qquad \frac{|P(z)S(z) - Q(z)R(z)|}{|S(z)|^2 - |R(z)|^2}$$

respectively.

Moreover, in our extended case, we consider the set

$$W'(z_i) = \{ f^{(n_i)}(z_i) : f \in \mathcal{E}\} \qquad (z_i \in \sigma).$$

Let $\pi \in \mathcal{P}$ be represented by (P, Q, R, S) such that $S \neq 0$ in D. To simplify the expression we shall denote, for $f \in \mathcal{B}$,

$$d_i f = f^{(n_i)}(z_i) \qquad (z_i \in \sigma).$$

With these symbols, we have

$$W'(z_i) = \left\{ d_i(Q/S) + \frac{d_i(PS - QR) \cdot \zeta}{(R(z_i)\zeta + S(z_i))S(z_i)} : \zeta \in \overline{D} \right\}.$$

Thus, $W'(z_i)$ is a nondegenerate compact disc in $\mathbb{C}$ of radius

$$\frac{|d_i(PS - QR)|}{|S(z_i)|^2 - |R(z_i)|^2}.$$

We state the following properties of the extremal solutions.

Theorem 5. *Assume the hypothesis* (H). *Let* $\pi \in \mathcal{P}$ *and* $f \in \mathcal{E}$. *The following are equivalent:*

(a) $f(z) \in \partial W(z)$ *for some* $z \in D \setminus \sigma$.

(b) $f(z) \in \partial W(z)$ *for all* $z \in D \setminus \sigma$.

(c) $f^{(n_i)}(z_i) \in \partial W'(z_i)$ *for some* $z_i \in \sigma$.

(d) $f^{(n_i)}(z_i) \in \partial W'(z_i)$ *for all* $z_i \in \sigma$.

(e) *There exists a* $\zeta \in \partial D$ *such that* $f = \pi(\zeta)$.

If one of these conditions is satisfied, ζ *in* (e) *is uniquely determined by* π *and* f.

Note. Let $\pi \in \mathcal{P}$ and $\zeta \in \partial D$. We observe that

(i) in the case where $\sigma = \{z_1, z_2, \cdots, z_k\}$ is finite, $\pi(\zeta)$ is a Blaschke product of degree $n = \sum_{i=1}^{k} n_i$; and

(ii) in the case where σ is infinite, by virtue of the argument of Nevanlinna [24], $\pi(\zeta)$ is an inner function.

§6. A SUFFICIENT CONDITION FOR NEVANLINNA PARAMETRIZATION

We continue to deal with extended interpolation in the open unit disc D and consider the propblem (EI). Let $\mathcal{B}$, $\mathcal{E}$ and σ be as in the preceding §5 and assume $\mathcal{E}$ has at least two elements. As mentioned in §5, each Nevanlinna parametrization has only one representative of the form $(P, Q, R, 1)$. In this section, we shall give a sufficient condition for a quadruple $(P, Q, R, 1)$, where P, Q and R are holomorphic in D, to represent a Nevanlinna parametrization of $\mathcal{E}$ [41].

We start with the following

Proposition 2. *Let* P, Q *and* R *be holomorphic functions in* D *and assume* $|R| < 1$ *in* D. *If for at least two values* ζ_1 *and* ζ_2 *in* D *the function*

$$f_\nu = \frac{P\zeta_\nu + Q}{R\zeta_\nu + 1} \qquad (\nu = 1, 2)$$

satisfy the conditions (EI), *then for any* $g \in \mathcal{B}$ *the function*

$$f = \frac{Pg + Q}{Rg + 1}$$

is holomorphic in D *and satisfies* (EI).

For three holomorphic functions P, Q and R in D and for each $z \in D$, consider two sets

$$\delta(z) = \left\{ \frac{P(z)\zeta + Q(z)}{R(z)\zeta + 1} : \zeta \in \overline{D} \right\}$$

and

$$W(z) = \{f(z) : f \in \mathcal{E}\}.$$

By a comparison of these two discs, we have

Lemma. *Suppose $\delta(z) \subset W(z)$ for all $z \in D$.*
(a) *Then we have*

$$\frac{Pg+Q}{Rg+1} \in \mathcal{E} \qquad (\forall g \in \mathcal{B}).$$

(b) *Moreover, if there is a point $z_0 \in D \setminus \sigma$ such that $\delta(z_0) = W(z_0)$, then we have $\delta(z) = W(z)$ for all $z \in D$.*

Now we present a sufficient condition for parametrization.

Theorem 6. *If $\delta(z) \subset W(z)$ for any $z \in D \setminus \sigma$ and $\delta(z_0) = W(z_0)$ for at least one $z_0 \in D \setminus \sigma$, then the quadruple $(P, Q, R, 1)$ represents a Nevanlinna parametrization of $\mathcal{E}$.*

The proofs of Lemma and Theorem 6 are found in [41].

§7. FINITELY CONNECTED DOMAINS.

In this section, we consider the extended interpolation problem in a finitely connected domain R in the complex plane $\mathbb{C}$, whose boundary ∂R consists of $m+1$ pairwise disjoint analytic simple closed curves γ_i $(i = 0, 1, \cdots, m)$.

In 1979, Abrahamse investigated the Pick interpolation theorem for such domains [1], by means of kernel functions depending on the periods. According to his method and studying the derivative of such kernel functions, we generalized his theorem to our extended interpolation [40]. Note that, in $\mathbb{C}$, interpolation problems may be given explicitly w.r.t.the canonical coordinate as in §3, not in terms of ideals and germs.

1. We consider the harmonic measure $d\omega$ on ∂R for a fixed point $z^* \in R$ and the Banach spaces $L^2 = L^2(\partial R, d\omega)$ of the complex-valued measurable functions f on ∂R with the norm

$$\|f\|_2 = \left(\int_{\partial R} |f|^2 \, d\omega \right)^{\frac{1}{2}},$$

(see [14]).

Let $\Lambda = \{\lambda = (\lambda_1, \cdots, \lambda_m) : \lambda_i \in \mathbb{C}, |\lambda_i| = 1 (i = 1, \cdots, m)\}$ be the m-torus. With Abrahamse, we consider m pairwise disjoint analytic cuts δ_i $(i = 1, \cdots, m)$, which starts from a point of γ_i and terminates at a point of γ_0 in such a way that the domain $R_0 = R \setminus (\bigcup_{i=1}^{m} \delta_i)$ is simply connected.

For $\lambda \in \Lambda$, let H^2_λ denote the set of complex-valued functions f in R such that

(i) f is holomorphic in R_0 ;
(ii) for each $t \in \delta_i \cap R$, $f(z)$ tends to $f(t)$ when $z \in R_0$ tends to t from the left side of δ_i and $f(z)$ tends to $\lambda_i f(t)$ when $z \in R_0$ tends to t from the right side of δ_i ; and
(iii) $|f|^2$ has a harmonic majorant in R.

It is well known ([14]) that any function f in H^2_λ admits nontangential limits $f^*(t)$ at almost all $t \in \partial R$ (w.r.t. $d\omega$). Via $f \mapsto f^*$, the space H^2_λ can be viewed as a closed subspace of the Hilbert space L^2 and H^2_λ is a Hilbert space with the inner product

$$\langle f, g\rangle = \int_{\partial R} f^* \overline{g^*}\, d\omega \qquad (f, g \in H^2_\lambda).$$

For $\lambda \in \Lambda$ and $\zeta \in R$, the mapping $f \mapsto f(\zeta)$ is a bounded linear functional on H^2_λ and ,by Riesz representation theorem, we have a unique $k_{\lambda\zeta} \in H^2_\lambda$ such that

$$f(\zeta) = \langle f, k_{\lambda\zeta}\rangle \qquad (\forall f \in H^2_\lambda).$$

Write

$$k_\lambda(z,\zeta) = k_{\lambda\zeta}(z) \qquad \text{and} \qquad k_\lambda(\ ,\zeta) = k_{\lambda\zeta}.$$

The kernel function $k_\lambda(z,\zeta)$ is holomorphic w.r.t.$(z,\overline{\zeta})$ in $R_0 \times R_0$ and

$$k_\lambda(z,\zeta) = \langle k_\lambda(\ ,\zeta), k_\lambda(\ ,z)\rangle = \overline{k_\lambda(\zeta, z)}.$$

Roughly speaking, we may say that $k_\lambda(z,\zeta)$ can be continued analytically across the boundary as a function of two variables $(z,\overline{\zeta})$.

2. Let $z_1, z_2, \cdots, z_k$ be k distinct points in a domain R. For each z_i, let $c_{i0}, \cdots, c_{in_i-1}$ be a sequence of n_i complex numbers. Our present problem (EI) is to find a holomorphic function f in R, satisfying $|f| \le 1$ and the conditions

$$\text{(EI)} \qquad f(z) = \sum_{\alpha=0}^{n_i-1} c_{i\alpha}(z-z_i)^\alpha + O((z-z_i)^{n_i}) \qquad (i = 1, \cdots, k).$$

For each element λ of the m-torus Λ, with k_λ introduced above, we define the following matrices for $i, j = 1, \cdots, k$:

$$C_i = \begin{bmatrix} c_{i0} & & & \\ c_{i1} & c_{i0} & & \\ \vdots & \ddots & \ddots & \\ c_{in_i-1} & \cdots & c_{i1} & c_{i0} \end{bmatrix}, \qquad C = \begin{bmatrix} C_1 & & \\ & \ddots & \\ & & C_k \end{bmatrix},$$

$$\Gamma_{ij}^{(\lambda)} = \mathbf{M}\,(k_\lambda; z_i, z_j; n_i, n_j)\,, \qquad A_{ij}^{(\lambda)} = \Gamma_{ij}^{(\lambda)} - C_i \cdot \Gamma_{ij}^{(\lambda)} \cdot C_j^*,$$

$$A_\lambda = \begin{bmatrix} A_{11}^{(\lambda)} & \cdots & A_{1k}^{(\lambda)} \\ \cdots & \cdots & \cdots \\ A_{k1}^{(\lambda)} & \cdots & A_{kk}^{(\lambda)} \end{bmatrix}.$$

In terms of these matrices A_λ $(\lambda \in \Lambda)$, we have

Theorem 7. *The problem* (EI) *admits a solution f with $|f| \leq 1$ in R if and only if the matrix A_λ is positive semidefinite for each $\lambda \in \Lambda$. The solution is unique if and only if the determinant of A_λ is zero for some $\lambda \in \Lambda$.*

We point out that, in the special case in which $m = 0$, the above statement is essentially the same as Theorem 1 in §3, but its proof gives a different approach from that of Theorem 1.

§8. AN EXTENSION OF THE HEINS THEOREM

In this section, let us generalize the Heins theorem for Pick's interpolation in doubly connected domains in [19] to our extended interpolation.

As Heins' theorem concerns some transformations, the coordinate free formulation of the interpolation problem on Riemann surfaces given in §1 would simplify the expression concerning transformations in our extended case.

Let Z be a simply connected Riemann surface of hyperbolic type and let be given an extended interpolation problem $(\mathcal{I}, \mathfrak{c})$ on Z, where

$$\mathcal{I} = (\mathcal{I}_z)_{z \in Z}\,,\ \mathcal{I}_z \text{ is a nonzero ideal of } \mathcal{O}_z;$$
$$\mathfrak{c} = (\mathfrak{c}_z)_{z \in Z}\,,\ \mathfrak{c}_z \in \mathcal{O}_z/\mathcal{I}_z\,.$$

As in §1, let $\mathcal{B}$ denote the set of functions f, holomorphic and $|f| \leq 1$ in Z, and $\mathcal{E}$ be the set of all solutions of the problem $(\mathcal{I}, \mathfrak{c})$ in $\mathcal{B}$. We assume $\sigma = \{z \in Z : \mathcal{I}_z \neq \mathcal{O}_z\} \neq \emptyset$. Moreover, consider an analytic automorphism T of Z and a Möbius transformation U :

$$U(w) = \lambda\,\frac{w + a}{1 + \bar{a}w} \qquad (|\lambda| = 1\,,\ |a| < 1)\,.$$

Our present problem is to find a function $f \in \mathcal{B}$ satisfying the interpolation condition

$$f_z + \mathcal{I}_z = \mathfrak{c}_z \qquad (\forall z \in Z)$$

and moreover the condition

$$f \circ T = U \circ f\,.$$

For the analytic automorphism T of Z, we have the ring homomorphisms, given in §1, $T_z^* : \mathcal{O}_{T(z)} \longrightarrow \mathcal{O}_z$ and $T_z^* : \mathcal{O}_{T(z)}/\mathcal{I}_{T(z)} \longrightarrow \mathcal{O}_z/\mathcal{I}_z$, provided that $T_z^*(\mathcal{I}_{T(z)}) \subset \mathcal{I}_z$. Here, for the Möubius transformation U of functions, we introduce the following notation.

For an element $\mathfrak{c}_z \in \mathcal{O}_z/\mathcal{I}_z$, whenever $1+\bar{a}\mathfrak{c}_z$ is a unit of $\mathcal{O}_z/\mathcal{I}_z$, we can define $U_z(\mathfrak{c}_z) \in \mathcal{O}_z/\mathcal{I}_z$ by

$$U_z(\mathfrak{c}_z) = \lambda \frac{\mathfrak{c}_z + a}{1+\bar{a}\mathfrak{c}_z}.$$

Note that if $f \in \mathcal{E}$ i.e. $\mathfrak{c}_z = f_z + \mathcal{I}_z$ and $z \in \sigma$, then $1+\bar{a}\mathfrak{c}_z$ is a unit and $U_z(\mathfrak{c}_z) = (U \circ f)_z + \mathcal{I}_z$.

In order to solve our present problem, it would be indispensable to impose on the transformations T and U the following consistency conditions (1) and (2):

$$T_z^*(\mathcal{I}_{T(z)}) = \mathcal{I}_z \qquad (\forall z \in Z)\,; \tag{1}$$

$$T_z^*(\mathfrak{c}_{T(z)}) = U_z(\mathfrak{c}_z) \qquad (\forall z \in \sigma)\,. \tag{2}$$

Theorem 8. *Suppose that the consistency conditons* (1) *and* (2) *are fulfilled. If* $\mathcal{E} \neq \emptyset$ *then there exists a function* $f \in \mathcal{E}$ *satisfying*

$$f \circ T = U \circ f\,.$$

Note that if $\mathcal{E} \neq \emptyset$ and (1) is satisfied, then $1+\bar{a}\mathfrak{c}_z$ is a unit for any $z \in \sigma$.

The proof of this theorem, which is based on Heins' method and given in [41], will be sketched here.

By (1) and (2), we can see, for $f \in \mathcal{E}$, $U^{-1} \circ f \circ T \in \mathcal{E}$, and $U \circ f \circ T^{-1} \in \mathcal{E}$. Consider the set $W(z) = \{f(z) : f \in \mathcal{E}\}$ $(z \in Z)$. Then we have

$$W(z) = U^{-1}(W(T(z))) \qquad (\forall z \in Z)\,. \tag{3}$$

In the case where $\mathcal{E}$ has only one element f, we have $f = U^{-1} \circ f \circ T$, which proves the theorem.

Now, assume $\mathcal{E}$ has at least two elements. Then, as was mentioned in §5, we may take a Nevanlinna parametrization π of $\mathcal{E}$ represented by a quadruple (P, Q, R, S). Taking account of

$$U^{-1}(w) = \frac{w - \lambda a}{\lambda - \bar{a}w}$$

and setting

$$\begin{bmatrix} P_0 & Q_0 \\ R_0 & S_0 \end{bmatrix} = \begin{bmatrix} 1 & -\lambda a \\ -\bar{a} & \lambda \end{bmatrix} \begin{bmatrix} P & Q \\ R & S \end{bmatrix},$$

we have by (3)

$$W(z) = U^{-1}\left(\left\{\frac{P(T(z))\zeta + Q(T(z))}{R(T(z))\zeta + S(T(z))} : |\zeta| \le 1\right\}\right)$$
$$= \left\{\frac{P_0(T(z))\zeta + Q_0(T(z))}{R_0(T(z))\zeta + S_0(T(z))} : |\zeta| \le 1\right\}.$$

By virtue of Theorem 6, with the functions

$$\widehat{P}(z) = \frac{P_0(T(z))}{S_0(T(z))}, \quad \widehat{Q}(z) = \frac{Q_0(T(z))}{S_0(T(z))} \quad \text{and} \quad \widehat{R}(z) = \frac{R_0(T(z))}{S_0(T(z))},$$

the quadruple $(\widehat{P}, \widehat{Q}, \widehat{R}, 1)$ represents a Nevanlinna parametrization of $\mathcal{E}$. By Theorem 4, there exists a unique Möbius transformation $\widehat{U}$ such that

$$\pi(g) = \widehat{\pi}(\widehat{U} \circ g) \qquad (\forall g \in \mathcal{B}).$$

We conclude that, for $f \in \mathcal{E}$,

$$f \circ T = U \circ f \iff \exists g \in \mathcal{B} \text{ s.t. } f = \pi(g), \quad g \circ T = \widehat{U} \circ g.$$

The existence of a fixed point ζ of $\widehat{U}$ in $\overline{D}$ asserts the existence of a function g satisfying $g \circ T = \widehat{U} \circ g$. □

We can apply the above Theorem 7 to the extended interpolation problem on the annuli.

Let X be a doubly connected Riemann surface biholomorphically equivalent to an annulus $\{z \in \mathbb{C} : 1 < |z| < \rho\}$ $(1 < \rho < \infty)$ and let $(\mathcal{I}, \mathfrak{c})$, where $\mathcal{I} = (\mathcal{I}_x)_{x \in X}$ and $\mathfrak{c} = (\mathfrak{c}_x)_{x \in X}$, be an extended interpolation problem on X. Choose a holomorphic mapping $\varphi : D \longrightarrow X$ of the open unit disc D onto X such that (D, φ) is a universal covering of X. For each $z \in D$, setting

$$\widetilde{\mathcal{I}}_z = \varphi_z^*(\mathcal{I}_{\varphi(z)}), \quad \widetilde{\mathcal{I}} = (\widetilde{\mathcal{I}}_z)_{z \in D} \quad \text{and} \quad \widetilde{\mathfrak{c}}_z = \varphi_z^*(\mathfrak{c}_{\varphi(z)}), \quad \widetilde{\mathfrak{c}} = (\widetilde{\mathfrak{c}}_z)_{z \in D},$$

we have an extended interpolation problem $(\widetilde{\mathcal{I}}, \widetilde{\mathfrak{c}})$ on D.

Corollary. *A necessary and sufficient condition that there exists a holomorphic function f on X satisfying*

$$|f| \le 1 \text{ on } X \qquad \text{and} \qquad f_x + \mathcal{I}_x = \mathfrak{c}_x \qquad (\forall x \in X)$$

is that there exists a holomorphic function $\widetilde{f}$ on D satisfying

$$|\widetilde{f}| \le 1 \text{ on } D \qquad \text{and} \qquad \widetilde{f}_z + \widetilde{\mathcal{I}}_z = \widetilde{\mathfrak{c}}_z \qquad (\forall z \in D).$$

Now, suppose X be an annulus in the complex plane $\mathbb{C}$ and (D,φ) be a universal covering of X as before.

Let $\sigma = \{x_1, x_2, \cdots, x_k\}$ be a set of k distinct points in X and, for each point $x_i \in \sigma$, let $(c_{i0}, \cdots, c_{in_i-1})$ be n_i-tuples of complex numbers. Let $T : D \longrightarrow D$ be a generator of the covering transformation group of (D,φ). For each $x_i \in \sigma$, choose a point z_i in D such that $\varphi(z_i) = x_i$. Then the inverse set $\widetilde{\sigma} = \varphi^{-1}(\sigma)$ is given by

$$\widetilde{\sigma} = \{T^m(z_i) : i = 1, \cdots, k;\ m \in \mathbb{Z}\}\,.$$

By Corollary, one can see that to solve the extended interpolation problem (EI) in X given by the data

$$\sigma = \{x_1, x_2, \cdots, x_k\} \quad \text{and} \quad \{(c_{i0}, \cdots, c_{in_i-1})\} \qquad (i = 1, \cdots, k)$$

is equivalent to solve the extended interpolation problem $(\widetilde{\text{EI}})$ in D given by the data

$$\widetilde{\sigma} = \{T^m(z_i)\} \quad \text{and} \quad \{(\widetilde{c}_{i0}^{(m)}, \cdots, \widetilde{c}_{in_i-1}^{(m)})\} \qquad (i = 1, \cdots, k;\ m \in \mathbb{Z})\,,$$

defined by

$${}^t(\widetilde{c}_{i0}^{(m)} \ \cdots \ \widetilde{c}_{in_i-1}^{(m)}) = \Omega(\varphi; T^m(z_i); n_i) \cdot {}^t(c_{i0} \ \cdots \ c_{in_i-1})\,,$$

where $\Omega(\varphi; T^m(z_i); n_i)$ is the transformation matrix of φ defined in §2.

Since $\widetilde{\sigma}$ is infinite, as mentioned in §3, the existence of solutions of the problem $(\widetilde{\text{EI}})$ is guaranteed by the semipositivity of a countably infinite number of Hermitian matrices.

On the other hand, we gave in §7 a generalization of the Abrahamse theorem, by which, in order to confirm the existence, one must, even in doubly connected domains, check the semipositivity of as many Hermitian matrices A_λ as the uncountable set $\Lambda = \{\lambda \in \mathbb{C} : |\lambda| = 1\}$.

References

[1] M. B. Abrahamse, *The Pick interpolation theorem for finitely connected domains*, Michigan Math. J. **26** (1979), 195–203.

[2] V. M. Adamyan, D. Z. Arov and M. G. Krein, *Infinite Hankel matrices and generalized problems of Carathéodory-Fejér and I.Schur*, Functional Anal. Appl. **2** (1968), 269–281.

[3] L. Ahlfors, *Conformal Invariants. Topics in Geometric Function Theory*, McGraw-Hill, New York, 1973.

[4] J. A. Ball, I. Gohberg and L. Rodman, *Interpolation of Rational Matrix Functions*, Oper.Theory Adv. Appl.**45**, Birkhäuser Verlag, Basel, 1990.

[5] S. Bergman, *The Kernel Function and Conformal Mapping, 2nd ed.*, Mathematical Surveys, No 5 Amer. Math. Soc., 1970.

[6] C. Carathéodory, *Über den Variabilitätsbereich der Fourier'schen Konstanten von positiven harmonischen Funktionen*, Rend. Circ. Mat. Palermo **32** (1911), 193–217.

[7] A. Denjoy, *Sur une classe de functions analytiques*, C. R. Acad. Sci.(Paris) **188** (1929), 140–142; 1084–1086.

[8] H. Dym, *J Contractive Matrix Functions, Reproducing Kernel Hilbert Spaces and Interpolation*, CBMS Regional Conference Series **71**, Amer. Math. Soc., Providence, Rhode Island, 1989.

[9] H. Dym, *On reproducing kernel spaces, J unitary matrix functions, interpolation and displacement rank*, Oper. Theory Adv. Appl. **41** (1989), 173–239, Birkhäuser Verlag, Basel.

[10] A. M. Džrbašjan, *Multiple interpolation in the* H^p *classes,* $0 < p \leq \infty$, Soviet Math. Dokl. **18** (1977), 837–841.

[11] J. P. Earl, *On the interpolation of bounded sequences by bounded functions*, J. London Math. Soc. (2) **2** (1970), 544–548.

[12] J. P. Earl, *A note on bounded interpolation in the unit disc*, J. London Math. Soc. (2) **13** (1976), 419–423.

[13] C. Faios and A. E. Frazho, *The Commutative Lifting Approach to Interpolation Problems*, Oper.Theory Adv. Appl.**44**, Birkhäuser Verlag, Basel, 1990.

[14] S. D. Fisher, *Function Theory on Planer Domains*, Wiley, New York, 1983.

[15] P. R. Garabedian, *Schwarz's lemma and the Szegö kernel function*, Trans. Amer. Math. Soc. **67** (1949), 1–35.

[16] J. B. Garnett, *Two remarks on interpolation by bounded analytic functions*, Lect. Notes in Math. (Springer) **604** (1977), 32–40.

[17] J. B. Garnett, *Bounded Analytic Functions*, Academic Press, New York, 1981.

[18] F. Hartogs, *Zur Theorie der analytischen Funktionen mehrerer unabhängiger Veränderlichen, insbesondere über die Darstellung derserben durch Reihen, welche nach Potenzen einer Veränderlichen fortschreiten*, Math. Ann. **62** (1906), 1–88.

[19] M. H. Heins, *Extremal problems for functions analytic and single-valued in a doubly-connected region*, Amer. J. Math. **62** (1940), 91–106.

[20] M. H. Heins, *Nonpersistence of the grenzkreis phenomenon for Pick-Nevanlinna interpolation on annuli*, Ann. Acad. Sci. Fenn. A. I. **596** (1975).

[21] D. E. Marshall, *An elementary proof of Pick-Nevanlinna interpolation theorem*, Michigan Math. J. **21** (1974), 219–223.

[22] V. M. Martirosjan, *Effective solution of an interpolation problem with nodes of bounded multiplicity*, Soviet Math. Dokl. **25** (1982), 441–445.

[23] R. Nevanlinna, *Über beschränkte Funktionen die in gegebenen Punkten vorgeschriebene Werte annehmen*, Ann. Acad. Sci. Fenn. **13** (1919), No 1.

[24] R. Nevanlinna, *Über beschränkte analytische Funktionen*, Ann. Acad. Sci. Fenn. Ser A **32** (1929), No 7.

[25] K. Øyma, *Extremal interpolatory functions in H^∞*, Proc. Amer. Math. Soc. **64** (1977), 272–276.

[26] K. Øyma, *Interpolation in H^p-spaces*, Proc. Amer. Math. Soc. **76** (1979), 81–88.

[27] K. Øyma, *An interpolation theorem for H_E^∞*, Pacific J. Math. **109** (1683), 457–462.

[28] G. Pick, *Über die Beschränkungen analytischer Funktionen, welche durch vorgegebene Funktionswerte bewirkt werden*, Math. Ann. **77** (1916), 7–23.

[29] W. Rudin, *Analytic functions of class H_p*, Trans. Amer. Math. Soc. **78** (1955), 46–66.

[30] A. M. Russakovskiĭ, *The problem of multiple interpolation in the classe of functions that are analytic in a half-plane and hence indicators not exceeding a given one*, Soviet Math. Dokl. **27** (1983), 433–436.

[31] S. Saitoh, *Theory of Reproducing Kernels and Its Applications*, Pitman Research Notes in Math. **189**, Longman, Harlow, 1988.

[32] I. Schur, *Über Potenzreihen, die im Innern des Einheitskreises beschränkt sind*, J. Reine Angew. Math. **147** (1917), 205–232.

[33] A. Stray, *Two applications of Schur-Nevanlinna algorithm*, Pacific J. Math. **91** (1980), 223–232.

[34] A. Stray, *On a formula by V. M. Adamjan, D. Z. Arov and M. G. Krein*, Proc. Amer. Math. Soc. **83** (1981), 337–340.

[35] A. Stray, *Minimal interpolation by Blaschke products*, J. London Math. Soc. (2) **32** (1985), 488–496.

[36] A. Stray, *Minimal interpolation by Blaschke products II*, Bull. London Math. Soc. **20** (1988), 329–332.

[37] A. Stray and K. O. Øyma, *On interpolations with minimal norm*, Proc. Amer. Math. Soc. **68** (1978), 75–119.

[38] S. Takahashi, *Extension of the theorems of Carathéodory-Toeplitz-Schur and Pick*, Pacific J. Math. **138** (1989), 391–399.

[39] S. Takahashi, *Nevanlinna parametrizations for the extended interpolation problem*, Pacific J. Math. **146** (1990), 115–129.

[40] S. Takahashi, *Extended interpolation problem in finitely connected domains*, Oper. Theory Adv. Appl. **59** (1992), 305–327, Birkhäuser Verlag, Basel.

[41] S. Takahashi,, *A sufficient condition for Nevanlinna parametrization and an extension of Heins theorem*, Nagoya Math. J. (to appear).

[42] O. Toeplitz, *Über die Fourier'sche Entwicklung positive Funktionen*, Rend. Circ. Math. Palermo **32** (1911), 191–192.

[43] H. Widom, *Extremal polynomials associated with a system of curves in the complex plane*, Advances in Math. **3** (1969), 127–232.

17 The Nehari problem for the weighted Szegő kernels

Masahiro Uehara

Department of Mathematics, Kagawa Medical University,
1750-1 Ikenobe Miki-cho Kagawa 761-0793, Japan
uehara@kms.ac.jp

The weighted kernels of Szegő type were systematically examined by Nehari ([Ne-1]) in 1950, in order to solve a number of extremal problems. Nehari derived some basic properties of the weighted Szegő kernels and interesting relations among those kernels and other domain functions. In there, he proposed a simple but important problem concerning the zero points of the weighted Garabedian kernels which are the adjoint L-kernels for the weighted Szegő kernels ([Ne-1]). We call this "the Nehari problem".

1. NOTATION AND BASIC FACTS

Now, we shall introduce the weighted kernels of Szegő type, related notations and present some basic facts ([Ne-1],[Sa],[Ue-1], [Ue-2]). Let D be a p-connected bounded domain on the complex plane, whose boundary ∂D consists of p -disjoint analytic Jordan curves $C_j (j = 1, \cdots, p)$. Let $H_2(D)$ denote the analytic Hardy class on D. For a positive continuous function $\lambda(z)$ on ∂D, let $H^2_\lambda(\partial D)$ the weighted Szegő space defined as the family of all analytic functions $f(z) \in H_2(D)$ equipped with norm:

$$\left\{ \int_{\partial D} |f(z)|^2 \lambda(z) \, |dz| \right\}^{1/2} < \infty ,$$

where $f(z)$ mean Fatou's nontangential boundary values at ∂D. There exist a reproducing kernel and its adjoint L-kernel for the space $H^2_\lambda(\partial D)$, that is, the weighted Szegő kernel $K_\lambda(z, \bar{t})$ and the weighted Garabedian kernel $L_\lambda(z,t)$. The kernel $K_\lambda(z, \bar{t})$ is analytic in $(z, \bar{t})$ for $(z,t) \in D \times D$, the kernel $L_\lambda(z,t)$ has one simple pole at $z = t$ with residue $1/2\pi$, and is analytic in (z,t) for $(z,t) \in D \times D$ except for the diagonal set of $D \times D$. The kernels $K_\lambda(z, \bar{t})$ and $L_\lambda(z,t)$ are extended continuously onto the boundary for fixed $t \in D$. These kernels are uniquely determined and are connected by the boundary relation

$$\overline{K_\lambda(z, \bar{t})} \lambda(z) \, |dz| = \frac{1}{i} L_\lambda(z,t) dz, z \in \partial D, t \in D .$$

As usual, we define the kernel $\ell_\lambda(z,t)$ by $\ell_\lambda(z,t) = L_\lambda(z,t) - \frac{1}{2\pi(z-t)}$, the regular term of the weighted Garabedian kernel.

The Hardy H_2 kernel and its conjugate kernel are the special cases of ours ([Sa]), that is, those are the cases that the weight functions are given by

$$\lambda(z) = \frac{1}{2\pi}\frac{\partial G(z,\zeta)}{\partial n} \quad \text{and} \quad \lambda(z) = \frac{1}{2\pi}\Big/\frac{\partial G(z,\zeta)}{\partial n},$$

respectively, where $G(z,\zeta)$ is the Green's function for the Laplace equation on D with pole at ζ, and $\partial/\partial n$ denotes the inner normal derivative with respect to D.

Further, we shall denote the higher order derivatives of an analytic function $F_\lambda(z,t)$ in (z,t) defined on $D \times D$ for each nonnegative integers m and n as follows:

$$\frac{\partial^{m+n} F_\lambda(z,t)}{\partial z^m \partial t^n} = F_\lambda^{mn}(z,t) \quad \text{and} \quad F_\lambda^{0n}(z,t) = F_\lambda^{[n]}(z,t).$$

In ohter words, the expression $F_\lambda^{[n]}(z,t)$ denotes the n-th derivative with respect to the second variable of $F_\lambda(z,t)$. Especially, in case of $\lambda(z) = 1$ or $n = 0$, we remove 1 or [0], respectively.

Both kernels $K_\lambda^{[n]}(z,\bar{t})$ and $L_\lambda^{[n]}(z,t)$ are still connected by the boundary relation

$$\overline{K_\lambda^{[n]}(z,\bar{t})}\lambda(z)\,|dz| = \frac{1}{i} L_\lambda^{[n]}(z,t)dz, z \in \partial D, t \in D.$$

Applying the argument principle to this identity, we have the following:

Lemma 1 ([Ue-1]). The total number of zeros of the kernels $K_\lambda^{[n]}(z,\bar{t})$ and $L_\lambda^{[n]}(z,t)$ as functions of z cannot exceed $n+p-1$ in D.

Remark. It is well-known ([Be]) that the Szegő kernel $K(z,\bar{t})$ has exactly $p-1$ zeros in D by counting multiplicity and the Garabedian kernel $L(z,t)$ has no zeros on $\overline{D}$.

We introduced two concepts for weight function $\lambda(z)$.

Definition 1 ([Ue-Sa]). We say that a weight function $\lambda(z)$ belongs to the class W_0, if there exists a nonvanishing regular function $P(z)$ in D, continuous on $\bar{D}$ and $|P(z)|^2 = \lambda(z)$ on ∂D.

Definition 2 ([Ue-1]). We say that a weight funciton $\lambda(z)$ belongs to the calss $W_n (n \geq 1)$, if $\lambda(z)$ belongs to W_0 and the corresponding function $P(z)$ satisfies $P(a) \neq 0, P'(a) = \cdots = P^{(n)}(a) = 0$ for some point $a \in D$.

In the sequel, we shall set $Q(z) = 1/P(z)$. We defined a new concept for the zero point of a function $F(z,t)$ defined on $D \times D$:

Definition 3 ([Ue-1]). For a function $F(z,t)$ defined on $D \times D$, we say that the point $z = a$ is a zero point of the first kind of the function $F(z,t)$, if there exists a point $z = a \in D$ satisfying $F(a,t) = 0$ for all $t \in D$.

Now, we shall prepare the following three lemmas:

Lemma 2 ([Ne-1], [Ue-1]). For a simply connected domain D, any weight function $\lambda(z)$ always belongs to W_0. Moreover, if $\lambda(z)$ belongs to W_0, then the following identities hold for the weighted kernels of Szegő type:

$$K_\lambda(z,\bar{t}) = Q(z)\overline{Q(t)}K(z,\bar{t}),\ L_\lambda(z,t) = P(z)Q(t)L(z,t)\,,$$

$$\ell_\lambda(z,t) = -P(z)\left\{\frac{Q(z)-Q(t)}{2\pi(z-t)} - Q(t)\ell(z,t)\right\}.$$

Lemma 3 ([Ue-1]). If $\lambda(z)$ belongs to W_n, then the following identities hold for a fixed point $a \in D$:

$$K_\lambda^{[n]}(z,\bar{a}) = Q(z)\overline{Q(a)}K^{[n]}(z,\bar{a}),\ L_\lambda^{[n]}(z,a) = P(z)Q(a)L^{[n]}(z,a)\,,$$

$$\ell_\lambda^{[n]}(z,a) = -P(z)\left\{\frac{n!\,(Q(z)-Q(a))}{2\pi(z-a)^{n+1}} - Q(a)\ell^{[n]}(z,a)\right\}.$$

Lemma 4 ([Ue-1]). The following identities hold on any simply connected domain D:

$$\frac{\partial}{\partial t}\left\{\frac{L(z,t)}{K(t,\bar{u})}\right\} = 2\pi\frac{L(z,t)^2}{K(z,\bar{u})} \quad \text{and} \quad \frac{\partial}{\partial \bar{t}}\left\{\frac{K(z,\bar{t})}{K(u,\bar{t})}\right\} = 2\pi\frac{K(z,\bar{t})^2}{L(z,u)}\,.$$

2. THE NEHARI PROBLEM

According to Lemma 1, there is an interesting problem which was proposed by Nehari (for the case of $n = 0$) in 1950 ([Ne-1]): What conditions have to be imposed on $\lambda(z)$ and D in order to exclude the existence of zeros of the kernel $L_\lambda^{[n]}(z,t)$ in its domain? We call this "the Nehari problem". Why did we consider this problem? There are two main reasons as follows: Firstly, what are the essential differences between the classical Szegő kernel and the weighted kernel? Secondly, it comes from a generalization of the Ahlfors function. The Ahlfors function $F(z) = \frac{K(z,\bar{a})}{L(z,a)}$ is characterized as follows. The function $F(z)$ gives the maximum of $\mathrm{Re} f'(a) \geq 0, a \in D$ among all the functions $f(z)$, which are regular, single valued in D, and $|f(z)| \leq 1$ on D, and the extremal value is given by $F'(a) = 2\pi K(a, \bar{a})$. It is well known that the Ahlfors function $F(z)$ has many important properties and applications. If $L_\lambda(z,t) \neq 0$ on D, then we can get the representation of an interesting extremal function which gives a general form of the Ahlfors function as follows ([Ne-1]):

Let $\hat{B}(D)$ denote the class of analytic funtions $f(z)$ which belong to $H_2(D)$, $f(a) = 0$ at $a \in D$ and $|f(z)| \leq \frac{1}{\lambda(z)}$ on ∂D. Then the function $F_\lambda(z) = \frac{K_\lambda(z,\bar{a})}{L_\lambda(z,a)}$ maximizes $\mathrm{Re} f'(a)$ among all the functions $f(z)$ of $\hat{B}(D)$ and the extremal value is given by $F'_\lambda(a) = 2\pi K_\lambda(a, \bar{a})$.

It will be very difficult to give the perfect solution for the Nehari problem.

Theorem 1. Suppose that the kernel $L^{[n]}(z,a)$ has no zeros on $\overline{D}$ for a fixed point $a \in D$. Then the kernel $L^{[n]}_\lambda(z,a)$ doesn't vanish on $\overline{D}$ if and only if there exists a weight function $\mu(z)$ belonging to W_n such that two kernels $L^{[n]}_\mu(z,a)$ and $L^{[n]}_\lambda(z,a)$ coincide.

Proof. Construct a weight function $\mu(z) \in W_n$ as follows;

$$P(z) = \frac{L^{[n]}_\lambda(z,a)}{L^{[n]}(z,a)}, z \in D;\ |P(z)|^2 = \mu(z), z \in \partial D\,.$$

Then from the residue theorem and the boundary relation of two kernels $K^{[n]}_\lambda(z,\bar{a})$ and $L^{[n]}_\lambda(z,a)$, we have for each function $f(z) \in H^2_\lambda(\partial D)$,

$$\int_{\partial D} f(z)\overline{L^{[n]}_\lambda(z,a)}\frac{1}{\mu(z)}\,|dz| = \frac{1}{i}\int_{\partial D} f(z)K^{[n]}_\lambda(z,\bar{t})\frac{L^{[n]}(z,a)}{L^{[n]}_\lambda(z,a)}dz = 0\,.$$

By the orthogonal property of the kernel $L^{[n]}_\mu(z,a)$ for every function $f(z) \in H^2_\lambda(\partial D)$ and the above identity, we have

$$\int_{\partial D} f(z)\overline{\left\{L^{[n]}_\lambda(z,a) - L^{[n]}_\mu(z,a)\right\}}\frac{1}{\mu(z)}\,|dz| = 0\,.$$

By putting $f(z) = L_\lambda^{[n]}(z,a) - L_\mu^{[n]}(z,a)$, we obtain $L_\lambda^{[n]}(z,a) = L_\mu^{[n]}(z,a)$ on $\overline{D}$. Thus the necessity condition of this theorem is proved.

Since the sufficiency of the condition follows from Lemma 3, the proof of Theorem 1 is herewith complete.

It seems that the class W_n plays a central role in solving the Nehari problem.

Corollary ([Ue-Sa]). The kernel $L_\lambda(z,a)$ does not vanish on $\overline{D}$ if and only if there exists $\mu(z)$ belonging to W_0 such that the kernel $L_\mu(z,a)$ is identical with the kernel $L_\lambda(z,a)$ on $\overline{D}$.

There arises a problem([Ue-1]): To determine the conditions on D in order that the kernel $L^{[n]}(z,a)$ has no zeros in D for a point a of D.

If the kernel $L^{[n]}(z,a)$ has no zeros in D, then we can give the representation of a general form of the Ahlfors function as follows:

Let $B_n(D)$ denote the class of all analytic functions $f(z)$ which belong to $H_2(D), f^{(m)}(a) = 0 (m = 0,1,\cdots,n)$ at some fixed $a \in D$ and $|f(z)| \leq 1$ on D. Then the function $F_n(z) = \frac{K^{[n]}(z,\bar{a})}{L^{[n]}(z,a)}$ maximizes $\mathrm{Re} f^{(2n+1)}(a)$ among all the functions $f(z)$ of the class $B_n(D)$ and the extremal value is given by $F_n^{(2n+1)}(a) = 2\pi \frac{(2n+1)!}{(n!)^2} K^{nn}(a,\bar{a})$.

In the sequel, we shall assume $\lambda(z)$ belongs to W_0. We have obtained several results([Ue-1], [Ue-2]).

Theorem 2. The kernel $L_\lambda^{[n]}(z,t)$ has no zeros of the first kind in its domain.

Theorem 3. For a disk D, the kernel $K_\lambda^{[n]}(z,\bar{t})$ has n zeros of the first kind at $z = a_k (k = 1,\cdots,n)$ in D if and only if the regular function $P(z)$ is represented by a product of n Szegő kernels

$$P(z) = c \prod_{k=1}^{n} K(z,\overline{a_k}), a_k \in D\,,$$

where c is a nonzero constant.

Proof. Without loss of generality, we can assume that D is the unit disk and all the zeros of $K_\lambda^{[n]}(z,\bar{t})$ are simple. Since for each $k, K_\lambda^{[n]}(a_k,\bar{t}) = \frac{1}{P(a_k)} \frac{\partial^n}{\partial t^n} \left\{ \frac{K(a_k,\bar{t})}{P(t)} \right\} = 0$, we have the expression of the kernel $K(t,\overline{a_k}) = P(t)H_k(t)$, where $H_k(t)$ is a polynomial of degree $n-1$ and has no zeros in D. Therefore, we have

$$P(z) = \frac{1}{2\pi H_k(z)\,(1-\overline{a_k}z)}, (k=1,\cdots,n) .$$

In general, we have

$$P(z) = \frac{c}{(2\pi)^n\,(1-\overline{a_1}z)\cdots(1-\overline{a_n}z)} = cK\left(z,\overline{a_1}\right)\cdots K\left(z,\overline{a_n}\right) ,$$

where c is a nonzero constant. Thus we finish the proof of the necessity condition.

The converse statement follows from simple calculation, and so the proof of the theorem is herewith complete.

It follows easily from Lemma 4 that for a simply connected domain D, the kernel $K_\lambda^{[1]}(z,\bar{t})$ has a zero point of the first kind at $z=a$ if and only if the regular function $P(z)$ is given by the Szegő kernel $P(z)=cK(z,\bar{a})$, where c is a nonzero constant. Furthermore, we obtain the following:

Theorem 4. For a simply connected domain D, the kernel $K_\lambda^{[2]}(z,\bar{t})$ has two zeros of the first kind at $z=a$ and $z=b$ in D if and only if D is a disk and $P(z)=cK(z,\bar{a})K(z,\bar{b})$, where c is a nonzero constant.

Proof. If the kernel $K_\lambda^{[2]}(z,\bar{t})$ has a zero point of the first kind at $z=a$ in D, then we have

$$K_\lambda^{[2]}(a,\bar{t}) = \frac{1}{P(a)}\frac{\partial^2}{\partial\bar{t}^2}\left\{\frac{K\left(a,\bar{t}\right)}{\overline{P(t)}}\right\} = 0, \ i.e., P(t) = \frac{K(t,\bar{a})}{\bar{A}t+\bar{B}},$$

where $\bar{A}t+\bar{B}$ does not vanish in D. By Lemma 4, the kernel $K_\lambda^{[2]}(z,\bar{t})$ can be written in the form

$$\begin{aligned} K_\lambda^{[2]}(z,\bar{t}) &= \frac{\bar{A}z+\bar{B}}{K(z,\bar{a})}\frac{\partial^2}{\partial\bar{t}^2}\left\{\frac{K\left(z,\bar{t}\right)}{K\left(a,\bar{t}\right)}(A\bar{t}+B)\right\} \\ &= \frac{4\pi K\left(z,\bar{t}\right)\left(\bar{A}t+\bar{B}\right)}{K\left(z,\bar{a}\right)L(z,a)}\frac{\partial}{\partial\bar{t}}\left\{K\left(z,\bar{t}\right)(A\bar{t}+B)\right\} . \end{aligned}$$

Since $K_\lambda^{[2]}(b,\bar{t})=0$, we have

$$\bar{c}K\left(b,\bar{t}\right)(A\bar{t}+B) = 1, \ i.e., cK\left(t,\bar{b}\right) = \frac{1}{\bar{A}t+\bar{B}},$$

where c is a nonzero constant. This shows that $P(z) = cK(z,\bar{a})K(z,\bar{b})$ and D must be a disk. This proves the necessity condition for this theorem.

Since the converse follows easily from Theorem 3, the proof of Theorem 4 is herewith complete.

3. FURTHER RESULT

In 1956 Ozawa ([Oz]) obtained an important result for the regular term of the Garabedian kernel. He proved that $\ell(z,t) = 0$ if and only if D is a disk. We obtained a general form of this fact as follows:

Theorem 5 ([Ue-2]). For a disk $D, \ell_\lambda^{[n]}(z,t) = 0$ on $D \times D$ if and only if the regular function $P(z)$ is given by the form

$$P(z) = c\prod_{k=1}^{n} K(z,\overline{a_k}) \,,$$

where $a_k \in D(k = 1, \cdots, n)$ and c is a nonzero constant.

Proof. Without loss of generality, we assume that D is the unit disk. By Lemma 2, we have the expansion of $\ell_\lambda^{[n]}(z,t)$ around $z = t$ in the form

$$\ell_\lambda^{[n]}(z,t) = -\frac{1}{2\pi Q(z)}\sum_{m=0}^{\infty}\frac{(-1)^m}{(m+1)!}\sum_{k=0}^{n}{}_nC_k\frac{m!}{(m-k)!}Q^{(m+n-k+1)}(t)(t-z)^{m-k}\,.$$

If $\ell_\lambda^{[n]}(z,t) = 0$ on $D \times D$, then by letting z tend to t, we obtain

$$\ell_\lambda^{[n]}(t,t) = -\frac{1}{2\pi Q(t)}\sum_{m=0}^{n}\frac{(-1)^m}{(m+1)!}\frac{n!}{(n-m)!}Q^{(n+1)}(t) = -\frac{Q^{(n+1)}(t)}{2\pi(n+1)Q(t)} = 0\,.$$

Since the kernel $\ell_\lambda^{[n]}(t,t)$ is the regular function of t, the function $Q(t)$ reduces to a polynomial of order n such as $Q(t) = \sum_{k=0}^{n}\alpha_k t^k$ for suitable constants α_k. Thus by Lemma 2, the kernel $K_\lambda(z,\bar{t})$ can be written in the form

$$K_\lambda(z,\bar{t}) = \frac{Q(z)}{2\pi\,(1-z\bar{t})}\sum_{k=0}^{n}\overline{\alpha_k}\bar{t}^k = \frac{Q(z)}{2\pi}\left\{\Phi_{n-1}(\bar{t}) + \frac{C_0(z)}{1-z\bar{t}}\right\}\,,$$

where $\Phi_{n-1}(\bar{t})$ is a polynomial of $\bar{t}$ of order $n-1$ and $C_0(z) = \overline{Q\,(1/\bar{z})}$. Then we have

$$K_\lambda^{[n]}(z,\bar{t}) = \frac{n! z^n Q(z) C_0(z)}{2\pi\,(1 - z\bar{t}\,)^{n+1}} = \frac{n! Q(z)}{2\pi\,(1 - z\bar{t}\,)^{n+1}} \sum_{k=0}^{n} \overline{\alpha_k} z^{n-k} .$$

In this case, the kernel $K_\lambda^{[n]}(z,\bar{t}\,)$ has the first kind zeros at $z = a_k\,(k = 1,\cdots,n)$ in D, which are the solutions of the equation $\sum_{k=0}^{n} \overline{\alpha_k} z^{n-k} = 0$. Thus by Theorem 3, we have the proof of a necessity condition of this theorem.

Conversely, if the identity $P(z) = c \prod_{k=1}^{n} K(z,\overline{a_k})$ holds, then by Lemma 2, we have

$$L_\lambda(z,t) = \frac{Q(t)}{2\pi Q(z)(z-t)} = \frac{1}{2\pi Q(z)} \left\{ \Psi_{n-1}(t) + \frac{C_1(z)}{z-t} \right\} ,$$

where $\Psi_{n-1}(t)$ is a polynomial of t of order $n-1$ and $C_1(z) = Q(z)$. Thus we obtain $L_\lambda^{[n]}(z,t) = \frac{n!}{2\pi(z-t)^{n+1}}$, which shows $\ell_\lambda^{[n]}(z,t) = 0$. This completes the proof of Theorem 5.

Theorems 1,3,4 and 5 give partial answers for the Nehari problem under some conditions. There are many ways to attack the Nehari problem. We have tried to solve this problem from a very general way and have obtained some interesting results ([Ue-2]). There is an important identity, which was proved by Nehari in 1952 ([Ne-2]),

$$K_\lambda(z,\bar{t}\,)K_{\lambda^{-1}}(z,\bar{t}\,) = K(z,\bar{t}\,)^2 + \sum_{j=1}^{p-1} c_j F_j'(z) ,$$

where each $\mathrm{Re}F_j(z)(j = 1,\cdots,p-1)$ is the harmonic measure function of the boundary component C_j and c_j are connstants. Using this identity, we are expected to achieve our purpose, but we have not yet attained the perfect solution to the problem.

4. ACKNOWLEDGMENT

Before I close this paper, I would like to express my sincere gratitude to Professor Saburou Saitoh for his kind advice regarding this paper and for his invitation to the Congress of ISAAC'97.

REFERENCES

[Be] S.Bergman, The Kernel Function and Conformal Mapping, Amer. Math. Soc. Providence, New York,1970, 2nd ed.

[Be-Sc] S.Bergman and M.Schiffer, Kernel Functions and Elliptic Differential Equations, Academic Press, New York,1953.

[Ep] B.Epstein, Orthogonal families of analytic function, The Macmillan Comp., New York,1965.

[Ne-1] Z.Nehari, A class of domain functions and some allied extremal problems, Trans. Amer. Math. Soc.,**69**, pp.161–178,1950.

[Ne-2] Z.Nehari, On weighted kernels, J.d'Analyse Math.,**2**, pp.126–149, 1952.

[Oz] M.Ozawa, Some estimations on the Szegő kernel function, Kōdai Math. Sem. Report,**8**, pp.71–78,1956.

[Sa] S.Saitoh, Theory of reproducing kernels and its applications, Pitman Research Notes in Math.,**189**, Longman, London,1988.

[Ue-1] M.Uehara, On the weighted Szegő L-kernel, Math. Japon.,**38**, pp. 1167–1174,1993.

[Ue-2] M.Uehara, On the weighted Szegő kernels, Math. Japon.,**42**, pp. 459–469,1995.

[Ue-Sa] M.Uehara and S.Saitoh, Some remarks for the weighted Szegő kernel functions, Math. Japon.,**29**, pp.887–891,1984.

18 FAY'S TRISECANT FORMULA AND HARDY H^2 REPRODUCING KERNELS

Akira Yamada

Department of Mathematics and Informatics,
Tokyo Gakugei University, Japan
yamada@u-gakugei.ac.jp

Abstract: By means of Riemann's theta function and Klein's prime form, we can express many important conformal invariants defined on a planar regular region. Fay's trisecant formula is the key to obtain various identities and inequalities among them. Also, we give a short proof of the trisecant formula and discuss its application to an analogue of the Pick-Nevanlinna extremal problems for Hardy H^2 spaces.

1. INTRODUCTION

Let R be a planar regular region with n boundary components. Its Schottky double $\hat{R}$ is a compact Riemann surface of genus $g = n - 1$ admitting an anti-conformal involution ϕ fixing the boundary ∂R of R. For simplicity we adopt the notation that $\bar{z} = \phi(z)$ for $z \in \hat{R}$ and $\bar{R} = \phi(R)$. Note that $\bar{R}$ is not the closure of R. Let C denote a general (not necessarily symmetric) compact Riemann surface of genus g.

For definitions, notation and relevant properties of theta functions $\theta(z)$ and prime-forms $E(x, y)$ used in this report, we refer the readers to the excellent lecture note by Fay [2].

Fay's trisecant formula is perhaps one of the most important tool for applications of theta functions to the theory of conformal invariants. In fact, almost all of the results included in this report are deduced from the trisecant formula.

To introduce it, we briefly review the work of Hejhal and Fay on an beautiful identity relating the Szegö kernel and the Bergman kernel.

In 1972 by using theta functions D. Hejhal [3] solved an open problem mentioned in the book of Sario and Oikawa [7]: If $g > 0$ then, for all $z \in R$,

$$C_B(z)^2 < \pi K(z, \bar{z}) \tag{1}$$

where $C_B(z)$ is the analytic capacity and $K(z, \overline{w})$ is the Bergman kernel. To prove this he found the following identity: for all $x, y \in C$ and $e \in \mathbb{C}^g$ with $\theta(e) \neq 0$

$$\frac{\theta(y - x - e)\theta(y - x + e)}{\theta^2(e)E(x, y)^2} = \omega(x, y) + \sum_{i,j=1}^{g} \frac{\partial^2 \log \theta}{\partial z_i \partial z_j}(e)u_i(x)u_j(y), \tag{2}$$

where $\omega(x,y)=\frac{d^2}{dxdy}\log E(x,y)dxdy$ is the bilinear normalized differential of the second kind on a compact Riemann surface C of genus g and $\{u_i(x)\}_{i=1}^g$ are the normalized differentials of the first kind with some canonical homology basis. Specializing $x=z,\ y=\bar{z}$ and $e=0$ in (2) we have

$$\left(\frac{\theta(z-\bar{z})}{\theta(0)iE(z,\bar{z})}\right)^2=-\omega(z,\bar{z})-\sum_{i,j=1}^{g}\frac{\partial^2\log\theta}{\partial z_i\partial z_j}(0)u_i(z)\overline{u_j(z)}. \tag{3}$$

By direct calculation using the definition of theta functions, the matrix $(\frac{\partial^2\log\theta}{\partial z_i\partial z_j}(0))$ is easily seen to be positive definite. Noting that $\frac{\theta(z-\bar{z})}{\theta(0)iE(z,\bar{z})}=C_B(z)$, $-\omega(z,\bar{z})=\pi K(z,\bar{z})$, Hejhal obtained (1) from (3) immediately.

In 1973 J. D. Fay obtained a formula more powerful than (2) in his lecture note [2] (trisecant formula):
for $x,y,a,b\in C$ and $e\in\mathbb{C}^g$ with $\theta(e)\neq 0$

$$\begin{aligned}\frac{\theta(x-a-e)\theta(y-b-e)}{\theta^2(e)E(x,a)E(y,b)}&-\frac{\theta(x-b-e)\theta(y-a-e)}{\theta^2(e)E(x,b)E(y,a)}\\&=\frac{\theta(x+y-a-b-e)E(x,y)E(b,a)}{\theta(e)E(x,a)E(x,b)E(y,a)E(y,b)}.\end{aligned}\tag{4}$$

Hejhal's formula (2) is obtained from (4) as follows. Dividing both sides of (4) by $\theta(x-a-e)/E(x,a)$ and then letting $y\to b$ using L'Hospital's rule, we have, after suitably permuting the variables,

$$\begin{aligned}&\frac{\theta(x-a+e)\theta(x-b-e)E(a,b)}{\theta(e)\theta(e+b-a)E(x,a)E(x,b)}\\&\qquad=\omega_{b-a}(x)+\sum_{j=1}^{g}\left[\frac{\partial\log\theta}{\partial z_j}(e+b-a)-\frac{\partial\log\theta}{\partial z_j}(e)\right]u_j(x),\end{aligned}\tag{5}$$

where $\omega_{b-a}(x)$ is the normalized differential of the third kind with simple poles of residue $-1,1$ at a,b respectively. Again differentiating on b at $b=a$, we have Hejhal's formula (2). Thus, (4) $\Longrightarrow$ (5) $\Longrightarrow$ (2). In fact, Hejhal deduced (2) by first proving (5).

In conclusion we can say that Fay has found a formula stronger than Hejhal's but Hejhal has found the meaning of the formula. We emphasize that the key point of the proof was the positive definiteness of the matrix $(\frac{\partial^2\log\theta}{\partial z_i\partial z_j}(0))$.

2. A PROOF OF THE TRISECANT FORMULA

In view of its importance we give here a short proof of Fay's trisecant formula.

For $a,x,y\in C$ and $e\in\mathbb{C}^g$ with $\theta(e)\neq 0$, put

$$\Lambda_a^e(x,y)=\frac{\theta(x-a-e)\theta(x-y+e)}{\theta(e)\theta(y-a-e)}\frac{E(a,y)}{E(x,a)E(x,y)}.$$

Observe that $\Lambda_a^e(x,y)$ is an Abelian differential in x of the third kind on C with simple poles of residue -1 and 1 at a and y respectively. Furthermore, note that $\Lambda_a^e(x,y)$ is invariant under addition of periods to e by the periodicity of theta functions, so that we see that it is in fact a function of $e\in J(C)=\mathbb{C}^g/(2\pi iI,\tau)$ the Jacobian variety of C. Considering singularity it is clear that the differential

$\Lambda_a^e(x,y)$ is expressible as a linear combination of $\omega_{y-a}(x)$ and $u_i(x)$ $(i = 1,\ldots,g)$. More precisely we have the following

Lemma 2.1 (Fay). *Let $u(\mathcal{A})$ be the $g \times g$ matrix $(u_i(a_j))$ where $\mathcal{A} = \sum_1^g a_i$ is the divisor of the function $\theta(\cdot - a - e)$ in C. Then*

$$\Lambda_a^e(x,y) = \frac{1}{\det(u(\mathcal{A}))} \begin{vmatrix} \omega_{y-a}(x) & \omega_{y-a}(a_1) & \ldots & \omega_{y-a}(a_g) \\ u_1(x) & u_1(a_1) & \ldots & u_1(a_g) \\ \vdots & \vdots & \ddots & \vdots \\ u_g(x) & u_g(a_1) & \ldots & u_g(a_g) \end{vmatrix}. \tag{6}$$

Proof. For fixed a and y, let us denote by $\Lambda(x)$ the right-hand side of (6). Since the index of speciality $i(\mathcal{A}) = 0$ by Riemann's vanishing theorem, we see easily that the determinant $\det(u(\mathcal{A}))$ does not vanish, and so $\Lambda(x)$ is a well-defined differential in x of the third kind with simple poles of residue -1 and 1 at a and y respectively. Also it is clear that $\Lambda(x)$ vanishes on $a_1,\ldots,a_g$. Thus, $\Lambda_a^e(x,y) - \Lambda(x)$ is a differential in x which is holomorphic everywhere on C and vanishes at the divisor $\mathcal{A}$. Since $i(\mathcal{A}) = 0$, we conclude that $\Lambda_a^e(x,y) = \Lambda(x)$ as desired. □

Now we proceed to prove the trisecant formula (4). As in the above Lemma let $\mathcal{A}$ be the divisor of the function $\theta(\cdot - a - e)$ in C. Then Riemann's vanishing theorem asserts that $e = \mathcal{A} - a - \Delta$ in $J(C)$ where Δ is the Riemann divisor class. By continuity we may assume without loss of generality that $b \notin \mathcal{A}$. Then by Riemann's vanishing theorem we see that $\theta(d) \neq 0$ with $d = \mathcal{A} - b - \Delta$. In view of the identity $\omega_{y-a}(x) + \omega_{a-b}(x) = \omega_{y-b}(x)$, Lemma 2.1 implies that

$$\Lambda_a^e(x,y) + \Lambda_b^d(x,a) = \Lambda_b^d(x,y)$$

for all $x, y, a, b \in C$. Since $d = a - b + e$ in $J(C)$, the above identity is, by suitably renaming the variables, easily seen to be the same as the trisecant formula, which completes the proof.

3. RELATION BETWEEN CONFORMAL INVARIANTS AND THETA FUNCTIONS

Let R be a planar regular region with boundary components $\Gamma_0,\ldots,\Gamma_{n-1}$.

We fix a symmetric canonical homology basis $\{A_i, B_j\}$ $(i,j = 1,\ldots,g)$ on R such that $B_j = \Gamma_j$ $(j = 1,\ldots,g)$ and the cycles $\{A_i\}$ $(i = 1,\ldots,g)$ satisfy the relations in $H_1(\hat{R},\mathbb{Z})$:

$$\phi(A_i) = -A_i, \ \phi(B_j) = B_j, \quad (i,j = 1,\ldots,g)$$

Let $u_1,\ldots,u_g$ be the normalized differentials of the first kind on $\hat{R}$ such that $\int_{A_i} u_j = 2\pi\sqrt{-1}\delta_{ij}$ (Kronecker delta), then

$$\phi^* u_j = \overline{u_j}, \quad (j = 1,\ldots,g). \tag{7}$$

The period matrix τ of $\hat{R}$ is by definition the $g \times g$ matrix $(\int_{B_i} u_j)$ $(i,j = 1,\ldots,g)$. It is well-known that τ is hermitian with $\operatorname{Re}\tau < 0$. In our case, however, from symmetry (7) we see easily that τ is a real symmetric matrix.

Remark 3.1. Here it should be pointed out that our choice of the canonical homology basis is different from Fay's lecture note [2]: we interchanged the A_i cycles with B_j cycles. Thus some of Fay's formulas must be modified suitably according as the transformation law of theta functions for the change of the homology basis [2, p. 7].

We record some symmetry relations because of its importance [2, Chap. VI].

Proposition 3.1 (Fay). *On $\hat{R}$ the following symmetry holds: for all $x, y \in \hat{R}$ and $e \in \mathbb{C}^g$*

$$\overline{\theta(x-y+e)} = \theta(\bar{x}-\bar{y}+\bar{e}),\ \overline{E(x,y)} = E(\bar{x},\bar{y}),\ \overline{\omega(x,y)} = \omega(\bar{x},\bar{y}), \tag{8}$$

where $\bar{e}$ is defined by $(\overline{e_1}, \dots, \overline{e_g})$ if $e = (e_1, \dots, e_g)$.

There is a simple relation between the prime-forms and Green's function $g(x,y)$ on R.

Lemma 3.1. $|\frac{E(x,y)}{E(x,\bar{y})}| = \exp(-g(x,y))$, *for all $x, y \in R$.*

Proof. See Lemma 2.2 of [11]. □

Proposition 3.2. *Let $K(x,\bar{y})$ and $\hat{K}(x,\bar{y})$ be the Bergman kernel and the Szegö kernel on R. Then we have*

$$K(x,\bar{y}) = -\frac{1}{\pi}\omega(x,\bar{y}),\ \hat{K}(x,\bar{y}) = \frac{1}{2\pi i}\frac{\theta(x-\bar{y})}{\theta(0)E(x,\bar{y})}.$$

Proof. See [3] or [2]. However, by using Schiffer's identity $K(x,\bar{y}) = -\frac{2}{\pi}\frac{\partial^2 g(x,y)}{\partial x \partial y}$ [6, p.21], the first identity is an easy consequence of Lemma 3.1, since we have $\omega(x,y) = \frac{d^2}{dxdy}\log E(x,y)$ [2, p.20]. □

As seen in Introduction, many conformal invariants are expressed by theta functions and prime-forms.

Let $C_B(a)$ and $C_\beta(a)$ be the analytic capacity and the logarithmic capacity of R at $a \in R$ respectively. More explicitly these constants are defined by $C_B(a) = \sup\{|f'(a)| \mid f \in H^\infty(R) \text{ and } ||f||_\infty \le 1\}$ and $C_\beta(a) = \lim_{z\to a}\exp(-g(z,a))/|z-a|$, where $H^\infty(R)$ is the space of bounded holomorphic functions on R.

Proposition 3.3. $C_B(x) = \frac{\theta(x-\bar{x})}{\theta(0)iE(x,\bar{x})}$, $C_\beta(x) = \frac{1}{iE(x,\bar{x})}$.

Proof. Since $C_B(x) = 2\pi\hat{K}(x,\bar{x})$ [1, p.118], the first identity is clear from Proposition 3.2. By noting that, near $x = y$, the expansion $E(x,y) = y - x +$ higher order terms holds [2, p.19], the second identity follows from Lemma 3.1. □

Remark 3.2. Our choice of a canonical homology basis permits us to assume that the vector $x - \bar{x} \in \mathbb{C}^g$ is purely imaginary for all $x \in R$. Then from the definition of theta function it is easy to see that $\theta(x-\bar{x}) < \theta(0)$ for all $x \in R$ when $g > 0$. Thus Proposition 3.3 gives immediately the trivial inequality $C_B(x) < C_\beta(x)$.

Remark 3.3. Now Hejhal's identity (3) is rewritten as

$$C_B(x)^2 = \pi K(x,\bar{x}) - \sum_{i,j=1}^{g}\frac{\partial^2 \log\theta}{\partial z_i \partial z_j}(0)u_i(x)\overline{u_j(x)}.$$

Hence the inequality $C_B(x)^2 < \pi K(x,\bar{x})$ holds at once as stated in the Introduction.

In 1972 N. Suita [8] conjectured, when $g > 0$, the inequality

$$C_\beta(x)^2 < \pi K(x, \bar{x}), \ \forall x \in R, \tag{9}$$

which is called the *Suita conjecture*. As far as the author knows, this problem is still open for planar domains with connectivity greater than two. Although $C_\beta(x)$ is expressed by the prime-form as $C_\beta(x) = \frac{1}{iE(x,\bar{x})}$, there seems to be no useful identity between $\frac{1}{E(x,\bar{x})^2}$ and $\omega(x, \bar{x})$ by the multi-valuedness of the prime-form.

S. Saitoh considered an analogous problem for Hardy H^2 kernel and posed an open problem [6, p.37], in our context, to prove that the matrix

$$\left(\frac{\partial^2 \log \theta}{\partial z_i \partial z_j}(e_0)\right) \text{ is negative definite.} \tag{10}$$

This is called the *Saitoh conjecture*. The constant e_0 is determined from the critical points of Green's function of R whose definition will be given in the next section.

Although we were not able to prove the above conjecture for $n \geq 3$, we shall show in the next section its relative version such that the matrix $\left(\frac{\partial^2 \log \theta}{\partial z_i \partial z_j}(a - \bar{a} + e) - \frac{\partial^2 \log \theta}{\partial z_i \partial z_j}(e)\right)$ is positive definite for e in some open set in $\hat{T}_0$ (Theorem 4.2).

4. WEIGHTED HARDY H² KERNELS AND THETA FUNCTIONS

We collect here the main results of the paper [11]. For proofs of these results see [11].

In connection with the extremal problems on the generalized Tchebycheff polynomials, H. Widom [9] studied multi-valued analytic functions on R. He obtained some result on the ranges of the extremal quantities with the help of the following Lemma [9]. For $a \in R$ let P_a be the set of positive differentials on $R \cup \partial R$ which is holomorphic except for a simple pole at a.

Lemma 4.1 (Widom). *Let $\{z_j\}_{j=1}^g$ be the zeroes of a positive differential in P_a. Then the inequality*

$$\sum_{j=1}^{g} g(z_j, a) \leq \sum_{j=1}^{g} g(z_j^*, a)$$

holds, where the set $\{z_j^*\}_{j=1}^g$ is the critical points of $g(\cdot, a)$ in R. The equality holds if and only if $\{z_j\}_{j=1}^g = \{z_j^*\}_{j=1}^g$.

On the other hand, by using theta functions, J. D. Fay [2] found a parameterization of the positive differentials on R by the set $\hat{T}_0 = \{z \in \mathbb{C}^g | \sqrt{-1}\, z \in \mathbb{R}^g\}$ and remarked that $\theta(e) > 0$ for all $e \in \hat{T}_0$ where $\theta(z)$ is the Riemann theta function with respect to the above canonical homology basis on $\hat{R}$.

We reformulate Lemma 4.1 in the context of theta functions. To this end, it is convenient to introduce the constants $e_0 = \sum_{j=1}^g z_j^* - a - \Delta \in \hat{T}_0$ and $e_1 = \sum_{j=1}^g \overline{z_j^*} - a - \Delta \in \hat{T}_0$ where Δ is the Riemann divisor class. Given $a \in R$ these constants are determined modulo $2\pi i \mathbb{Z}^g$ and have the following extremal property.

Theorem 4.1. *For $a \in R$ and $e \in \hat{T}_0$, we have*

$$\exp\left(-\sum_{j=1}^{g} g(z_j^*, a)\right) \leq \theta(a - \bar{a} + e)/\theta(e) \leq \exp\left(\sum_{j=1}^{g} g(z_j^*, a)\right).$$

The equality occurs on the second (resp. first) inequality if and only if $e = e_0$ (resp. $e = e_1$).

Theorem 4.1 implies that, for fixed a, the function $F(e) = \theta(a-\bar{a}+e)/\theta(e)$ $(e \in \hat{T}_0)$ attains its maximum at $e = e_0 \in \hat{T}_0$. Hence it is trivial that the first derivatives $(\frac{\partial \log F}{\partial z_j}(e_0))$ vanish. As for the second derivatives, we naturally expect that the matrix $(\frac{\partial^2 \log F}{\partial z_i \partial z_j}(e_0))$ is positive definite.

Remark 4.1. Since our parameter space $\hat{T}_0$ is purely imaginary, the second derivatives have opposite sign as usual. Hence it is not negative definite but positive definite.

To show indeed $(\frac{\partial^2 \log F}{\partial z_i \partial z_j}(e_0))$ is positive definite, for fixed $a \in R$ let us introduce four subsets T_+, T_{++}, T_- and T_{--} of $\hat{T}_0$ by

$$T_+ = \mathcal{P} - a - \Delta \text{ and } T_{++} = \mathcal{P}_0 - a - \Delta$$
$$T_- = \overline{\mathcal{P}} - a - \Delta \text{ and } T_{--} = \overline{\mathcal{P}_0} - a - \Delta$$

where $\mathcal{P}$ (resp. $\mathcal{P}_0$) is the set of zero divisors δ in $R \cup \partial R$ of a positive (resp. strictly positive) differential ω with a simple pole at a satisfying $\delta + \overline{\delta} - a - \bar{a} = \text{div}(\omega)$. In view of this notation, $e_0 \in T_{++}$ and $e_1 \in T_{--}$. It is easily seen that T_{++} is an open subset of $\hat{T}_0$ with its closure equals T_+. Similar assertion holds for the sets T_- and T_{--}.

Then, we have

Theorem 4.2. *For $e \in \hat{T}_0$ and $a \in R$ let Φ be the $g \times g$ matrix $(\frac{\partial^2 \log F}{\partial z_i \partial z_j}(e))$. Then, Φ is a real matrix and the following hold.*

1. *For any $e \in T_+$ the matrix Φ is positive semi-definite. Φ is positive definite for $e \in T_+$ if and only if $e \in T_{++}$.*
2. *For any $e \in T_-$ the matrix Φ is negative semi-definite. Φ is negative definite for $e \in T_-$ if and only if $e \in T_{--}$.*

The proof of the positive definiteness of Φ is obtained by the following scheme:

1. Construct a weighted Hardy H^2-differential space X with weight coming from a strictly positive differential,
2. Orthogonal decomposition: $X = Y \oplus \{\text{Schottky differentials}\}$ for some Hilbert space Y,
3. Identity among the kernel functions: $k_X = k_Y + \sum c_{ij} u_i \overline{u_j} \implies (c_{ij}) > 0$ where k_X (resp. k_Y) denotes the kernel function of the space X (resp. Y),
4. The matrix $\{c_{ij}\}$ coincides with the desired matrix Φ.

Now we give an outline of proof of Theorem 4.2.

Definition 4.1 (c.f. [6, Chapter III]). For $e \in T_{++}$ let

$$\omega_e(x) = \frac{i\theta(x-a-e)\theta(x-\bar{a}+e)E(a,\bar{a})}{\theta(e)\theta(a-\bar{a}+e)E(x,a)E(x,\bar{a})} \tag{11}$$

be the strictly positive differential on R with simple poles of residue $-i, i$ at $a, \bar{a}$. Denote by $H^2_e(R)$ the Hilbert space of holomorphic functions $f(z)$ on R such that the function $|f(z)|^2$ admits a harmonic majorant on R. For $f, g \in H^2_e(R)$ the inner product is defined by $\langle f, g\rangle_e = \frac{1}{2\pi}\int_{\partial R} f\bar{g}\omega_e$ where $f\bar{g}$ in the integrand is evaluated by its non-tangential boundary value on ∂R. Also, denote by $H^{2,1}_e(R)$ the Hilbert space of holomorphic differentials $\xi(z)$ on R such that $\xi(z)/dz \in H_e^2(R)$. For $\xi, \eta \in H_e^{2,1}(R)$ the inner product is defined by $\langle \xi, \eta \rangle_{e,1} = \frac{1}{2\pi} \int_{\partial R} \xi\bar{\eta}/\omega_e$.

By the residue theorem we obtain explicit representations of the kernel functions of the spaces $H_e^2(R)$ and $H_e^{2,1}(R)$.

Lemma 4.2. *The Hilbert spaces $H_e^2(R)$ and $H_e^{2,1}(R)$ possess the reproducing kernels $R_e(x, \bar{y})$ and $\hat{R}_e(x, \bar{y})$ respectively given by:*

$$R_e(x,\bar{y}) = \frac{\theta(a-\bar{a}+e)\theta(x-\bar{y}+e)}{\theta(x-\bar{a}+e)\theta(a-\bar{y}+e)} \frac{E(x,\bar{a})E(\bar{y},a)}{E(\bar{a},a)E(x,\bar{y})}, \quad x, y \in R \tag{12}$$

$$\hat{R}_e(x,\bar{y}) = \frac{\theta(x-\bar{a}+e)\theta(a-\bar{y}+e)\theta(x-\bar{y}-e)E(a,\bar{a})}{\theta^2(e)\theta(a-\bar{a}+e)E(x,\bar{y})E(x,\bar{a})E(\bar{y},a)}, \quad x, y \in R. \tag{13}$$

In particular,

$$\hat{R}_e(x,\bar{y}) = -R_e(\bar{y},x)\omega_e(x)\omega_e(\bar{y}). \tag{14}$$

We shall call $R_e(x, \bar{y})$ the *Hardy H^2 kernel* and $\hat{R}_e(x, \bar{y})$ the *conjugate Hardy H^2 kernel* associated with $e \in T_{++}$ (c.f. [6, Chapter III]). Note that $R_{e_0}(x, \bar{y}) = K_a(x, \bar{y})$ and $\hat{R}_{e_0}(x, \bar{y}) = \hat{K}_a(x, \bar{y})$ where $K_a(x, \bar{y})$ and $\hat{K}_a(x, \bar{y})$ are the classical Hardy H^2 kernel and conjugate Hardy H^2 kernel respectively [6, p.30].

Let $H_{e,a}^2(R)$ be the subspace $\{f \in H_e^2(R) | f(a) = 0\}$ of $H_e^2(R)$. For $f \in H_{e,a}^2(R)$ define the mapping l: $H_{e,a}^2(R) \to H_e^{2,1}(R)$ by $l(f) = f\omega_e$. It follows at once that the mapping l is a complex linear isometry from $H_{e,a}^2(R)$ into $H_e^{2,1}(R)$. Identifying the spaces $H_{e,a}^2(R)$ and $l(H_{e,a}^2(R))$ via l we may regard $H_{e,a}^2(R)$ as a subspace of $H_e^{2,1}(R)$.

Lemma 4.3. *The following orthogonal decomposition holds:*

$$H_e^{2,1}(R) = H_{e,a}^2(R) \oplus \Gamma(\hat{R}) \tag{15}$$

where $\Gamma(\hat{R})$ denotes the space of holomorphic differentials on R which extend to Abelian differentials of the first kind on $\hat{R}$.

Thus, by means of the orthogonal decomposition (15) we have

$$\hat{R}_e(x,\bar{y}) = (R_e(x,\bar{y}) - 1)\omega_e(x)\omega_e(\bar{y}) + \sum_{i,j=1}^{g} c_{ij}u_i(x)u_j(\bar{y}) \tag{16}$$

where (c_{ij}) $(i, j = 1, \ldots, g)$ is a positive definite matrix.

To determine the matrix (c_{ij}) explicitly, we need the following theorem derived from applying Fay's trisecant formula three times. To state it we introduce the convenient notation $(a, \bar{b}) = \frac{\theta(a-\bar{b}-e)}{\theta(e)E(a,\bar{b})}$ similar to the Szegö kernel.

Theorem 4.3. *For $a, b, x, y \in \hat{R}$ and $e \in \mathbb{C}^g$ with $\theta(e) \neq 0$, we have*

$$\frac{(a,x)(x,y)(y,b)}{(a,b)} + \frac{(a,y)(y,x)(x,b)}{(a,b)} - \frac{(a,x)(x,b)(a,y)(y,b)}{(a,b)^2} = \sum_{i,j=1}^{g} \left\{ \frac{\partial^2 \log\theta}{\partial z_i \partial z_j}(a\text{-}b\text{-}e) - \frac{\partial^2 \log\theta}{\partial z_i \partial z_j}(e) \right\} u_i(x)u_j(y). \tag{17}$$

From (12),(13),(16) and (17) we find that $c_{ij} = \frac{\partial^2 \log\theta}{\partial z_i \partial z_j}$(a-b-e) $- \frac{\partial^2 \log\theta}{\partial z_i \partial z_j}(e)$. Hence for all $e \in T_{++}$ the matrix $\frac{\partial^2 \log\theta}{\partial z_i \partial z_j}(a - \bar{a} + e) - \frac{\partial^2 \log\theta}{\partial z_i \partial z_j}(e)$ is positive definite. This completes an outline of the proof of Theorem 4.2.

5. A REMARK ON THE SAITOH CONJECTURE

From (12) and (13) we have, for all $x, y \in R$,

$$R_e(x,\bar{y})\hat{R}_e(x,\bar{y}) = \frac{\theta(x-\bar{y}+e)\theta(x-\bar{y}-e)}{\theta^2(e)(iE(x,\bar{y}))^2}, \tag{18}$$

and so by (2) and (7)

$$R_e(x,\bar{y})\hat{R}_e(x,\bar{y}) = \pi K(x,\bar{y}) - \sum_{i,j=1}^{g} \frac{\partial^2 \log\theta}{\partial z_i \partial z_j}(e) u_i(x)\overline{u_j(y)}. \tag{19}$$

In particular the Saitoh conjecture (10) implies

$$\pi K(a,\bar{a}) < \hat{R}_{e_0}(a,\bar{a}), \tag{20}$$

since $R_{e_0}(a,\bar{a}) = 1$. Although we cannot prove the inequality (20) (the weak Saitoh conjecture) except for the case $g = 1$ [10, 5], it is of some interest to see that the following is an easy consequence of Theorem 4.1.

Proposition 5.1. *If $g > 0$ then for all $a \in R$*

$$C_\beta(a)^2 < \hat{R}_{e_0}(a,\bar{a}).$$

Proof. From (18) and Proposition 3.3 it suffices to show that

$$\frac{\theta(a-\bar{a}+e_0)\theta(a-\bar{a}-e_0)}{\theta^2(e_0)} > 1,$$

or equivalently, $F(e_0)/F(e_0 - a + \bar{a}) > 1$ with $F(e) = \theta(a - \bar{a} + e)/\theta(e)$, $e \in \hat{T}_0$. But this is clear, since the function $F(e)$ takes a maximum only at $e = e_0$ by Theorem 4.1. □

Of course, if both the Suita and the Saitoh conjectures are proved then the above inequality becomes trivial.

6. REPRESENTATION OF THE HARDY H^2 EXTREMAL FUNCTIONS

As a counterpart of the trisecant formula or its generalization, we give explicit representations of some extremal functions in $H_e^2(R)$.

First we need an analogue of Bergman's formula on minimum integral [1, p.26].

Proposition 6.1. *Let $\mathcal{H}$ be a Hilbert space with inner product $\langle\cdot,\cdot\rangle$ and assume that the set $\{x_j\}_{j=1}^n$ is linearly independent in $\mathcal{H}$. Then, for any $(b_j)_{j=1}^n \in \mathbb{C}^n$, the element $f_n \in \mathcal{H}$ defined by*

$$f_n = -\frac{1}{D_n}\begin{vmatrix} 0 & x_1 & \dots & x_n \\ b_1 & \langle x_1, x_1\rangle & \dots & \langle x_n, x_1\rangle \\ \vdots & \vdots & \ddots & \vdots \\ b_n & \langle x_1, x_n\rangle & \dots & \langle x_n, x_n\rangle \end{vmatrix} \tag{21}$$

is the unique element of $\mathcal{H}$ *with the minimum norm* Λ_n *satisfying* $\langle f_n, x_j\rangle = b_j$ *for all* $j = 1, \ldots, n$*, where* D_n *denotes* $\det(\langle x_j, x_i\rangle)$ *and* Λ_n *is given by*

$$\Lambda_n^2 = -\frac{1}{D_n}\begin{vmatrix} 0 & \overline{b_1} & \ldots & \overline{b_n} \\ b_1 & \langle x_1, x_1\rangle & \ldots & \langle x_n, x_1\rangle \\ \vdots & \vdots & \ddots & \vdots \\ b_n & \langle x_1, x_n\rangle & \ldots & \langle x_n, x_n\rangle \end{vmatrix}. \tag{22}$$

Proof. First note that D_n is positive, since $\{x_j\}_{j=1}^n$ is linearly independent. From (21) the inner product $\langle f_n, x_j\rangle$ is given by

$$\langle f_n, x_j\rangle = -\frac{1}{D_n}\begin{vmatrix} 0 & \langle x_1, x_j\rangle & \ldots & \langle x_n, x_j\rangle \\ b_1 & \langle x_1, x_1\rangle & \ldots & \langle x_n, x_1\rangle \\ \vdots & \vdots & \ddots & \vdots \\ b_n & \langle x_1, x_n\rangle & \ldots & \langle x_n, x_n\rangle \end{vmatrix}, \tag{23}$$

and so expanding the determinant in the first column we see that $\langle f_n, x_j\rangle = b_j$ $(j = 1, \ldots, n)$. To show that f_n has the desired extremal property assume that $g \in \mathcal{H}$ satisfies $\langle g, x_j\rangle = b_j$ $(j = 1, \ldots, n)$. Then $f_n - g \perp \{x_j\}_{j=1}^n$. Since f_n is a linear combination of $\{x_j\}_{j=1}^n$, we have $f_n - g \perp f_n$ which implies $\|g\|^2 = \|f_n\|^2 + \|g - f_n\|^2$. Thus it is clear that f_n is uniquely extremal. Again, since f_n is a linear combination of $\{x_j\}_{j=1}^n$, from (23) the norm of f_n is given by

$$\Lambda_n^2 = \langle f_n, f_n\rangle = -\frac{1}{D_n}\begin{vmatrix} 0 & \langle x_1, f_n\rangle & \ldots & \langle x_n, f_n\rangle \\ b_1 & \langle x_1, x_1\rangle & \ldots & \langle x_n, x_1\rangle \\ \vdots & \vdots & \ddots & \vdots \\ b_n & \langle x_1, x_n\rangle & \ldots & \langle x_n, x_n\rangle \end{vmatrix},$$

which completes the proof. □

Let $\mathcal{H}_K$ be the reproducing kernel Hilbert space on E with the reproducing kernel $K(x, y)$. For simplicity we write $K_a = K(\cdot, \bar{a})$.

Corollary 6.1. *Let* $\{a_j\}_{j=1}^\infty$ *be a sequence in* E*. Assume that the set* $\{K_{a_j}\}_{j=1}^\infty$ *is linearly independent in* $\mathcal{H}_K$ *and that an orthonormal sequence* $\{F_n\}_{n=1}^\infty$ *is obtained from* $\{K_{a_j}\}_{j=1}^\infty$ *by the Gram-Schmidt orthogonalization process. Then, for any* $n \geq 1$*,* F_n *is the unique extremal function for the problem*

$$\sup_{\|f\|\leq 1} \{\operatorname{Re} f(a_n) \mid f \in \mathcal{H}_K,\ f(a_j) = 0\ (j = 1, \ldots, n-1)\}.$$

The extremal value is given by $\sqrt{D_n/D_{n-1}}$ *with* $D_n = \det(K(a_i, \overline{a_j}))$ $(i, j = 1, \ldots, n)$.

Proof. Given $n \geq 1$ setting $b_j = 0$ $(j = 1, \ldots, n-1)$, $b_n = 1$ and $x_j = K_{a_j}$ $(n = 1, \ldots, n)$, from Proposition 6.1 we obtain the extremal function f_n defined by (21). By the reproducing property of K_{a_j}, it is easy to see that the function $h = f_n/\|f_n\|$ uniquely solves the extremal problem stated in this Corollary. From (21) and (22) the function h is given by

$$h = \frac{1}{\sqrt{D_n D_{n-1}}}\begin{vmatrix} \langle x_1, x_1\rangle & \langle x_2, x_1\rangle & \ldots & \langle x_n, x_1\rangle \\ \vdots & \vdots & \ddots & \vdots \\ \langle x_1, x_{n-1}\rangle & \langle x_2, x_{n-1}\rangle & \ldots & \langle x_n, x_{n-1}\rangle \\ x_1 & x_2 & \ldots & x_n \end{vmatrix}. \tag{24}$$

By the well-known formula describing the explicit form of the element obtained by the Gram-Schmidt orthogonalization process, we find that the extremal function h coincides with F_n. □

Fay's general addition-theorem is this.

Theorem 6.1 (Fay). *If $e \in \mathbb{C}^g$ with $\theta(e) \neq 0$, then*

$$\theta(\sum_1^n x_i - \sum_1^n y_i - e)\frac{\prod_{i<j} E(x_i, x_j)E(y_j, y_i)}{\theta(e)\prod_{i,j} E(x_i, y_j)} = \det\left(\frac{\theta(x_i - y_j - e)}{\theta(e)E(x_i, y_j)}\right) \tag{25}$$

for all $x_1, \dots, x_n, y_1, \dots, y_n \in C$.

Remark 6.1. When $n = 2$ the above identity coincides with the trisecant formula.

Proof. See [2, p.33]. This theorem, however, is easily deduced from its special case the trisecant formula. We sketch a proof as follows. Let $A = (a_{ij})$ be an $n \times n$ matrix. Recall that by virtue of Jacobi's formula we have the identity

$$|A_{n-2}| \cdot |A| = \begin{vmatrix} \Delta_{n-1,n-1} & \Delta_{n-1,n} \\ \Delta_{n,n-1} & \Delta_{n,n} \end{vmatrix} \tag{26}$$

where $A_{n-2} = (a_{ij})_{i,j=1}^{n-2}$ and Δ_{ij} denotes the (i, j)-cofactor of the matrix A. Let I_n be the identity (25). We show that I_n holds for all $n \in \mathbb{N}$ by induction on n. Case $n = 1$ is trivial. Case $n = 2$ is the trisecant formula as remarked above. Assume next that $n > 2$ and that for all $k < n$ the identity I_k holds. We have to prove the identity I_n. To this end set $a_{ij} = \frac{\theta(x_i - y_j - e)}{\theta(e)E(x_i, y_j)}$ and apply Jacobi's formula (26) to the matrix A. Then the determinant $|A_{n-2}|$ and the cofactor Δ_{ij} can be calculated by using the identity I_{n-2} and I_{n-1} respectively. Moreover, the right-hand side of (26) can be simplified by using the trisecant formula. Thus we find that $|A|$ is given by the left-hand side of (25), which proves the identity I_n. □

With this preparation we now turn to the representation of the extremal functions in the space $H_e^2(R)$ $(e \in \hat{T}_0)$. Note that the point $a \in R$ is a fixed reference point implicit in the definition of $H_e^2(R)$.

Theorem 6.2. *Let $\{a_j\}_{j=1}^n$ be distinct points in $R \setminus \{a\}$. Let $f_n \in H_e^2(R)$ be the extremal function attaining the value*

$$\sup_{\|f\| \leq 1} \{\operatorname{Re} f(b) \mid f \in H_e^2(R),\ f(a_j) = 0\ (j = 1, \dots, n)\}. \tag{27}$$

Then f_n is given by

$$f_n(x) = A_n \frac{\theta(x - \bar{b} + \sum_1^n a_i - \sum_1^n \overline{a_i} + e)}{\theta(x - \bar{a} + e)} \frac{E(x, \bar{a}) \prod_1^n E(x, a_i)}{E(x, \bar{b}) \prod_1^n E(x, \overline{a_i})}, \quad x \in R,$$

where A_n is a constant depending only on $a, b, a_1, \dots, a_n$. If $\Gamma_n(b)$ denotes the extremal value (27), then $\Gamma_n^2(b)$ is given by

$$\frac{\theta(b - \bar{b} + \sum_1^n a_i - \sum_1^n \overline{a_i} + e)\theta(e)\theta(a - \bar{a} + e)}{\theta(\sum_1^n a_i - \sum_1^n \overline{a_i} + e)|\theta(b - \bar{a} + e)|^2} \prod_1^n \left|\frac{E(b, a_i)}{E(b, \overline{a_i})}\right|^2 \frac{|E(b, \bar{a})|^2}{iE(b, \bar{b})iE(a, \bar{a})}. \tag{28}$$

Proof. Setting $a_{n+1} = b$, from Corollary 6.1 and (24), we have

$$f_n(x) = \frac{1}{\sqrt{D_n D_{n+1}}} \begin{vmatrix} R_e(a_1,\overline{a_1}) & R_e(a_1,\overline{a_2}) & \dots & R_e(a_1,\overline{a_{n+1}}) \\ \vdots & \vdots & \ddots & \vdots \\ R_e(a_n,\overline{a_1}) & R_e(a_n,\overline{a_2}) & \dots & R_e(a_n,\overline{a_{n+1}}) \\ R_e(x,\overline{a_1}) & R_e(x,\overline{a_2}) & \dots & R_e(x,\overline{a_{n+1}}) \end{vmatrix}$$

with $D_k = \det(R_e(a_i,\overline{a_j}))$ $(i,j = 1,\dots,k)$. Since $R_e(x,\bar{y}) = \frac{(\bar{a},a)(\bar{y},x)}{(\bar{a},x)(\bar{y},a)}$ by (12), we have

$$D_n = \frac{(\bar{a},a)^n \det(\overline{a_j},a_i)}{\prod_1^n(\bar{a},a_i)\prod_1^n(\overline{a_i},a)},$$

and by Theorem 6.1

$$\det(\overline{a_j},a_i) = \theta\Big(\sum_1^n a_i - \sum_1^n \overline{a_i} + e\Big)\frac{\prod_{i<j} E(a_j,a_i)E(\overline{a_i},\overline{a_j})}{\theta(e)\prod_{i,j} E(\overline{a_i},a_j)}.$$

Hence

$$D_n = \frac{(\bar{a},a)^n}{\prod_1^n(\bar{a},a_i)\prod_1^n(\overline{a_i},a)} \frac{\theta(\sum_1^n a_i - \sum_1^n \overline{a_i} + e)}{\theta(e)} \frac{\prod_{i<j} E(a_j,a_i)E(\overline{a_i},\overline{a_j})}{\prod_{i,j} E(\overline{a_i},a_j)}. \tag{29}$$

Since $\Gamma_n^2(b) = D_{n+1}/D_n$ by Corollary 6.1, the exact form of Γ_n is computed from (29) at once. Also, substituting x for a_{n+1} in D_{n+1}, we find that f_n has the form stated in the theorem. □

From extremality observe that the sequence $\{\Gamma_n(b)\}_{n=1}^\infty$ is monotone decreasing. Furthermore we have

Proposition 6.2.

$$\Gamma_{n-1}^2(b) - \Gamma_n^2(b) =$$
$$\frac{\theta(e)\theta(a-\bar{a}+e)|\theta(b-\overline{a_n}+e')|^2 E(a_n,\overline{a_n})|E(b,\bar{a})|^2}{\theta(e')\theta(a_n-\overline{a_n}+e')|\theta(b-\bar{a}+e)|^2 E(a,\bar{a})|E(b,\overline{a_n})|^2} \prod_1^{n-1}\left|\frac{E(b,a_i)}{E(b,\overline{a_i})}\right|^2$$

with $e' = \sum_1^{n-1} a_i - \sum_1^{n-1} \overline{a_i} + e$.

Proof. From (28) we have

$$\Gamma_{n-1}^2(b) - \Gamma_n^2(b) = \frac{\theta(e)\theta(a-\bar{a}+e)|E(b,\bar{a})|^2}{|\theta(b-\bar{a}+e)|^2 iE(a,\bar{a})iE(b,\bar{b})} \prod_1^{n-1}\left|\frac{E(b,a_i)}{E(b,\overline{a_i})}\right|^2 \cdot I$$

where

$$I = \frac{\theta(b-\bar{b}+e')}{\theta(e')} - \frac{\theta(b+a_n-\bar{b}-\overline{a_n}+e')}{\theta(a_n-\overline{a_n}+e')}\left|\frac{E(b,a_n)}{E(b,\overline{a_n})}\right|^2.$$

Applying the trisecant formula and symmetry (8) we have

$$I = -\frac{|\theta(b-\overline{a_n}+e')|^2 E(b,\bar{b})E(a_n,\overline{a_n})}{\theta(e')\theta(a_n-\overline{a_n}+e')|E(b,\overline{a_n})|^2},$$

so our Proposition is proved. □

References

1. S. Bergman, *The kernel function and conformal mapping*, Amer. Math. Soc. Providence, R.I. 1750, 1970.
2. J. D. Fay, *Theta functions on Riemann surfaces*, Lecture Notes in Mathematics **352**, Springer-Verlag, 1973.
3. D. A. Hejhal, *Theta functions, kernel functions and Abelian integrals*, Amer. Math. Soc. Memoir **129**, 1972.
4. A. H. Read, *A converse of Cauchy's theorem and applications to extremal problems*, Acta Math. **160** (1959), 1-22.
5. S. Saitoh, *The Bergman norm and the Szegö norm*, Trans. Amer. Math. Soc. **249** (1979), 261–279.
6. ———, *Theory of reproducing kernels and its applications*, Pitman Research Notes in Mathematics Series **189**, Longman Scientific & Technical, 1988.
7. L. Sario and K. Oikawa, *Capacity functions*, Die Grundlehren der mathematischen Wissenschaften, Band 149 Springer-Verlag New York Inc., New York, 1969.
8. N. Suita, *Capacities and kernels on Riemann surfaces*, Arch. Rational Mech. Anal. **46** (1972), 212–217.
9. H. Widom, *Extremal polynomials associated with a system of curves in the complex plane*, Adv. in Math. **2** (1969), 127–232.
10. A. Yamada, *Theta functions and domain functions*, RIMS Kokyuroku **323** (1978), 84–101 (in Japanese).
11. ———, *Positive differentials, theta functions and Hardy H^2 kernels*, Proc. Amer. Math. Soc., (to appear).

International Society for Analysis, Applications and Computation

1. H. Florian et al. (eds.): *Generalized Analytic Functions.* Theory and Applications to Mechanics. 1998 ISBN 0-7923-5043-X
2. H.G.W. Begehr et al. (eds.): *Partial Differential and Integral Equations.* 1999 ISBN 0-7923-5482-6
3. S. Saitoh, D. Alpay, J.A. Ball and T. Ohsawa (eds.): *Reproducing Kernels and their Applications.* 1999 ISBN 0-7923-5618-7

KLUWER ACADEMIC PUBLISHERS – DORDRECHT / BOSTON / LONDON